This authoritive volume shows how modern dynamical systems theory can help us in understanding the evolution of cosmological models. It also compares this approach with Hamiltonian methods and numerical studies. A major part of the book deals with the spatially homogeneous (Bianchi) models and their isotropic subclass, the Friedmann–Lemaître models, but certain classes of inhomogeneous models (for example 'silent universes') are also examined. The analysis leads to an understanding of how special (high symmetry) models determine the evolution of more general families of models; and how these families relate to real cosmological observations.

This is the first book to relate modern dynamical systems theory to both cosmological models and cosmological observations. It provides an invaluable reference for graduate students and researchers in relativity, cosmology and dynamical systems theory.

T0172102

Dynamical Systems in Cosmology

Dynamical Systems in Cosmology

Edited by
J. WAINWRIGHT

Department of Applied Mathematics,
University of Waterloo, Waterloo,
Ontario N2L 3G1, Canada

G. F. R. ELLIS

Department of Mathematics and Applied Mathematics,
University of Cape Town
Rondebosch 7000, South Africa

CAMBRIDGE
UNIVERSITY PRESS

CAMBRIDGE UNIVERSITY PRESS
Cambridge, New York, Melbourne, Madrid, Cape Town, Singapore, São Paulo

Cambridge University Press
The Edinburgh Building, Cambridge CB2 2RU, UK

Published in the United States of America by Cambridge University Press, New York

www.cambridge.org
Information on this title: www.cambridge.org/9780521554572

First published 1997
This digitally printed first paperback version 2005

A catalogue record for this publication is available from the British Library

Library of Congress Cataloguing in Publication data

Dynamical systems in cosmology / edited by J. Wainwright, G. F. R. Ellis
p. cm.
Includes bibliographical references and index.
ISBN 0 521 55457 8
1. Cosmology–Mathematical models.   2. Differential dynamical systems.
I. Wainwright, J. (John)   II. Ellis, George Francis Rayner.
QB981.D96 1996
523.1′01′5118–dc20   96–28592   CIP

ISBN-13 978-0-521-55457-2 hardback
ISBN-10 0-521-55457-8 hardback

ISBN-13 978-0-521-67352-5 paperback
ISBN-10 0-521-67352-6 paperback

Contents

Contents

Contributors

M. Bruni
SISSA, via Beirut 2-4, 34013 Trieste, Italy

P.K.S. Dunsby
Department of Mathematics and Applied Mathematics, University of Cape Town, Rondebosch 7000, South Africa

G.F.R. Ellis
Department of Mathematics and Applied Mathematics, University of Cape Town, Rondebosch 7000, South Africa

C.G. Hewitt
Faculty of Mathematics, University of Waterloo, Waterloo, Ontario N2L 3G1, Canada

D.W. Hobill
Department of Physics, University of Calgary, Calgary, Alberta T2N 1N4, Canada

S. Matarrese
Dipartimento di Fisica "Galileo Galilei" Università di Padova, via Marzolo 8, 35131 Padova, Italy

O. Pantano
Dipartimento di Fisica "Galileo Galilei" Università di Padova, via Marzolo 8, 35131 Padova, Italy

S.T.C. Siklos
Newnham College, Cambridge, U.K.

R. Tavakol
Queen Mary & Westfield College, London, U.K.

C. Uggla
Department of Physics, University of Stockholm, Stockholm, Sweden

J. Wainwright
Department of Applied Mathematics, University of Waterloo, Waterloo, Ontario N2L 3G1, Canada

Preface

This book had its origins in a workshop held in Cape Town from June 27 to 2 July 1994, with participants from South Africa, USA, Canada, UK, Sweden, Germany, and India. The meeting considered in depth recent progress in analyzing the evolution and structure of cosmological models from a dynamical systems viewpoint, and the relation of this work to various other approaches (particularly Hamiltonian methods). This book is however not a conference report. It was written by some of the conference participants, based on what they presented at the workshop but altered and extended after reflection on what was learned there, and then extensively edited so as to form a coherent whole. This process has been very useful: a considerable increase in understanding has resulted, particularly through the emphasis on relating the results of the qualitative analysis to possible observational tests. Apart from describing the development of the subject and what is presently known, the book serves to delineate many areas where the answers are still unknown. The intended readers are graduate students or research workers from either discipline (cosmological modeling or dynamical systems theory) who wish to engage in research in the area, tackling some of these unsolved problems.

The role of the two editors has been somewhat different. One (GFRE) has played a standard editorial role in terms of reviewing and editing material; he also initiated and organized the meeting, which took place in South Africa shortly after the transition to democracy there through the remarkable elections of 1994. The meeting was a suitable scientific celebration of those events. The other editor (JW) has been responsible for the main burden of shaping the volume, giving it its overall coherence. We thank the authors for their cooperation in this process, and for reading and commenting on other chapters, which has resulted in a far more useful volume than if it had been just a record of the workshop proceedings. We also thank Richard Matzner, one of the participants in the workshop, for his contributions concerning cosmological observations, which were very useful to us in writing Chapter 3. We would like to thank Alan Coley, Malcolm MacCallum and David Siegel for their comments on some of the chapters; Henk van Elst for discussions concerning evolution and constraint equations, and for detailed comments on the manuscript; and research students Robert Bridson and Derek Harnett at the University of Waterloo for reading parts of the manuscript and checking calculations. We are particularly indebted to Conrad Hewitt and Claes Uggla for continuing discussions over the past eighteen months.

We thank the Foundation for Research and Development (South Africa) and the Research Committee of the University of Cape Town for financial support of the meeting; our secretary Di Loureiro for her unfailing good humour, efficiency, and helpfulness in its organization; and graduate students Bruce Bassett, Tim Gebbie, Nico Christodoulides, and particularly Conrad Mellin, for the assistance they gave to conference participants. One of us (J. Wainwright) would like to thank the Department of Mathematics and Applied Mathematics at the University of Cape Town and the School of Mathematical Sciences at Queen Mary and Westfield College for their hospitality during several visits while the book was being edited.

As regards the preparation of the manuscript we thank Ann Puncher and Helen Warren at the University of Waterloo for their excellent and efficient work, and unfailing good humour when faced with successive rounds of changes. We also thank Lorraine Kritzer for help with the figures.

J Wainwright   G F R Ellis
Waterloo   *Cape Town*

Introduction

G. F. R. ELLIS
University of Cape Town

J. WAINWRIGHT
University of Waterloo

The cosmological models proposed by A. Einstein and W. de Sitter in 1917, based on Einstein's theory of general relativity, initiated the modern study of cosmology. The concept of an expanding universe was introduced by A. Friedmann and G. Lemaître in the 1920s, and gained credence in the 1930s because of Hubble's observations of galaxies showing a systematic increase of redshift with distance, together with Eddington's proof of the instability of the Einstein static model.† Since the 1940s the implications of following an expanding universe back in time have been systematically investigated, with an emphasis on four distinct epochs in the history of the universe:

(1) The *galactic epoch*, which is the period of time extending from galaxy formation to the present. This is the epoch that is most accessible to observation. During this period, matter in a cosmological model is usually idealized as a pressure–free perfect fluid, with galaxy clusters or galaxies acting as the particles of the fluid. The cosmic background radiation has negligible dynamic effect in this period.

(2) The *pre–galactic epoch*, during which matter is idealized as a gas, with the particles being the gas molecules, atoms, nuclei, or elementary particles at different times. The epoch is divided into a *post–decoupling period*, when matter and radiation evolve essentially independently, and a *pre–decoupling period*, when matter is ionized and is strongly interacting with radiation through Thomson scattering. The observed cosmic microwave background radiation is interpreted as evidence for the existence of this pre–decoupling period. In addition big–bang nucleosynthesis, i.e., the formation of ^{4}He and other light isotopes, occurs during this period, and has observational implications. Matter dominates dynamically at late times in this epoch, but the cosmic background radiation, which to a good approximation can

---

† We refer to Ellis (1989) for a discussion of the early history of relativistic cosmology and detailed references.

be represented as an ideal relativistic gas, is dynamically dominant in the earlier part of the epoch.

The physics of these two epochs is relatively well understood, and together they constitute the *standard or hot big–bang model* of the universe. This model has provided a very useful framework for explaining various facts about the universe and analyzing observations (e.g. Peebles *et al.* 1991). Additionally two earlier periods have been under intensive study, although the physics at those times is not so well understood:

(3) The conjectured *inflationary epoch*, during which the universe under-goes accelerated expansion. In most of the models for this epoch, which occurs before the pre-galactic epoch just mentioned, and after the Planck time, the source is taken to be a scalar field (often termed the inflaton), self–interacting through a potential which may or may not give rise to a mass term. The prototype for these fields is the Higgs boson, which arises in the standard model of particle physics (e.g. Kolb & Turner 1990). Scalar fields have not been observed in nature to date, so the physical status of the entire inflationary scenario is open to question. Nevertheless the idea provides important links to modern theoretical physics, and provides possible explanations for aspects of the universe that are otherwise puzzling. It also has observational consequences as regards density fluctuations that can be related to anisotropies of the cosmic background radiation.

(4) The presumed *quantum gravity epoch*, which occurs before the Planck time. In this epoch, classical general relativity may no longer be assumed to be valid.

It is usually assumed that the evolution of the universe is governed by the Einstein field equations during epochs (1) and (2), and usually also in epoch (3), although sometimes an inflationary epoch is ascribed to effects of modified gravitational theories. It is also usually assumed, to the extent of becoming dogma, that the space–time geometry is spatially homogeneous and isotropic.

The systematic study of expanding, spatially homogeneous and isotropic universe models, first undertaken by Friedmann and Lemaître, was placed on a firm geometric footing by H. P. Robertson and A. G. Walker. We will refer to these models of the universe, whose evolution is governed by the Einstein field equations, as *Friedmann–Lemaître (FL) universes*; the underlying geometry is given by the *Robertson–Walker (RW) metric*. The standard assumption, then, is that the universe can be described by a FL model, with perturbations about that model being adequate to explain the formation of galaxies and other structures.

The extent to which this assumption is necessitated by observations is open to debate. For example, it is clear that in the galactic epoch, the universe is not homogeneous on all scales. Thus the validity of the isotropic and spatially homogeneous RW metric entails the assumption that on *some* scale the galaxies are distributed homogeneously in space. In recent years, however, structures of increasingly large scale in the distribution of galaxies and deviations from a uniform Hubble flow on an increasingly large scale have been detected, which raises the question of whether the RW metric is in fact viable in this epoch (and if it is, what is the associated averaging scale needed to make sense of that model as a description of reality?). The best evidence is the high degree of isotropy of the cosmic background radiation, which lends strong plausibility to the FL models in the epoch since decoupling (provided we introduce a weak Copernican Principle: namely the assumption that observers elsewhere in the universe will see a high degree of isotropy in this radiation; see Section 3.2.5).

The FL models are clearly very special within the class of all cosmological models, and so a priori are highly unlikely; and various types of possible behaviour are completely suppressed by assuming the RW metric. Perturbation theory can probe the space of general models in a neighbourhood of the FL models, but cannot fully describe the non–linear behaviour inherent in the Einstein field equations. For example some models which are very different from a FL model at early times, evolve to become close to a FL model for a period, and then diverge strongly from a FL model in the future; and they can be fully compatible with the observational evidence available to us. Developing an understanding of the range of such models provides a parametrized set of test cases against which to compare observations.

For instance, one can examine what dynamics of the early universe are required to lead to the presently observed galactic epoch, without assuming the RW metric at earlier times. Misner's chaotic cosmology program studied this in the 1960s (e.g. Barrow 1982a). In the 1980s inflation was introduced to provide another physical answer (e.g. Narlikar 1993, pages 179–92; Kolb & Turner 1990, Chapter 8). Early analytical work on inflation had an element of inconsistency, because it only considered almost–FL models, which already assumed some of the behaviour inflation was intended to produce. To answer this necessitates the study of inflation in more general models.

Thus for a variety of reasons it is desirable to analyze the evolution of cosmological models that are more general than the FL models, in that they exhibit both anisotropy and inhomogeneity. It is primarily this problem that has led researchers, over the past 25 years, to apply the theory of dynamical systems to theoretical cosmology.

In general the theory of dynamical systems is used to study physical sys-

tems that evolve in time. It is assumed that the state of the physical system
at an instant in time can be described by an element **x** of a state space X,
which may be finite dimensional ($\mathbb{R}^n$, or a non–trivial differentiable manifold) or infinite dimensional (a function space). The evolution of the system
is described by an autonomous differential equation on X, written symbolically

$$\frac{d\mathbf{x}}{dt} = \mathbf{f}(\mathbf{x}), \quad \mathbf{x} \in X \tag{0.1}$$

where $\mathbf{f} : X \to X$. If X is finite dimensional then (0.1) represents an
autonomous system of *ordinary differential equations*, while if X is a function
space, the map **f** would involve taking spatial derivatives and (0.1) represents
an autonomous system of *partial differential equations*.

The theory of dynamical systems has its origins in the work of Poincaré
at the end of the nineteenth century (e.g. Hirsch 1984). He proposed that
instead of trying to find particular exact solutions of a differential equation
(0.1), one should use topological and geometrical methods to determine
properties of the set of all solutions, viewed as orbits (trajectories) in a state
space. It was some years later, however, in the late 1920s, that Birkhoff
and others began the formal mathematical development of the theory of
dynamical systems, by introducing concepts such as the flow associated with
a differential equation, and the fundamental concept of an ω–limit set.

As the title suggests, the main focus of this book is the application of
the theory of dynamical systems to the Einstein field equations in a cosmological setting. The goal in applying these methods is to obtain qualitative
information about the evolution of general classes of cosmological models,
both spatially homogeneous and inhomogeneous, at the expense of making
highly idealized assumptions about the physical processes in the universe.
In this way one hopes to shed light on the question of what is the range of
initial conditions and subsequent evolution that is compatible with current
observations of the universe. The second aim of the book is to give a glimpse
of how the other approaches to studying cosmological models more general
than FL interact with and complement the dynamical systems approach. In
the authors' opinion this interaction has not been fully explored. A related
issue is formalism. In order to apply the theory of dynamical systems it is
necessary to write the Einstein field equations as first–order evolution equations (i.e. an autonomous system of ordinary or partial differential equations,
depending on the context). This has been done in a variety of ways, and it
is thus useful to describe and compare the different approaches.

We now summarize the methods that have been used for analyzing cosmological models more general than FL, with a view to placing the various contributions to the book in a broader context.

(1) *Topological methods.* By this we primarily mean the global analysis leading to the singularity theorems that were proved in the 1960s by R. Penrose and S. Hawking (and are discussed in Hawking & Ellis 1973). These theorems assert that subject to very general conditions, which seem to be satisfied in the physical universe, a cosmological model begins at a space–time singularity (representing a breakdown of classical physics). These results do not depend on any symmetry or other simplifying assumptions on the geometry, or on a detailed analysis of the field equations; rather they come from geometrically and physically motivated inequalities, together with a careful analysis of properties of various causal regions in a space–time and their boundaries, based on generic properties of key equations (specifically the Raychaudhuri and conservation equations).

(2) *Qualitative methods.* By this we mean methods of analyzing the Einstein field equations with a view to deducing the evolutionary behaviour of general classes of cosmological models. The bulk of the results obtained to date concern spatially homogeneous models. Three different approaches have been used, which we refer to as follows:

 (i) piecewise approximation methods,
 (ii) Hamiltonian methods,
(iii) dynamical systems methods.

The first approach was developed by the Russian school of physicists. The evolution of a cosmological model in time is approximated as a sequence of time periods during each of which certain terms in the field equations can be neglected, leading to a simpler system of equations (e.g. Belinskii *et al.* 1970, Doroshkevich *et al.* 1973, Lukash 1974 and Zel'dovich & Novikov 1983, Section 22.7). This heuristic approach can in fact be placed on a sound foundation through the analysis of the so–called heteroclinic sequences in the dynamical systems approach (see Chapters 6–9).

The use of Hamiltonian methods in cosmology originated with C. W. Misner and his coworkers in the 1960s (Misner 1969a,b), and has been extensively developed *inter alia* by Jantzen, Rosquist and Uggla. The Einstein field equations were reduced to a time–dependent Hamiltonian system for a particle (the 'universe point') in two dimensions. The system was analyzed heuristically by approximating the time–dependent potentials by moving 'potential walls', which are assumed to 'reflect' the moving particle instantaneously. This approach, which has been used primarily to analyze the dynamics near the big–bang, will be discussed in Chapter 10.

The third approach is based on the fact that the Einstein field equations for spatially homogeneous cosmologies can be written as an autonomous system of first–order differential equations (more briefly, an autonomous differential equation on $\mathbb{R}^n$). The solution curves partition $\mathbb{R}^n$ into orbits, thereby defining a dynamical system on $\mathbb{R}^n$. This approach, which was initiated by Collins (1971) and developed extensively by Bogoyavlensky and collaborators (Bogoyavlensky 1985), Wainwright and collaborators, and others, will be discussed in Chapter 5–8, and is one of the principal topics of this book.

The theory of dynamical systems can also be applied to autonomous systems of first–order partial differential equations, with an infinite dimensional function space as the state space (e.g. Hale 1988, Temam 1988a,b), but the subject is still in its infancy compared with the finite dimensional case. It is not clear how much progress can be made analytically, and indeed one expects that numerical methods will have to play a major role. Nevertheless a start has been made in applying these ideas to the Einstein field equations, in particular to the class of cosmologies that we shall refer to as G_2 cosmologies, which permit one spatial degree of freedom (Hewitt & Wainwright 1990). These developments are discussed in Chapter 12.

(3) *Numerical methods*. With the advent of modern computers, numerical methods are being increasingly used to investigate various aspects of the evolution of inhomogeneous cosmologies (e.g. Anninos *et al.* 1991, Berger & Moncrief 1993, Shinkai & Maeda 1994), with emphasis on the case of one spatial degree of freedom (partial differential equations with one time and one space variable). Numerical experiments are also needed in connection with dynamical systems methods, to explore the state space in situations where the qualitative techniques cannot be fully implemented. This is discussed in Chapter 11 in connection with chaotic behaviour in the Bianchi models.

(4) *Perturbation methods*. The study of perturbations of the FL models in connection with the growth of density perturbations dates back to E.M. Lifshitz in the late 1940s. Perturbation theory in effect enables one to study inhomogeneous models by reducing partial differential equations to ordinary differential equations. From a dynamical systems perspective the method enables one to explore the infinite–dimensional state space associated with the partial differential equations by investigating the finite dimensional state space associated with the approximating ordinary differential equations. This matter is discussed in Chapter 14, using the geometrical approach to cosmological perturbations recently introduced by Ellis & Bruni (1989). A related approach is discussed in Chapter 13, where it

is shown that for a class of inhomogeneous models without symmetry, the so–called silent universes, the spatial dependence decouples from the time evolution, so that the evolution equations reduce from partial to ordinary differential equations, which facilitates their qualitative analysis.

(5) *Exact solutions.* One can expect to find exact solutions of the Einstein field equations only by imposing strong restrictions on the space–time geometry, primarily by assuming the existence of symmetries, or by imposing restrictions on the Weyl tensor or on the kinematical quantities. Despite the special properties inherent in their very existence, exact solutions can (and often do) play an important role in the evolution of whole classes of models, by acting as asymptotic or intermediate states. We shall illustrate this role of exact solutions when we give a concise overview of the known solutions in Chapter 9 (spatially homogeneous) and Chapter 13 (spatially inhomogeneous), using the results from the dynamical systems analysis.

We have attempted to make this book relatively self–contained by providing introductory and background material in Part one. This Part also serves to motivate some of the subsequent developments. In Chapter 1 we give the geometrical background that is needed for studying cosmological models. Much of this material has been presented elsewhere in greater detail, and we refer the reader to Ellis (1971, 1973) and MacCallum (1973). Because of their importance in cosmology, we give an introduction to the FL models in Chapter 2. In order to illustrate the use of qualitative methods and dimensionless variables in a simple setting, we analyze the evolution of the FL models assuming pressure–free matter as the source. In Chapter 3 we review cosmological observations, and assess how compelling is the evidence that the universe is close to a FL universe in different epochs. Finally, Chapter 4 contains an introduction to some of the main ideas in the theory of dynamical systems, with emphasis on asymptotic behaviour.

Dynamical systems methods have been used primarily to study spatially homogeneous cosmologies, and thus a major portion of the book (Part two) is devoted to this topic. In Part three, however, we have attempted to give a glimpse of the potential for using dynamical systems methods to study inhomogeneous models. This Part contains three chapters each of which illustrates a different way in which an autonomous system of ordinary differential equations can be used to probe the infinite–dimensional state space of the inhomogeneous models.

On assessing the utility of dynamical systems methods, it is clear that despite their mathematical sophistication, they are limited from a physical point of view because of the strong simplifying assumptions that are usu-

ally made about the matter content. Indeed most of the results that are
described in this book concern cosmological models whose matter content is
that of a perfect fluid with a linear equation of state, with possible inclusion
of a cosmological constant. While the cases of dust ($p = 0$) and radiation
($p = \frac{1}{3}\mu$) are important from a cosmological point of view, we acknowledge
that less idealised matter descriptions are needed at some epochs; this has
been pursued to some extent by various authors. Some efforts have also
been made to use dynamical systems methods to study evolution during
an inflationary epoch. Their applicability in exploring the consequences of
other gravitational theories is also of interest; space and time limits pre-
clude us from pursuing these topics. It is also clear that these methods have
so far achieved only a limited success in analyzing inhomogeneous models
(which are described by infinite dimensional dynamical systems in general).
Despite the above limitations, we hope the progress reported in this book
will encourage others to extend the use of dynamical systems methods to a
broader range of problems than those considered here (and to complete the
analyses of these problems).

Part one

Background

1

Geometry of cosmological models

G. F. R. ELLIS
University of Cape Town

S. T. C. SIKLOS
Cambridge University

J. WAINWRIGHT
University of Waterloo

In this chapter we give a summary of the geometrical background that is needed to classify and analyze the evolution of cosmological models. In particular, we focus on the description of space–time curvature and the kinematics of a fluid, symmetry properties of space–times, and methods for performing a $1 + 3$ decomposition of the Einstein field equations into evolution and constraint equations. In Sections 1.5 and 1.6 we specialize these methods to two important special cases, the Bianchi cosmologies and the G_2 cosmologies.

1.1 Cosmological models

A cosmological model $(\mathcal{M}, \mathbf{g}, \mathbf{u})$ represents the universe at a particular averaging scale. It is defined by specifying *the space–time geometry*, determined by a Lorentzian metric $\mathbf{g}$ defined on the manifold $\mathcal{M}$, and *a family of fundamental observers*, whose congruence of world–lines is represented by the 4–velocity field $\mathbf{u}$ (which will usually be the matter 4–velocity). We will assume that this congruence is expanding in some epoch. Together, these determine the kinematics of the universe. To determine its dynamic evolution, we also need to specify *the matter content*, represented by the stress–energy tensor $T_{ab}^{(A)}$ of each matter component (labelled by A), the equations governing the behaviour of each component and describing the interactions between them, and finally, *the interaction of the geometry and matter* – how matter determines the geometry, which in turn determines the motion of the matter. We will assume this interaction is governed by the *Einstein field equations* (EFE)

$$R_{ab} - \tfrac{1}{2}Rg_{ab} = T_{ab}, \quad T_{ab} = \sum_A T_{ab}^{(A)}, \tag{1.1}$$

which imply the conservation equations

$$T^{ab}{}_{;b} = 0. \tag{1.2}$$

That is, we will assume General Relativity is the correct classical theory of gravitation.

As discussed in the Introduction one of our goals is to investigate cosmological models which are close to a FL model during some epoch. However, models which are not close to FL are of interest as possible asymptotic states at early times and late times. We thus do not impose an assumption of 'close to FL' a priori.

The notation and conventions that are used in (1.1), (1.2) and the rest of the book are introduced systematically in this Section. We use geometrized units throughout, i.e. $c = 1$ and $8\pi G = 1$, where c is the speed of light in vacuum and G is the gravitational constant. The result of this choice is that the dimensions of all dynamical variables are expressed as integer powers of length (see, for example, Wald 1984, page 471).

1.1.1 Description of models

It is useful to describe cosmological models in terms of a basis of vector fields $\{e_a\}$ and the dual basis of 1–forms $\{\omega^a\}$, where $a = 0, 1, 2, 3$. The components of tensors and of the connection relative to such a basis will be labelled with indices $a, b, c, \dots$. For example for a vector field $\mathbf{X}$, we have

$$\mathbf{X} = X^a \mathbf{e}_a.$$

The components of the metric tensor $\mathbf{g}$ relative to this basis are given by

$$g_{ab} = \mathbf{g}(\mathbf{e}_a, \mathbf{e}_b), \tag{1.3}$$

and the line–element will be written symbolically as

$$ds^2 = g_{ab}\omega^a\omega^b. \tag{1.4}$$

In any coordinate chart, there is a natural basis: $\{e_a\}$ can be chosen to be a *coordinate basis* $\{\partial/\partial x^i\}$, with the dual basis being the coordinate 1–forms $\{dx^i\}$. We shall use indices $i, j, k, \dots$ to label coordinate components. The general basis vector fields $\mathbf{e}_a$ and 1–forms ω^a can be written in terms of a coordinate basis as follows

$$\mathbf{e}_a = e_a{}^i(x^j)\frac{\partial}{\partial x^i}, \quad \omega^a = \omega^a{}_i(x^j)dx^i. \tag{1.5}$$

It is thus natural to regard a vector field $\mathbf{X}$ as a differential operator which acts on scalars:

$$\mathbf{X}(f) = X^i \frac{\partial f}{\partial x^i} . \tag{1.6}$$

In particular,

$$\mathbf{e}_a(f) = e_a{}^i \frac{\partial f}{\partial x^i} .$$

In terms of a coordinate basis, (1.3) and (1.4) assume the form

$$g_{ij} = \mathbf{g}\left(\frac{\partial}{\partial x^i}, \frac{\partial}{\partial x^j}\right), \qquad ds^2 = g_{ij} dx^i dx^j . \tag{1.7}$$

Another special type of basis is an *orthonormal frame*, for which the four vectors $\mathbf{e}_a$ are mutually orthogonal and of unit length, with $\mathbf{e}_0$ timelike. The vector fields $\mathbf{e}_a$ thus satisfy

$$\mathbf{g}(\mathbf{e}_a, \mathbf{e}_b) = \eta_{ab}, \tag{1.8}$$

where

$$\eta_{ab} = \text{diag}(-1, 1, 1, 1) . \tag{1.9}$$

Relative to an orthonormal frame the line–element (1.4) assumes the form

$$ds^2 = \eta_{ab} \omega^a \omega^b . \tag{1.10}$$

Given any basis of vector fields $\mathbf{e}_a$, the commutators $[\mathbf{e}_a, \mathbf{e}_b]$ are vector fields and hence can be written as a linear combination of the basis vectors:

$$[\mathbf{e}_a, \mathbf{e}_b] = \gamma^c{}_{ab} \mathbf{e}_c. \tag{1.11}$$

The coefficients $\gamma^c{}_{ab}(x^j)$ are called the *commutation functions* of the basis. If $\{\mathbf{e}_a\}$ is a coordinate basis, then the commutation functions are zero.

We use the Levi–Civita connection ∇ to define covariant derivatives.† For vector fields $\mathbf{X}$ and $\mathbf{Y}$ the covariant derivative of $\mathbf{Y}$ with respect to $\mathbf{X}$ is a vector field denoted $\nabla_{\mathbf{X}} \mathbf{Y}$. The components of the connection relative to a basis $\{\mathbf{e}_a\}$ are the set of functions $\Gamma^c{}_{ab}$ defined by writing the vector fields $\nabla_{\mathbf{e}_b} \mathbf{e}_a$ as linear combinations of the basis vectors:

$$\nabla_{\mathbf{e}_b} \mathbf{e}_a = \Gamma^c{}_{ab} \mathbf{e}_c . \tag{1.12}$$

Since the connection is assumed to be (i) *torsion-free*,

$$\nabla_{\mathbf{X}} \mathbf{Y} - \nabla_{\mathbf{Y}} \mathbf{X} = [\mathbf{X}, \mathbf{Y}], \tag{1.13}$$

---

† We refer to Stewart (1990, pages 25–43), for a clear and concise introduction to connections, covariant derivatives and curvature.

and (ii) *metric*, $\nabla \mathbf{g} = 0$, it follows that the $\Gamma^c{}_{ab}$ are related to the derivatives of the metric components (1.3) and the commutation functions $\gamma^c{}_{ab}$, which are defined by (1.11). The formula is (Stewart 1990, page 43),

$$\Gamma_{abc} = \tfrac{1}{2}\left[\mathbf{e}_b(g_{ac}) + \mathbf{e}_c(g_{ba}) - \mathbf{e}_a(g_{cb}) + \gamma^d{}_{cb}g_{ad} + \gamma^d{}_{ac}g_{bd} - \gamma^d{}_{ba}g_{cd}\right]. \quad (1.14)$$

where $\Gamma_{abc} = g_{ad}\Gamma^d{}_{bc}$. The components of $\nabla_{\mathbf{X}}\mathbf{Y}$ relative to $\{\mathbf{e}_c\}$ are denoted by $Y^c{}_{;b}X^b$, and it follows that

$$Y^c{}_{;b} = \mathbf{e}_b(Y^c) + \Gamma^c{}_{ab}Y^a. \quad (1.15)$$

These quantities are the components of the $(1,1)$ tensor $\nabla\mathbf{Y}$, the covariant derivative of $\mathbf{Y}$. This formula generalizes in the usual way to arbitrary tensors (e.g. Kramer *et al.* 1980, page 40). We shall sometimes use the notation ∇_a for covariant derivatives (Wald 1984, page 31), for example

$$\nabla_b Y^c = Y^c{}_{;b},$$

and also to denote the gradient of a scalar field f

$$\nabla_a f = \mathbf{e}_a(f).$$

Relative to a coordinate basis the commutation functions are zero and equations (1.14) and (1.15) reduce to

$$\Gamma_{ijk} = \tfrac{1}{2}\left(\frac{\partial g_{ik}}{\partial x^j} + \frac{\partial g_{ji}}{\partial x^k} - \frac{\partial g_{kj}}{\partial x^i}\right),$$

which are the usual Christoffel symbols, and

$$Y^k{}_{;j} = \frac{\partial Y^k}{\partial x^j} + \Gamma^k{}_{ij}Y^i.$$

On the other hand, relative to an orthonormal basis, $g_{ab} = \eta_{ab}$, and hence the first three terms on the right side of (1.14) are zero (see (1.59) below).

The components of the Riemann tensor are given by the usual formula (Stewart 1990, page 34, but note the difference in sign convention),

$$R^a{}_{bcd} = \mathbf{e}_c(\Gamma^a{}_{bd}) - \mathbf{e}_d(\Gamma^a{}_{bc}) + \Gamma^a{}_{fc}\Gamma^f{}_{bd} - \Gamma^a{}_{fd}\Gamma^f{}_{bc} - \Gamma^a{}_{bf}\gamma^f{}_{cd}, \quad (1.16)$$

and the Ricci identity is

$$Y^a{}_{;dc} - Y^a{}_{;cd} = R^a{}_{bcd}Y^b.$$

The components of the Ricci tensor and the Ricci scalar are given by

$$R_{bd} = R^a{}_{bad}, \qquad R = R^a{}_a. \quad (1.17)$$

If we use a coordinate basis the 10 metric tensor components g_{ij} are the gravitational field variables, and the EFE (1.1) are second–order partial differential equations as follows from the coordinate versions of (1.16) and (1.17). The gauge freedom is the arbitrariness in the choice of local coordinates.

If we use an orthonormal frame, the 24 commutation functions $\gamma^c{}_{ab}$ are the basic variables, and the gauge freedom is an arbitrary Lorentz transformation, representing the freedom in choice of the orthonormal frame. The Jacobi identity for vector fields,

$$[[\mathbf{X}, \mathbf{Y}], \mathbf{Z}] + [[\mathbf{Z}, \mathbf{X}], \mathbf{Y}] + [[\mathbf{Y}, \mathbf{Z}], \mathbf{X}] = 0,$$

when applied to the frame vectors $\mathbf{e}_a$ yields a set of 16 identities

$$\mathbf{e}_{[c}\gamma^d{}_{ab]} - \gamma^d{}_{e[c}\gamma^e{}_{ab]} = 0. \tag{1.18}$$

These identities, in conjunction with the EFE, give first–order evolution equations for some† of the commutation functions, and also provide a set of constraints involving only spatial derivatives (see Section 1.4.2). We shall refer to the resulting formalism as the *orthonormal frame formalism*. Since it leads directly to first–order evolution equations, it is a natural choice when applying dynamical systems methods.

1.1.2 The source terms

In order to be able to incorporate a variety of source terms in the EFE, we use the standard decomposition of the stress–energy tensor T_{ab} with respect to a timelike vector field $\mathbf{u}$ (Ellis 1971, page 7):

$$T_{ab} = \mu u_a u_b + 2q_{(a}u_{b)} + p(g_{ab} + u_a u_b) + \pi_{ab}, \tag{1.19}$$

where

$$q_a u^a = 0, \quad \pi_{ab}u^b = 0, \quad \pi_a{}^a = 0, \quad \pi_{ab} = \pi_{ba}.$$

The following source terms are of interest, although most of our attention will be focussed on perfect fluids. In general one will have a mixture of different forms of matter (see Section 2.1 for a simple example).

(1) A *non–tilted perfect fluid* is described by its 4–velocity $\mathbf{u}$, which is the fundamental 4–velocity, its energy density μ and pressure p, with barotropic equation of state $p = p(\mu)$. The stress–energy tensor is

$$T_{ab} = \mu u_a u_b + p(g_{ab} + u_a u_b), \quad u_a u^a = -1, \tag{1.20}$$

† The gauge freedom can be used to specifiy some of the commutation functions.

so that

$$q_a = 0, \quad \pi_{ab} = 0,$$

in the general form (1.19). We will usually work with an equation of state of the form

$$p = (\gamma - 1)\mu,$$

where γ is a constant. From a physical point of view, the most important values are $\gamma = 1$ (dust) and $\gamma = \frac{4}{3}$ (radiation). The value $\gamma = 0$ corresponds to a cosmological constant and the value $\gamma = 2$ is sometimes considered, corresponding to a 'stiff fluid'. We will thus assume that γ satisfies

$$0 \le \gamma \le 2.$$

(2) A *tilted perfect fluid* has 4–velocity $\hat{u}$ which is *not* aligned with the fundamental 4–velocity u. The stress–energy tensor is

$$T_{ab} = \hat{\mu}\hat{u}_a\hat{u}_b + \hat{p}(g_{ab} + \hat{u}_a\hat{u}_b), \quad \hat{u}_a\hat{u}^a = -1. \tag{1.21}$$

Following van Elst & Uggla (1996), we decompose $\hat{u}$ into a component parallel to u and one orthogonal to u, writing

$$\hat{u}^a = \Gamma(u^a + v^a),$$

where $u^a v_a = 0$ and

$$\Gamma = (1 - v^2)^{-\frac{1}{2}}, \quad v^2 = v^a v_a.$$

It follows that the various terms in (1.19) are given by

$$\begin{aligned}
\mu &= \hat{\mu} + \Gamma^2 v^2(\hat{\mu} + \hat{p}), \\
p &= \hat{p} + \tfrac{1}{3}\Gamma^2 v^2(\hat{\mu} + \hat{p}), \\
q_a &= \Gamma^2(\hat{\mu} + \hat{p})v_a, \\
\pi_{ab} &= \Gamma^2(\hat{\mu} + \hat{p})\left[v_a v_b - \tfrac{1}{3}v^2(g_{ab} + u_a u_b)\right].
\end{aligned} \tag{1.22}$$

(3) A *pure source–free magnetic field* is described by a stress–energy tensor of the form

$$T_{ab} = F_{ac}F_b{}^c - \tfrac{1}{4}F^{cd}F_{cd}g_{ab}, \tag{1.23}$$

where†

$$F_{ab} = \eta_{ab}{}^{cd}H_c u_d,$$

---

† We use the convention $\eta^{0123} = 1/\sqrt{-g}$, where $g = \det(g_{ab})$, for the totally antisymmetric permutation tensor η^{abcd}. See Ellis (1973, page 2) for the identities satisfied by η^{abcd}.

H_d being the magnetic field relative to the fundamental 4–velocity **u**. It follows that

$$\mu = \tfrac{1}{2} H_a H^a, \quad p = \tfrac{1}{3}\mu,$$

$$q_a = 0, \quad \pi_{ab} = -H_a H_b + \tfrac{1}{3}(H_c H^c)(g_{ab} + u_a u_b).$$

Maxwell's equations are

$$F^{ab}{}_{;b} = 0, \quad F_{[ab;c]} = 0.$$

(4) A *scalar field*† is described by a stress–energy tensor of the form

$$T_{ab} = \nabla_a \phi \nabla_b \phi - \left[\tfrac{1}{2}\nabla_c \phi \nabla^c \phi + V(\phi)\right] g_{ab}, \tag{1.24}$$

where the potential $V(\phi)$ has to be specified. If $\nabla_a \phi$ is timelike, we can define a unit timelike vector field **u** normal to the surfaces $\{\phi = const\}$:

$$u^a = \frac{\nabla^a \phi}{(-\nabla_c \phi \nabla^c \phi)^{\frac{1}{2}}}.$$

Then T_{ab} has the algebraic form of a perfect fluid, with

$$\mu = -\tfrac{1}{2}\nabla_a \phi \nabla^a \phi + V(\phi), \quad p = -\tfrac{1}{2}\nabla_a \phi \nabla^a \phi - V(\phi),$$

$$q_a = 0, \quad \pi_{ab} = 0.$$

The equation of motion for the scalar field is the Klein–Gordon equation

$$\nabla^a \nabla_a \phi - V'(\phi) = 0,$$

which is a consequence of (1.24) and the conservation equation (1.2).

We note that imperfect fluids are of interest in the context of the early universe. A detailed discussion of the stress–energy tensor and the thermo-dynamics of such fluids is, however, beyond the scope of this book. We refer to Section 8.6.2 for a brief summary of the qualitative analyses that have been given in the literature, and some of the basic references.

1.1.3 Characterizing models

We are interested, on the one hand, in specific exact solutions of the EFE and, on the other, in the properties of general classes of solutions. Simple exact solutions can be analyzed in detail as idealized cosmological models in their own right. They can also act as asymptotic states of general classes

---

† In the context of the early universe a scalar field is used as the source of inflation and as such is often referred to as the 'inflaton'.

of models or as background solutions for perturbation analyses, thereby creating models of greater generality which may describe physical reality more accurately. On the other hand, by analyzing the evolution of general classes of models qualitatively, one can gain insight into what initial conditions in the early universe can evolve into the present observable universe.

In all of these endeavours one needs to be able to characterize a cosmological model $(\mathcal{M}, \mathbf{g}, \mathbf{u})$ invariantly. We shall use the following features to characterize models: *symmetry* properties, *kinematic* properties, and properties of the *Weyl conformal curvature tensor*.

Traditionally symmetry properties have referred to *isometries*, and their use in cosmology dates back to Robertson (1935, 1936) and Walker (1936). Over the past ten years, however, it has been realized that, in addition, *similarities* (also known as *homothetic transformations*) play an important role in the analysis of cosmological models. We discuss symmetry properties in some depth in Section 1.2.

The kinematic quantities associated with a timelike congruence **u** were first introduced by Raychaudhuri, Schücking and Ehlers (Ehlers 1961†). We refer to Ellis (1971, pages 8–11) for a discussion of these quantities in a cosmological context, and here give a brief review.

A unit timelike vector field **u** determines a projection tensor h_{ab} according to

$$h_{ab} = g_{ab} + u_a u_b, \tag{1.25}$$

which at each point projects into the 3–space orthogonal to **u**. It follows that

$$h_a{}^c h_c{}^b = h_a{}^b, \quad h_a{}^b u_b = 0, \quad h_a{}^a = 3.$$

One can decompose the covariant derivative $u_{a;b}$ into its irreducible parts according to

$$u_{a;b} = \sigma_{ab} + \omega_{ab} + \tfrac{1}{3}\Theta h_{ab} - \dot{u}_a u_b, \tag{1.26}$$

where σ_{ab} is symmetric and trace–free, ω_{ab} is antisymmetric, and $u^a \sigma_{ab} = 0 = u^a \omega_{ab}$. It follows that

$$
\begin{aligned}
\Theta &= u^a{}_{;a}, \\
\dot{u}_a &= u_{a;b} u^b, \\
\sigma_{ab} &= u_{(a;b)} - \tfrac{1}{3}\Theta h_{ab} + \dot{u}_{(a} u_{b)} \\
\omega_{ab} &= u_{[a;b]} + \dot{u}_{[a} u_{b]}.
\end{aligned}
\tag{1.27}
$$

The scalar Θ is the rate of *expansion scalar* and $\dot{u}_a$ is the *acceleration vector*.

† See Ehlers (1993) for a translation of this classic article.

The tensor σ_{ab} is the rate of *shear tensor*, with magnitude σ given by

$$\sigma^2 = \tfrac{1}{2}\sigma_{ab}\sigma^{ab}, \qquad (1.28)$$

and ω_{ab} is the *vorticity tensor*. It is convenient to define the vorticity vector

$$\omega^a = \tfrac{1}{2}\eta^{abcd}u_b\omega_{cd}, \qquad (1.29)$$

which satisfies $u^a\omega_a = 0$. The magnitude ω of the vorticity is given by

$$\omega^2 = \omega^a\omega_a = \tfrac{1}{2}\omega^{ab}\omega_{ab}.$$

If the vorticity is zero, we say that the vector field **u** is irrotational. It is sometimes convenient to combine the rate of shear tensor and rate of expansion scalar to give the rate of *expansion tensor* Θ_{ab},

$$\Theta_{ab} = \sigma_{ab} + \tfrac{1}{3}\Theta h_{ab}. \qquad (1.30)$$

Because we are working in a cosmological context we shall usually replace Θ by the *Hubble scalar H* defined by

$$H = \tfrac{1}{3}\Theta. \qquad (1.31)$$

The *Weyl conformal curvature tensor* (e.g. Kramer *et al.* 1980, page 50) is defined by

$$C^{ab}{}_{cd} = R^{ab}{}_{cd} - 2\delta^{[a}_{[c}R^{b]}_{d]} + \tfrac{1}{3}R\delta^a_{[c}\delta^c_{d]}. \qquad (1.32)$$

Through the Bianchi identities

$$R_{ab[cd;e]} = 0 \;\Leftrightarrow\; C_{abc}{}^d{}_{;d} = R_{c[a;b]} - \tfrac{1}{6}g_{c[a}\nabla_{b]}R, \qquad (1.33)$$

and the EFE, it is determined non–locally by the matter elsewhere in the universe, together with suitable boundary conditions, and represents the 'free gravitational field'. It is useful to define its *electric part* E_{ab} and *magnetic part* H_{ab} relative to **u** according to

$$E_{ac} = C_{abcd}u^bu^d, \qquad H_{ac} = {}^*C_{abcd}u^bu^d, \qquad (1.34)$$

where the dual is defined by

$${}^*C_{abcd} = \tfrac{1}{2}\eta_{ab}{}^{st}C_{stcd}.$$

These tensors are symmetric and trace–free and satisfy $E_{ab}u^b = 0 = H_{ab}u^b$. It also follows that

$$E_{ab} = 0 = H_{ab} \Leftrightarrow C_{abcd} = 0.$$

It can be shown that

$$C_{abcd}C^{abcd} = 8(E_{ab}E^{ab} - H_{ab}H^{ab}), \quad C_{abcd}{}^*C^{abcd} = 16E_{ab}H^{ab}. \qquad (1.35)$$

The algebraic structure of the Weyl tensor is characterized by its *Petrov type*, (e.g. Kramer *et al.* 1980, Chapter 4 or Wald 1984, pages 179–80).

1.2 Symmetries of space–time

The aim of this section is to summarize the main concepts relating to symmetry properties of space–times, in particular isometry and similarity groups, isotropy subgroups and group orbits, and the restrictions on their dimensions. This is the broader symmetry classification within which we locate the classification of Bianchi models and inhomogeneous models, considered in Sections 1.5 and 1.6. For generality we formulate the concepts in terms of an n–dimensional manifold with metric tensor, denoted $(\mathcal{M}, \mathbf{g})$, with space–time as a particular example. For brevity, we will refer to $(\mathcal{M}, \mathbf{g})$ as a manifold. For a more detailed discussion of this topic, with applications to cosmology, we refer to MacCallum (1973, section II); see also Kramer *et al.* (1980, Chapter 8), where further references may be found.

1.2.1 Killing vectors and isometry groups

An isometry of a manifold $(\mathcal{M}, \mathbf{g})$ is a mapping of $\mathcal{M}$ into itself that leaves the metric $\mathbf{g}$ invariant. The Lie derivative is used to give a formal description of this idea. A vector field $\boldsymbol{\xi}$ generates a one–parameter group of transformations and vice versa. The orbits of this group are the integral curves of $\boldsymbol{\xi}$. The requirement that these transformations be isometries is that the metric $\mathbf{g}$ has zero Lie derivative with respect to $\boldsymbol{\xi}$, i.e.

$$\mathcal{L}_{\boldsymbol{\xi}} g_{ab} = 0.$$

This is equivalent to the Killing equation

$$\xi_{a;b} + \xi_{b;a} = 0,$$

and $\boldsymbol{\xi}$ is called a *Killing vector field* ('KVF' for short). It is well known that:

(i) any linear combination (with constant coefficients) of KVFs is a KVF,
(ii) the commutator $[\boldsymbol{\xi}_1, \boldsymbol{\xi}_2]$ (if non–zero) of any two KVFs $\boldsymbol{\xi}_1, \boldsymbol{\xi}_2$ is a KVF,
(iii) a manifold admits at most a finite number of linearly independent KVFs.

The set of all KVFs thus forms a Lie algebra, with a basis $\boldsymbol{\xi}_\alpha$, $\alpha = 1, 2, \ldots, r$, where r is the dimension of the algebra. The commutator $[\boldsymbol{\xi}_\alpha, \boldsymbol{\xi}_\beta]$ can be expanded relative to the basis, giving the basic identity

$$[\boldsymbol{\xi}_\alpha, \boldsymbol{\xi}_\beta] = C^\mu{}_{\alpha\beta} \boldsymbol{\xi}_\mu, \tag{1.36}$$

where

$$C^\mu{}_{\alpha\beta} = -C^\mu{}_{\beta\alpha}. \tag{1.37}$$

The constants $C^\mu{}_{\alpha\beta}$ are called the *structure constants* of the Lie algebra. The Jacobi identities for the $\boldsymbol{\xi}_\alpha$ are equivalent to

$$C^\mu{}_{\nu[\gamma}C^\nu{}_{\alpha\beta]} = 0. \tag{1.38}$$

Under a constant change of basis

$$\boldsymbol{\xi}'_\alpha = A_\alpha{}^\beta \boldsymbol{\xi}_\beta,$$

$C^\mu{}_{\alpha\beta}$ transforms as a $(1,2)$ tensor. This freedom can be used to classify the structure constants, and hence the Lie algebras (see Section 1.4.1 for the case $r = 3$).

It is a standard result that the set of all isometries of a given manifold $(\mathcal{M}, \mathbf{g})$ forms a Lie group G_r of dimension r, called the *isometry group* of $(\mathcal{M}, \mathbf{g})$. Each one–dimensional subgroup of G_r defines a family of curves whose tangent field is a KVF. In this way the Lie group G_r generates the Lie algebra of KVFs. Conversely, the Lie algebra can be thought of as generating the Lie group, since each KVF determines a one–parameter group of isometries.†

The structure constants of the Lie algebra of KVFs determine the algebraic properties of the Lie group G_r of isometries, but they do not determine how the group acts on the manifold. The action is specified by giving the orbits of points in the manifold.

The *orbit* of a point $p \in \mathcal{M}$ under the group G_r is the set of all points into which p is mapped when all elements of G_r act on p. The KVFs $\boldsymbol{\xi}_\alpha$ at p are tangent to the orbit of p. We say that a group acts *simply transitively* on an orbit if the dimension of the orbit equals the dimension of the group G_r. In this case, the KVFs $\boldsymbol{\xi}_\alpha$ evaluated at p are linearly independent. Otherwise, the group is said to act *multiply transitively* on the orbit. In this case, the KVFs $\boldsymbol{\xi}_\alpha$, evaluated at p, are linearly dependent, and span a subspace of the tangent space at p of dimension $s < r$, say. This means that the orbit of p will have dimension s. Since the KVFs $\boldsymbol{\xi}_\alpha$ at p are linearly dependent, there will be a subspace of the Lie algebra of KVFs, of dimension $r - s$ which consists of KVFs that are zero at p. These KVFs will generate a subgroup of isometries which leave p fixed, called the *isotropy subgroup of p*. We will

---

† In order for the Lie algebra to generate a Lie group it is necessary that the transformations generated by the vector fields in the basis are complete. In practice, one usually has a Lie algebra, without knowledge of the completeness, in which case the algebra gives rise to a local group of transformations. We refer to Hall & Steele (1990) for a discussion of this point, and other references.

denote the dimension of the isotropy subgroup by $d = r - s$. The isotropy subgroup at p induces a group of linear transformations in the tangent space at p. In the case of space–time, a one–dimensional isotropy subgroup that leaves a timelike vector fixed gives a *rotational symmetry* at p.

We finally note the standard bounds on the dimension of the isometry group G_r and the isotropy subgroup in an n–dimensional space:

$$r \leq \tfrac{1}{2}n(n + 1),$$
$$d \leq \tfrac{1}{2}n(n - 1).$$

These bounds follow from the fact that a KVF is uniquely determined by the $n + \tfrac{1}{2}n(n - 1)$ quantities ξ^a and $\xi_{a;b}$ at a single point p. Isotropies at p are determined by KVFs that satisfy $\xi^a = 0$ at p.

1.2.2 Isometry–based classification of models

In this section we classify cosmological models $(\mathcal{M}, \mathbf{g}, \mathbf{u})$ based on the dimension s of the orbits of the isometry group G_r, and on the dimension d of the isotropy subgroup (see Ellis 1967, page 1191). The value of d determines the *isotropy properties* of the model, and the value of s determines the *homogeneity properties*. The classification itself is independent of the EFE; the specific examples mentioned are models with a perfect fluid and possibly a cosmological constant.

As regards isotropy, there are three main possibilities†, $d = 3, 1$ or 0.

(1) *Isotropic* ($d = 3$). The isometry group is at least a G_6, giving the FL family of models (see Chapter 2). The Weyl tensor is zero, as are the kinematic quantities of the preferred congruence, apart from the Hubble scalar H.

(2) *Locally rotationally symmetric*, LRS ($d = 1$). The isometry group is at least a G_3. These models have been classified using orthonormal frames and local coordinates (see Ellis 1967 for dust and Stewart & Ellis 1968 for non–zero pressure). A covariant classification is given in van Elst & Ellis (1996). At each point the kinematic quantities are rotationally symmetric about a preferred spacelike axis, and the Weyl tensor is of Petrov type D or O.

(3) *Anisotropic* ($d = 0$). There are no continuous isotropies or rotational symmetries (although there can still be discrete isotropies; see MacCallum & Ellis 1970).

---

† The value $d = 6$ is also possible and corresponds to a space–time of constant curvature, called de Sitter space–time (e.g. Kramer *et al.* 1980, pages 102–3).

As regards homogeneity, there are five possibilities, specified by the value of s, $0 \le s \le 4$.

(a) *Homogeneous space-time* ($s = 4$). The field equations reduce to purely algebraic equations, and all scalar invariants are constant. In particular, the density μ is constant, which implies that the expansion Θ is zero, assuming $\mu + p > 0$. Their only relevance in cosmology is as a non–expanding asymptotic state of an expanding model. The isotropy subcases are

$d = 3$: the Einstein static model (G_7) (see Section 2.1)
$d = 1$: the Gödel model (G_5) (Kramer *et al.* 1980, page 122)
$d = 0$: the Oszvath–Kerr models (G_4) (Kramer *et al.* 1980, pages 122–4)

Note: The above remarks apply when $\mu \ne 0$. The vacuum plane wave solutions are homogeneous space–times, admitting a G_6. They also admit a family of spacelike hypersurfaces whose normal congruence is expanding, and hence can be interpreted as Bianchi cosmologies with $\Theta \ne 0$ (Siklos 1981, 1984; see also Section 9.1.4). With this interpretation they play an important role as asymptotic states of non–vacuum Bianchi cosmologies (see Chapters 7 and 12).

(b) *Homogeneous on three-dimensional orbits* ($s = 3$). The case of interest in cosmology is *spacelike orbits*, in which case the models are called *spatially homogeneous*. The field equations reduce to ordinary differential equations. The isotropy subcases are

$d = 3$: the FL models (G_6) (see Section 2.1)
$d = 1$: spatially homogeneous LRS models (G_4)
$d = 0$: the Bianchi models (G_3) (see Section 1.5)

Note: There are two possibilities when $s = 3$, $d = 1$. If the G_4 has a subgroup G_3 which acts simply transitively on the three–dimensional orbits we obtain the *LRS Bianchi models*. Otherwise, we obtain the *Kantowski–Sachs models* (see Section 8.5.3).

(c) *Homogeneous on two-dimensional orbits* ($s = 2$). The case of interest in cosmology is *spacelike orbits*. The models evolve in time and are *spatially inhomogeneous* with one spatial degree of freedom. The field equations reduce to a system of partial differential equations with one time and one space variable. The isotropy subcases are

$d = 1$: the inhomogenous LRS models (G_3)
$d = 0$: the most important case is an Abelian G_2, discussed in Section 1.6 and Chapter 12.

Note: There are three possibilities when $s = 2$, $d = 1$. The G_3 acts multiply transitively on 2–spaces of constant curvature K, with

$\quad K > 0$ (spherical symmetry),

$\quad K = 0$ (plane symmetry),

$\quad K < 0$ (hyperbolic symmetry).

The well–known Lemaître–Tolman solutions (e.g. Krasiński 1996, section 16) occur when $K > 0$ and the pressure is zero.

(d) *Homogeneous on one–dimensional orbits* ($s = 1$). The case of interest in cosmology is *spacelike orbits*. The models evolve in time and are spatially inhomogeneous, with two spatial degrees of freedom. The isotropy subgroup is trivial at each point ($d = 0$).

(e) *Fully inhomogeneous* ($s = 0$). The models evolve in time and are spatially inhomogeneous, with three spatial degrees of freedom. The silent universes, analyzed in Chapter 13, belong to this class in general. Several families of exact solutions are known in this class, for example

$\quad$ the Szekeres solutions (Krasiński 1996, Part II),

$\quad$ the Stephani–Barnes solutions (Krasiński 1996, Part IV),

$\quad$ the Oleson (1971) solutions.

Although these solutions are general in that they admit no KVFs, they are very special as regards the Weyl tensor and the kinematic quantities. The second class is conformally flat ($C_{abcd} = 0$), in the first class the magnetic part of the Weyl tensor is zero ($H_{ab} = 0$) and the Weyl tensor is of Petrov type D, and in the third class the Weyl tensor is of Petrov type N. In each class the fluid vorticity is zero ($\omega = 0$) and in the first class the fluid acceleration is zero.

1.2.3 Homothetic vectors and similarity groups

A *similarity* (or homothetic transformation) of a manifold ($\mathcal{M}, \mathbf{g}$) is a mapping of $\mathcal{M}$ into itself which induces a *constant* scale transformation of the metric $\mathbf{g}$. An isometry is a special case of a similarity. Thus, while isometries preserve lengths and angles, similarities preserve angles but change lengths. If $\boldsymbol{\xi}$ is the generating vector field of a one–parameter group of similarities, this transformation property is equivalent to the equation

$$\mathcal{L}_{\boldsymbol{\xi}} g_{ab} = 2\kappa g_{ab}, \qquad (1.39)$$

where κ is a constant. The vector field $\boldsymbol{\xi}$ is called a *homothetic vector field* (HVF). If $\kappa \neq 0$, $\boldsymbol{\xi}$ is called a *proper* HVF, and if $\kappa = 0$, $\boldsymbol{\xi}$ is a KVF. We shall refer to a space–time $(\mathcal{M}, \mathbf{g})$ whose metric admits a proper HVF as a *self–similar space–time*.†

As with isometries, the set of all similarities of $(\mathcal{M}, \mathbf{g})$ forms a finite dimensional Lie group H_r, called the *similarity group* of $\mathcal{M}$, and the set of all homothetic vector fields forms a Lie algebra $\mathcal{H}_r$. An important result is that the group H_r admits a subgroup G_{r-1} of isometries (Eardley 1974). Equivalently, the Lie algebra $\mathcal{H}_r$ (assumed to admit at least one proper HVF, i.e. $\kappa \neq 0$) admits a basis consisting of one proper HVF and $r - 1$ KVFs.†

One can extend the isometry–based classification of cosmological models given in Section 1.2.2 to include similarities. Suppose that space–time admits an isometry group G_r, such that the dimension s of the orbits, assumed spacelike, satisfies $0 < s \leq 3$ (i.e. cases (b)–(d) of Section 1.2.2). One can, in addition, assume that the space–time admits a proper HVF, leading to the existence of a similarity group H_{r+1} with orbits of dimension $s + 1$. We shall refer to the cosmological model as *self–similar*. The simplest case is when $s = 3$, leading to a similarity group whose orbits are four–dimensional; in this case we shall refer to the cosmological model as *transitively self–similar*. This means that the geometry at two different points of space–time differs by at most a change in the overall length scale. In addition, although the model evolves in time, the evolution is fully specified by this symmetry property, and the EFE are purely algebraic when expressed in terms of expansion–normalized variables (see Section 5.2.1). This type of solution plays an important role in the analysis of the Bianchi models in Chapters 6 and 7. The other case that occurs in this book is $s = 2$, leading to a similarity group whose orbits are three–dimensional timelike hypersurfaces. The resulting models are spatially inhomogeneous and evolve in time, yet the EFE reduce to a system of ordinary different equations. These models arise in connection with the analysis of G_2 cosmologies in Chapter 12.

1.2.4 Inheritance of symmetry

Suppose that $\boldsymbol{\xi}$ is a HVF. Since $R^a{}_{bcd}$ is unchanged by the constant scaling $g_{ab} \to \lambda g_{ab}$, we have

$$\mathcal{L}_{\boldsymbol{\xi}} R^a{}_{bcd} = 0.$$

† This terminology was introduced by Eardley (1974, page 289), generalizing the earlier work of Cahill & Taub (1971) on spherically symmetric space–times. Carter & Henriksen (1989) have introduced the more general concept of 'kinematic self–similarity' in the context of relativistic fluid mechanics.

† We note that the orbits of G_{r-1} and H_r can only coincide if they are four–dimensional or three–dimensional and null (Hall & Steele 1990).

This implies that the Ricci tensor and Ricci scalar satisfy

$$\mathcal{L}_{\xi} R_{ab} = 0, \quad \mathcal{L}_{\xi} R = -2\kappa R,$$

and hence that the Einstein tensor satisfies

$$\mathcal{L}_{\xi} G_{ab} = 0.$$

It now follows from the EFE (1.1) that

$$\mathcal{L}_{\xi} T_{ab} = 0. \tag{1.40}$$

We say that the stress–energy tensor *inherits the symmetry* of the space–time†.

Equation (1.40) in turn restricts the physical quantities that describe the source. For a perfect fluid (1.20), it follows from (1.39) and (1.40) that

$$\mathcal{L}_{\xi} \mu = -2\kappa\mu, \quad \mathcal{L}_{\xi} p = -2\kappa p, \quad \mathcal{L}_{\xi} u^a = -\kappa u^a. \tag{1.41}$$

If follows immediately from the first two of these equations that if there is an equation of state of the form $p = p(\mu)$ and ξ is a proper HVF ($\kappa \neq 0$) then the equation of state must be of the form

$$p = (\gamma - 1)\mu, \quad \gamma = \text{constant}.$$

A detailed analysis of the contracted Bianchi identities (Wainwright 1985) shows that in addition $\gamma \neq 0$, and that if ξ is orthogonal to the fluid velocity **u** then $\gamma = 2$, and if ξ is parallel to **u**, then $\gamma = \frac{2}{3}$.

For a pure magnetic field (1.23), it follows that

$$\mathcal{L}_{\xi} H^a = -2\kappa H^a.$$

For a scalar field (1.24), the situation is complicated by the presence of the arbitrary potential $V(\phi)$. If ξ is a KVF and $V'(\phi) \neq 0$, then it follows from (1.39) and (1.40) that

$$\mathcal{L}_{\xi} \phi = 0.$$

If ξ is a proper HVF, then it follows that the potential is required to have an exponential form

$$V = V_0 e^{\alpha\phi},$$

---

† The form of the symmetry restriction satisfied by the stress–energy tensor depends on the position of the indices. For example, (1.39) and (1.40) imply that
$$\mathcal{L}_{\xi} T_a{}^b = -2\kappa T_a{}^b$$

where α and V_0 are non–zero constants, and that

$$\mathcal{L}_{\boldsymbol{\xi}}\phi = -\frac{2\kappa}{\alpha}.$$

The overall result is that for each source considered in Section 1.1.2, the symmetry of the space–time is inherited by the source variables in an appropriate sense[†], but that if $\boldsymbol{\xi}$ is a proper HVF, then the perfect fluid equation of state and the scalar potential are restricted. It should be noted, however, that if the stress–energy tensor is the sum of more than one component, *it does not follow in general that the separate components inherit the symmetry.* When we assume that a cosmological model $(\mathcal{M}, \mathbf{g}, \mathbf{u})$ admits an isometry or similarity group, we shall mean that $\mathbf{g}, \mathbf{u}$ and *each* matter component is invariant under the group.

1.3 Evolution and constraint equations

As shown by Trümper (see Hawking 1966, Ellis 1971 and Ellis 1973), in a space–time with a preferred timelike congruence $\mathbf{u}$, the information contained in the EFE can be displayed in an indirect manner by expressing the Ricci identities as applied to $\mathbf{u}$, namely[‡]

$$\nabla_c \nabla_d u_a - \nabla_d \nabla_c u_a = R_a{}^b{}_{cd} u_b,$$

and the Bianchi identities (1.33), or equivalently

$$\nabla_d C_{ab}{}^{cd} = -\nabla_{[a} R^c{}_{b]} - \tfrac{1}{6}\delta^c{}_{[a}\nabla_{b]}R,$$

in terms of the kinematic quantities (1.27) and the electric and magnetic parts of the Weyl tensor, as given by (1.34). One essentially projects these identities along directions parallel and orthogonal to $\mathbf{u}$. In both cases the Ricci tensor is expressed in terms of the stress energy tensor (1.19) using the EFE (1.1) i.e.

$$R_{ab} = T_{ab} - \tfrac{1}{2}T_c{}^c g_{ab}.$$

The result is a set of *evolution equations* giving the derivatives of $H = \tfrac{1}{3}\Theta$, σ_{ab}, ω_a, $E_{ab} + \tfrac{1}{2}\pi_{ab}$, H_{ab}, μ and q_a along the congruence $\mathbf{u}$, and a set of

---

[†] If the source variables are not determined uniquely by the eigenvectors and eigenvalues of the stress–energy tensor, for example an arbitrary electromagnetic field (Michalski & Wainwright 1975), an imperfect fluid (Coley & Tupper 1989), or a kinetic theory description of matter based on the Einstein–Liouville equations (Ellis *et al.* 1983), the source variables do not, in general, inherit the symmetry.

[‡] In this Section, we use the notation ∇_a for covariant derivatives.

constraint equations that contain only spatial derivatives (e.g. $h^a{}_b\nabla_a\omega^b$). A derivative along the congruence **u** is denoted by an overdot, e.g.,

$$\dot\mu = u^a\nabla_a\mu, \quad \dot\sigma_{ab} = u^c\nabla_c\sigma_{ab},$$

and, if applied to a tensor, involves a covariant derivative. We note that these evolution and constraint equations provide the basis for the description of the so–called silent universes given in Chapter 13.

1.3.1 Ricci identities

The results of decomposing the Ricci identities are evolution equations for H, σ_{ab} and ω_a, and three constraint equations:

$$\dot H = -H^2 + \tfrac{1}{3}(h^a{}_b\nabla_a\dot u^b + \dot u_a\dot u^a - 2\sigma^2 + 2\omega^2) - \tfrac{1}{6}(\mu+3p), \quad (1.42)$$

$$
\begin{aligned}
h^a{}_c h^b{}_d\dot\sigma^{cd} = {}& -2H\sigma^{ab} + h^{(a}{}_c h^{b)}{}_d\nabla^c\dot u^d + \dot u^a\dot u^b - \sigma^a{}_c\sigma^{bc} - \omega^a\omega^b \\
& -\tfrac{1}{3}(h^c{}_d\nabla_c\dot u^d + \dot u_c\dot u^c - 2\sigma^2 - \omega^2)h^{ab} - (E^{ab}-\tfrac{1}{2}\pi^{ab}), \quad (1.43)
\end{aligned}
$$

$$h^a{}_b\dot\omega^b = -2H\omega^a + \sigma^a{}_b\omega^b - \tfrac{1}{2}\eta^{abcd}(\nabla_b\dot u_c)u_d, \quad (1.44)$$

$$0 = h^a{}_c h^c{}_d\nabla_b\sigma^{cd} - 2h^a{}_b\nabla^b H - \eta^{abcd}(\nabla_b\omega_c + 2\dot u_b\omega_c)u_d + q^a, \quad (1.45)$$

$$0 = h^a{}_b\nabla_a\omega^b - \dot u_a\omega^a, \quad (1.46)$$

$$
\begin{aligned}
0 = {}& H_{ab} - 2\dot u_{(a}\omega_{b)} - h^c{}_{(a}h^d{}_{b)}\nabla_c\omega_d + \tfrac{1}{3}(2\dot u_c\omega^c + h^c{}_d\nabla_c\omega^d)h_{ab} \\
& - h^c{}_{(a}h^d{}_{b)}\eta_{cefg}(\nabla^e\sigma^f{}_d)u^g. \quad (1.47)
\end{aligned}
$$

These equations correspond to equations (61a), (68), (63), (69), (70) and (71) in Ellis (1973), respectively. We refer to that article for an outline of the derivation (see pages 24–30). The above equation for $\dot H$ is the well-known *Raychaudhuri equation*.

1.3.2 Bianchi identities

Firstly, the contracted Bianchi identities (1.2) with (1.19), lead to evolution equations for μ and q_a:

$$\dot\mu = -3H(\mu+p) - h^a{}_b\nabla_a q^b - 2\dot u_a q^a - \sigma^a{}_b\pi^b{}_a, \quad (1.48)$$

$$
\begin{aligned}
h^a{}_b\dot q^b = {}& -4Hq^a - h^a{}_b\nabla^b p - (\mu+p)\dot u^a - h^a{}_c h^b{}_d\nabla_b\pi^{cd} \\
& - \dot u_b\pi^{ab} - \sigma^a{}_b q^b + \eta^{abcd}\omega_b q_c u_d. \quad (1.49)
\end{aligned}
$$

These are equations (37) and (38) in Ellis (1973). For a perfect fluid ($q_a = 0$, $\pi_{ab} = 0$) they specialize to the well–known form

$$\dot{\mu} + 3H(\mu + p) = 0, \quad (\mu + p)\dot{u}_a + h_a{}^b \nabla_b p = 0. \tag{1.50}$$

The full Bianchi identities lead to evolution equations for $E_{ab} + \frac{1}{2}\pi_{ab}$ and H_{ab}, together with two constraints. For simplicity, we give these equations subject to the restriction that the source (1.19) is a perfect fluid:

$$
\begin{aligned}
h^a{}_c h^b{}_d \dot{E}^{cd} &= -3HE^{ab} + 3\sigma^{(a}{}_c E^{b)c} - \sigma^c{}_d E^d{}_c h^{ab} - \tfrac{1}{2}(\mu + p)\sigma^{ab} \\
&\quad + h^{(a}{}_c h^{b)}{}_d \eta^{cefg}(\nabla_e H^d{}_f + 2\dot{u}_e H^d{}_f - \omega_e E^d{}_f)u_g, \tag{1.51}
\end{aligned}
$$

$$
\begin{aligned}
h^a{}_c h^b{}_d \dot{H}^{cd} &= -3HH^{ab} + 3\sigma^{(a}{}_c H^{b)c} - \sigma^c{}_d H^d{}_c h^{ab} \\
&\quad - h^{(a}{}_c h^{b)}{}_d \eta^{cefg}(\nabla_e E^d{}_f + 2\dot{u}_e E^d{}_f + \omega_e H^d{}_f)u_g, \tag{1.52}
\end{aligned}
$$

$$0 = h^a{}_c h^b{}_d \nabla_b E^{cd} + 3\omega_b H^{ab} - \eta^{abcd}\sigma_{be} H^e{}_c u_d - \tfrac{1}{3} h^a{}_b \nabla^b \mu, \tag{1.53}$$

$$0 = h^a{}_c h^b{}_d \nabla_b H^{cd} - 3\omega_b E^{ab} + \eta^{abcd}\sigma_{be} E^e{}_c u_d - (\mu + p)\omega^a. \tag{1.54}$$

These equations correspond to equations (74b, d, a, c), respectively, in Ellis (1973). They are closely analogous to Maxwell's equations written in a 1+3 form for a general timelike reference congruence (Ellis 1973).

1.3.3 Spatial curvature

If the congruence **u** is irrotational, one can obtain expressions for the curvature of the 3–spaces orthogonal to the congruence. One obtains (Ellis 1973, page 34)

$$^3R_{abcd} = h_a{}^p h_b{}^q h_c{}^r h_d{}^s R_{pqrs} - \Theta_{ac}\Theta_{bd} + \Theta_{ad}\Theta_{bc}$$

where Θ_{ab} is the rate of expansion tensor (1.30). The trace–free spatial Ricci tensor is defined by

$$^3S_{ab} = {}^3R_{ab} - \tfrac{1}{3}{}^3R h_{ab}, \tag{1.55}$$

where

$$^3R_{ab} = h^{pq}\, {}^3R_{apbq}, \quad ^3R = h^{pq}\, {}^3R_{pq}.$$

By using (1.1), (1.19), (1.32) and (1.34), one obtains (when **u** is irrotational)

$$^3S_{ab} = E_{ab} + \tfrac{1}{2}\pi_{ab} - H\sigma_{ab} + \sigma_a{}^c \sigma_{cb} - \tfrac{2}{3}\sigma^2 h_{ab}, \tag{1.56}$$

and

$$^3R = -6H^2 + 2\sigma^2 + 2\mu. \tag{1.57}$$

When **u** is irrotational, (1.56) gives a convenient formula for calculating the electric part of the Weyl tensor E_{ab}.

1.3.4 Singularity types

In any cosmological model $(\mathcal{M}, g, u)$, the rate of expansion scalar Θ of the timelike congruence **u**, or equivalently the Hubble scalar $H = \frac{1}{3}\Theta$, can be used to define a scale factor ℓ, according to

$$H = \frac{\dot{\ell}}{\ell},$$

where $\dot{\ell} = \ell_{,a} u^a$. At a big–bang singularity the scale factor ℓ tends to zero, corresponding to a curvature singularity since in general the energy density, a Ricci invariant, will diverge as $\ell \to 0$. (If the source is a perfect fluid with $p = (\gamma - 1)\mu$, this follows from the contracted Bianchi identities (1.50) provided that $\gamma > 0$.)

One can use the rate of expansion tensor Θ_{ab}, as given by (1.30), to distinguish different types of big–bang singularity. We introduce an orthonormal frame $\{e_a\}$ with $e_0 = u$, and such that the spatial vectors $\{e_\alpha\}$, $\alpha = 1, 2, 3$, are eigenvectors of Θ_{ab}. Then

$$\Theta_{ab} = \text{diag}\,(0, \Theta_1, \Theta_2, \Theta_3),$$

where $H = \frac{1}{3}(\Theta_1 + \Theta_2 + \Theta_3)$. We now define scale factors ℓ_α in the expansion eigendirections according to

$$\Theta_\alpha = \frac{\dot{\ell}_\alpha}{\ell_\alpha}. \tag{1.58}$$

The different singularity types are distinguished by the behaviour of the length scales ℓ_α as $\ell \to 0$:

(i) *point* means that all three ℓ_α tend to 0,

(ii) *cigar* means that two of the ℓ_α tend to 0 and the third increases without bound,

(iii) *barrel* means that two of the ℓ_α tend to 0 and the third approaches a finite value,

(iv) *pancake* means that one of the ℓ_α tends to 0 and two approach finite values.

These names refer to the change of the shape of a spherical element of the cosmological fluid as the singularity is approached. They were introduced by Thorne (1967) in connection with Bianchi I models, but can be applied generally. It is an open question whether this is a complete set of possibilities for the case in which the limits of the ℓ_α either exist or are infinite.

1.4 Orthonormal frame formalism

The orthonormal frame formalism is essentially a $1 + 3$ decomposition of the EFE into evolution and constraint equations, relative to the timelike vector field $\mathbf{e}_0$ of an orthonormal frame $\{\mathbf{e}_a\}$. In a cosmological context $\mathbf{e}_0$ is usually chosen to be the fundamental 4–velocity $\mathbf{u}$ or, in models which admit an isometry group, to be normal to the spacelike group orbits. The rate of expansion scalar Θ of this congruence plays a significant role in the evolution equations. Because we are working in a cosmological context, we usually replace Θ by the Hubble scalar $H = \frac{1}{3}\Theta$ (see (1.31)). There is a significant difference between the evolution and constraint equations in Section 1.3 and those given in this section; in Section 1.3 the equations are tensorial, and involve covariant derivatives, while in this section the variables are scalars (for example, frame components of vectors and tensors), and hence covariant derivatives are not required.

The orthonormal frame formalism† was pioneered in the cosmological context by Ellis (1967), Ellis & MacCallum (1969) and MacCallum (1973). The formalism has recently been revisited and extended by van Elst & Uggla‡ (1996).

1.4.1 Commutation functions as variables

In the orthonormal frame formalism, the metric components satisfy

$$g_{ab} = \eta_{ab},$$

and the basic variables are the commutation functions $\gamma^c{}_{ab}$, as defined by (1.11). Equation (1.14) for the components of the connection simplifies to

$$\Gamma_{abc} = \tfrac{1}{2}(\gamma^d{}_{cb}\eta_{ad} + \gamma^d{}_{ac}\eta_{bd} - \gamma^d{}_{ba}\eta_{cd}). \tag{1.59}$$

The spatial frame vectors are denoted by $\{\mathbf{e}_\alpha\}$ with Greek indices assuming the values 1 to 3. Here and in what follows $\varepsilon_{\alpha\beta\gamma}$ denotes the alternating symbol ($\varepsilon_{123} = +1$), and Greek indices are raised and lowered with the spatial metric tensor $g_{\alpha\beta} = \delta_{\alpha\beta}$. We begin by decomposing the commutation functions into algebraically simple quantities, some of which have a direct

---

† Null tetrads were already in extensive use in the study of vacuum space–times and gravitational radiation (e.g. Newman & Penrose 1962). The orthonormal frame equations are analogous to the Newman–Penrose equations.

‡ We thank the authors for making their preprint available to us.

physical or geometrical interpretation. Firstly, the commutation functions $\gamma^c{}_{ab}$ with one index zero can be expressed in terms of the kinematic quantities (1.27) of the timelike congruence $\mathbf{u} = \mathbf{e}_0$ and the quantity§

$$\Omega^\alpha = \tfrac{1}{2}\varepsilon^{\alpha\mu\nu}e^i_\mu e_{\nu i;j}u^j, \tag{1.60}$$

which is the local angular velocity of the spatial frame $\{\mathbf{e}_\alpha\}$ with respect to a Fermi–propagated spatial frame. It follows from (1.11), (1.26), (1.27), (1.29), (1.31) and (1.60) that

$$\gamma^\alpha{}_{0\beta} = -\sigma_\beta{}^\alpha - H\delta_\beta{}^\alpha - \varepsilon^\alpha{}_{\beta\mu}(\omega^\mu + \Omega^\mu)$$

$$\gamma^0{}_{0\alpha} = \dot{u}_\alpha, \quad \gamma^0{}_{\alpha\beta} = -2\varepsilon_{\alpha\beta}{}^\mu\omega_\mu. \tag{1.61}$$

Here, $\sigma_{\alpha\beta}$, ω_α and $\dot{u}_\alpha$ are the non–zero orthonormal frame components of σ_{ab}, ω_a and $\dot{u}_a$. Secondly, the spatial components $\gamma^\mu{}_{\alpha\beta}$ are decomposed into a 2–index symmetric object $n_{\alpha\beta}$ and a 1–index object a_α as follows:

$$\gamma^\mu{}_{\alpha\beta} = \varepsilon_{\alpha\beta\nu}n^{\mu\nu} + a_\alpha\delta_\beta{}^\mu - a_\beta\delta_\alpha{}^\mu. \tag{1.62}$$

Finally, the choice $\mathbf{u} = \mathbf{e}_0$ in (1.19) means that the frame components of the source terms satisfy $q_0 = 0$, $\pi_{00} = 0$ and $\pi_{0\alpha} = 0$.

1.4.2 Evolution and constraint equations

The EFE (1.1), the Jacobi identities (1.18) and the contracted Bianchi identities (1.2) can now be written in terms of the basic variables

$$(H, \sigma_{\alpha\beta}, \dot{u}_\alpha, \omega_\alpha, \Omega_\alpha, n_{\alpha\beta}, a_\alpha) \tag{1.63}$$

and the source terms

$$(\mu, p, q_\alpha, \pi_{\alpha\beta}), \tag{1.64}$$

using (1.16), (1.17), (1.19), (1.42), (1.59), (1.61) and (1.62). We write the EFE in the form

$$R_{ab} = T_{ab} - \tfrac{1}{2}T_c{}^c g_{ab},$$

and calculate the 00–equation, the trace–free $\alpha\beta$–equation, the combination (00)+(11)+(22)+(33), and the 0α–equation to obtain (1.65)–(1.68) below, respectively.† In these equations ∂_a denotes the vector field $\mathbf{e}_a$ acting on a

---

§ In the literature there are two conventions for defining Ω^α, due to Ellis & MacCallum (1969) and MacCallum (1973), which differ only in sign. We are using the MacCallum 1973 convention.

† The resulting equations are essentially those given by MacCallum (1973, page 105); we have replaced θ, $\theta_{\alpha\beta}$ and $^3R_{\alpha\beta}$ by $H = \tfrac{1}{3}\theta$, $\sigma_{\alpha\beta} = \theta_{\alpha\beta} - \tfrac{1}{3}\theta\delta_{\alpha\beta}$ and $^3S_{\alpha\beta} = {}^3R_{\alpha\beta} - \tfrac{1}{3}{}^3R\delta_{\alpha\beta}$, where $^3R = {}^3R_\alpha{}^\alpha$.

scalar as a differential operator, i.e.

$$\eth_a(f) = e_a{}^i \frac{\partial f}{\partial x^i}.$$

Einstein field equations:

$$\eth_0 H = -H^2 - \tfrac{2}{3}\sigma^2 + \tfrac{2}{3}\omega^2 + \tfrac{1}{3}(\eth_\alpha + \dot{u}_\alpha - 2a_\alpha)\dot{u}^\alpha - \tfrac{1}{6}(\mu + 3p), \qquad (1.65)$$

$$
\begin{aligned}
\eth_0 \sigma_{\alpha\beta} = & -3H\sigma_{\alpha\beta} + 2\varepsilon^{\mu\nu}{}_{(\alpha}\sigma_{\beta)\mu}\Omega_\nu - 2\omega_{(\alpha}\Omega_{\beta)} + \tfrac{2}{3}\omega^\mu\Omega_\mu\delta_{\alpha\beta} \\
& + [\eth_{(\alpha} + \dot{u}_{(\alpha} + a_{(\alpha}]\dot{u}_{\beta)} - \tfrac{1}{3}(\eth^\mu + \dot{u}^\mu + a^\mu)\dot{u}_\mu\delta_{\alpha\beta} \\
& - \varepsilon^{\mu\nu}{}_{(\alpha}n_{\beta)\mu}\dot{u}_\nu - {}^3S_{\alpha\beta} + \pi_{\alpha\beta}, \qquad (1.66)
\end{aligned}
$$

$$\mu = 3H^2 - \sigma^2 + \omega^2 - 2\omega_\alpha\Omega^\alpha + \tfrac{1}{2}{}^3R, \qquad (1.67)$$

$$
\begin{aligned}
q_\alpha = & \; 2\eth_\alpha H - (\eth_\beta - 3a_\beta)\sigma_\alpha{}^\beta - \varepsilon_\alpha{}^{\mu\nu}\sigma_\mu{}^\beta n_{\beta\nu} \\
& + \varepsilon_\alpha{}^{\mu\nu}(\eth_\mu + 2\dot{u}_\mu - a_\mu)\omega_\nu - n_\alpha{}^\beta\omega_\beta, \qquad (1.68)
\end{aligned}
$$

where

$$
{}^3S_{\alpha\beta} = \eth_{(\alpha}a_{\beta)} - \tfrac{1}{3}\eth_\mu a^\mu\delta_{\alpha\beta} - (\eth_\mu - 2a_\mu)n_{\nu(\alpha}\varepsilon_{\beta)}{}^{\mu\nu} + b_{\alpha\beta} - \tfrac{1}{3}b_\mu{}^\mu\delta_{\alpha\beta}, \quad (1.69)
$$

$$
{}^3R = 4\eth_\mu a^\mu - 6a_\mu a^\mu - \tfrac{1}{2}b_\mu{}^\mu, \qquad (1.70)
$$

and

$$
b_{\alpha\beta} = 2n_\alpha{}^\mu n_{\mu\beta} - (n_\mu{}^\mu)n_{\alpha\beta}.
$$

Jacobi identities:

$$
\begin{aligned}
\eth_0 n_{\alpha\beta} = & \; -Hn_{\alpha\beta} + 2\sigma_{(\alpha}{}^\mu n_{\beta)\mu} + 2\varepsilon^{\mu\nu}{}_{(\alpha}n_{\beta)\mu}(\omega_\nu + \Omega_\nu) \\
& - [\eth_{(\alpha} + \dot{u}_{(\alpha}][\omega_{\beta)} + \Omega_{\beta)}] - (\eth_\mu + \dot{u}_\mu)\sigma_{\nu(\alpha}\varepsilon_{\beta)}{}^{\mu\nu} \\
& + (\eth_\mu + \dot{u}_\mu)(\omega^\mu + \Omega^\mu)\delta_{\alpha\beta}, \qquad (1.71)
\end{aligned}
$$

$$
\begin{aligned}
\eth_0 a_\alpha = & \; -Ha_\alpha - \sigma_\alpha{}^\beta a_\beta - (\eth_\alpha + \dot{u}_\alpha)H + \tfrac{1}{2}(\eth_\beta + \dot{u}_\beta)\sigma_\alpha{}^\beta \\
& - \tfrac{1}{2}\varepsilon_\alpha{}^{\mu\nu}(\eth_\mu + \dot{u}_\mu - 2a_\mu)(\omega_\nu + \Omega_\nu), \qquad (1.72)
\end{aligned}
$$

$$
\begin{aligned}
\eth_0 \omega_\alpha = & \; -2H\omega_\alpha + \sigma_\alpha{}^\beta\omega_\beta + \varepsilon_\alpha{}^{\mu\nu}\omega_\mu\Omega_\nu \\
& - \tfrac{1}{2}\varepsilon_\alpha{}^{\mu\nu}(\eth_\mu - a_\mu)\dot{u}_\nu + \tfrac{1}{2}n_\alpha{}^\beta\dot{u}_\beta, \qquad (1.73)
\end{aligned}
$$

$$
0 = (\eth_\beta - 2a_\beta)n_\alpha{}^\beta + \varepsilon_\alpha{}^{\mu\nu}\eth_\mu a_\nu - 2H\omega_\alpha - 2\sigma_\alpha{}^\beta\omega_\beta - 2\varepsilon_\alpha{}^{\mu\nu}\omega_\mu\Omega_\nu, \qquad (1.74)
$$

$$
0 = (\eth_\alpha - \dot{u}_\alpha - 2a_\alpha)\omega^\alpha. \qquad (1.75)
$$

Contracted Bianchi identities:

$$\partial_0 \mu = -3H(\mu + p) - \sigma_\alpha{}^\beta \pi_\beta{}^\alpha - (\partial_\alpha + 2\dot{u}_\alpha - 2a_\alpha)q^\alpha, \qquad (1.76)$$

$$\begin{aligned}\partial_0 q_\alpha =\ & -4Hq_\alpha - \sigma_\alpha{}^\beta q_\beta + \varepsilon_\alpha{}^{\mu\nu}(\omega_\mu - \Omega_\mu)q_\nu - \partial_\alpha p - (\mu + p)\dot{u}_\alpha \\ & -(\partial_\beta + \dot{u}_\beta - 3a_\beta)\pi_\alpha{}^\beta + \varepsilon_\alpha{}^{\mu\nu} n_\mu{}^\beta \pi_{\beta\nu}. \qquad (1.77)\end{aligned}$$

The notation ${}^3 S_{\alpha\beta}$ and ${}^3 R$ used in writing the field equations requires an explanation. In this section these quantities simply represent the parts of the field equations that only involve the quantities $n_{\alpha\beta}$, a_α and their spatial derivatives. If the timelike congruence $\mathbf{e}_0$ is irrotational ($\omega_\alpha = 0$), however, ${}^3 R$ represents the *scalar curvature* and ${}^3 S_{\alpha\beta}$ the *tracefree Ricci–tensor* of the 3–metric induced on the spacelike hypersurfaces orthogonal to $\mathbf{e}_0$ (see (1.55)).

By inspecting the preceding equations, one can see that the field equations and Jacobi identities provide evolution equations for all of the basic variables (1.63) except for $\dot{u}_\alpha$ and Ω_α. The whole system thus appears to be undetermined. Two facts should be noted. First, the timelike congruence $\mathbf{u} = \mathbf{e}_0$ has not been specified uniquely, which is related to the fact that the source has not been specified, and this is the source of arbitrariness in $\dot{u}_\alpha$. One could simply choose $\mathbf{u}$ to be tangent to a family of timelike geodesics in which case $\dot{u}_\alpha = 0$, although it is not necessarily desirable to make this choice. If the source is a perfect fluid with $p = (\gamma - 1)\mu$, and we choose $\mathbf{e}_0$ to be the fluid 4–velocity, then the commutator (1.78), when applied to μ in conjunction with (1.76) and (1.77), yields an evolution equation for $\dot{u}_\alpha$ (see Section 12.1.1). Second, once $\mathbf{u}$ is chosen, there is still freedom in the choice of spatial frame, namely a position–dependent rotation which involves three arbitrary functions. This freedom could be used to set $\Omega_\alpha = 0$, although it is not always convenient to do so.

In addition, we stress that the differential operators ∂_0 and ∂_α are not simply partial derivative operators, and indeed may 'evolve' with time. If one regards ∂_0 as given (one can, without loss of generality, write $\partial_0 = \partial/\partial t$), then the commutator $[\partial_0, \partial_\alpha]$ can be regarded as formally determining the time evolution of the spatial operators ∂_α. In addition, through equation (1.11), the spatial commutators $[\partial_\alpha, \partial_\beta]$ provide compatibility conditions for the spatial operators. If the space–time admits an isometry group, however, and one can choose an orthonormal frame that is invariant under the group, then one or more of the spatial operators will drop out of the equations (see Sections 1.5.3 and 1.6.2). In this case, the corresponding commutator(s) will play no role in determining the evolution of the basic variables (1.63).

With this understanding, we include the commutators $[\boldsymbol{\partial}_0, \boldsymbol{\partial}_\alpha]$ and $[\boldsymbol{\partial}_\alpha, \boldsymbol{\partial}_\beta]$, as given by (1.11), (1.61) and (1.62), in the full set of evolution equations and constraints.

Commutators:

$$[\boldsymbol{\partial}_0, \boldsymbol{\partial}_\alpha] = \dot{u}_\alpha \boldsymbol{\partial}_0 - \left[\sigma_\alpha{}^\beta + H\delta_\alpha{}^\beta + \varepsilon^\beta{}_{\alpha\mu}(\omega^\mu + \Omega^\mu)\right]\boldsymbol{\partial}_\beta, \qquad (1.78)$$

$$[\boldsymbol{\partial}_\alpha, \boldsymbol{\partial}_\beta] = -2\varepsilon_{\alpha\beta}{}^\mu \omega_\mu \boldsymbol{\partial}_0 + \left[\varepsilon_{\alpha\beta\nu}n^{\mu\nu} + a_\alpha \delta_\beta{}^\mu - a_\beta \delta_\alpha{}^\mu\right]\boldsymbol{\partial}_\mu. \qquad (1.79)$$

Of course, the commutators are needed if one wants to determine the coordinate components (1.5) of the orthonormal frame vectors, and through them the coordinate components of the metric tensor. We shall, however, emphasize using the formal evolution equations to obtain qualitative information about the dynamics of the cosmological models.

In analyzing cosmological models it is useful to be able to keep track of the Weyl curvature tensor (1.32). The equations of Section 1.3 enable one to express the electric and magnetic components of the Weyl tensor in terms of the commutation functions and the source terms. Equations (1.43) and (1.47) are valid in general, i.e. for an arbitrary timelike congruence $\mathbf{e}_0$ and arbitrary source terms. In applications we shall choose $\mathbf{e}_0$ to be irrotational ($\omega_{ab} = 0$), in which case we can use the simpler formula (1.56) for E_{ab}, and the formula (1.47) for H_{ab} simplifies considerably. On performing the $1 + 3$ decomposition, one obtains the following equations.

Weyl tensor (irrotational congruence $\mathbf{e}_0$):

$$E_{\alpha\beta} = H\sigma_{\alpha\beta} - (\sigma_\alpha{}^\mu \sigma_{\mu\beta} - \tfrac{2}{3}\sigma^2 \delta_{\alpha\beta}) + {}^3 S_{\alpha\beta} - \tfrac{1}{2}\pi_{\alpha\beta}, \qquad (1.80)$$

$$H_{\alpha\beta} = (\boldsymbol{\partial}_\mu - a_\mu)\sigma_{\nu(\alpha}\varepsilon_{\beta)}{}^{\mu\nu} - 3\sigma^\mu{}_{(\alpha}n_{\beta)\mu} + n_{\mu\nu}\sigma^{\mu\nu}\delta_{\alpha\beta} + \tfrac{1}{2}n_\mu{}^\mu \sigma_{\alpha\beta}. \qquad (1.81)$$

One can also express the evolution and constraint equations (1.51)–(1.54) for E_{ab} and H_{ab}, i.e. the full Bianchi identities, in terms of the orthonormal frame variables (van Elst & Uggla 1996).

1.5 Bianchi cosmologies

As described in Section 1.2.2, a Bianchi cosmology $(\mathcal{M}, \mathbf{g}, \mathbf{u})$ is a model whose metric admits a three–dimensional group of isometries acting simply transitively on spacelike hypersurfaces, which are surfaces of homogeneity in space–time. A Bianchi cosmology thus admits a Lie algebra of KVFs with basis $\boldsymbol{\xi}_\alpha$, $\alpha = 1, 2, 3$, and structure constants $C^\mu{}_{\alpha\beta}$:

$$[\boldsymbol{\xi}_\alpha, \boldsymbol{\xi}_\beta] = C^\mu{}_{\alpha\beta} \boldsymbol{\xi}_\mu. \tag{1.82}$$

The $\boldsymbol{\xi}_\alpha$ are tangent to the group orbits, which are called the surfaces of homogeneity. The fundamental 4–velocity **u** may be orthogonal to the group orbits, giving the family of *orthogonal or non–tilted models*, or it may not, leading to the family of *tilted models*. The dynamics of the latter are considerably more complex than the former.

1.5.1 Classification of Bianchi cosmologies

The Bianchi cosmologies can be classified by classifying the Lie algebras of KVFs, and hence the associated isometry group G_3. The problem becomes that of classifying the structure constants $C^\mu{}_{\alpha\beta}$, which transform as a rank $(1, 2)$ tensor under a change of basis of the Lie algebra, and which satisfy the algebraic restrictions (1.37) and (1.38).

One can decompose $C^\mu{}_{\alpha\beta}$ in a manner analogous to (1.62):

$$C^\mu{}_{\alpha\beta} = \varepsilon_{\alpha\beta\nu} \hat{n}^{\mu\nu} + \hat{a}_\alpha \delta_\beta{}^\mu - \hat{a}_\beta \delta_\alpha{}^\mu,$$

where $\hat{n}^{\mu\nu} = \hat{n}^{\nu\mu}$ and $\hat{a}_\alpha$ are constants. The hats distinguish these quantities from the corresponding quantities in (1.62), which are not constant in general. The identity (1.37) is satisfied, and (1.38) is equivalent to

$$\hat{n}^{\alpha\beta} \hat{a}_\beta = 0. \tag{1.83}$$

Following Ellis & MacCallum (1969) one first divides the Lie algebras into class A ($\hat{a}_\alpha = 0$) and class B ($\hat{a}_\alpha \neq 0$). One then classifies further by the signs of the eigenvalues of $\hat{n}^{\alpha\beta}$. In class B one may introduce a scalar h by the following formula (Collins & Hawking 1973b, page 322):

$$\hat{a}_\alpha \hat{a}_\beta = \tfrac{1}{2} h \varepsilon_{\alpha\mu\nu} \varepsilon_{\beta\sigma\tau} \hat{n}^{\mu\sigma} \hat{n}^{\nu\tau}.$$

In the eigenbasis of $\hat{n}^{\alpha\beta}$, with $\hat{a}_1 \neq 0$,

$$(\hat{n}^{\alpha\beta}) = \mathrm{diag}\,(\hat{n}_1, \hat{n}_2, \hat{n}_3), \quad (\hat{a}_\alpha) = (\hat{a}, 0, 0),$$

this identity reduces to

$$\hat{a}^2 = h \hat{n}_2 \hat{n}_3,$$

so that h is well defined if and only if $\hat{n}_2 \hat{n}_3 \neq 0$ in class B. Note that $h < 0$ in type VI_h and $h > 0$ in type VII_h. The labels given to the ten equivalence classes, as listed in Table 1.1, are based on a classification of

Table 1.1. *Classification of Bianchi cosmologies into ten group types using the eigenvalues of $\hat{n}^{\alpha\beta}$.*

Group class	Group type	$\hat{n}_1$	$\hat{n}_2$	$\hat{n}_3$	Contains RW?
A($\hat{a}=0$)	I	0	0	0	$k=0$
	II	+	0	0	–
	VI$_0$	0	+	–	–
	VII$_0$	0	+	+	$k=0$
	VIII	–	+	+	–
	IX	+	+	+	$k=+1$
B($\hat{a}\neq 0$)	V	0	0	0	$k=-1$
	IV	0	0	+	–
	VI$_h$	0	+	–	–
	VII$_h$	0	+	+	$k=-1$

Bianchi, as modified by Schücking and Behr (Estabrook *et al.* 1968, Ellis & MacCallum 1969). Our convention for labelling the signs of the $\hat{n}_\alpha$ is that of MacCallum (1973, see page 113). Following Collins & Hawking (1973b, page 321), we also indicate which group types are admitted by each of the three RW metrics ($k = 0, \pm 1$). We note that the 'missing' Bianchi type III corresponds to VI$_{-1}$ in this classification.

1.5.2 Choice of frame

Let **n** be the unit vector field normal to the orbits of the G_3. It follows that **n** is tangent to a geodesic congruence and

$$n_{[i,j]} = 0 .$$

This gives a natural choice for the time coordinate t:

$$n_i = -t_{,i} ,$$

so that the group orbits are given by $t = $ constant. It is often convenient to introduce a new time variable $\tilde{t}$ which is related to t by

$$dt = N(\tilde{t})d\tilde{t} . \qquad (1.84)$$

The function N is called the *lapse function*. The unit normal **n** of the group orbits is invariant under the group:

$$[\xi_\alpha, \mathbf{n}] = 0 . \qquad (1.85)$$

We can now choose a triad of spacelike vectors $\mathbf{e}_\alpha$ that are tangent to the group orbits, i.e. $g(\mathbf{n}, \mathbf{e}_\alpha) = 0$, and that commute with the Killing vector fields,

$$[\mathbf{e}_\alpha, \boldsymbol{\xi}_\beta] = 0.$$

We shall refer to a frame $\{\mathbf{n}, \mathbf{e}_\alpha\}$ chosen in this way as a *group–invariant frame*. It follows that the frame components of the metric and the commutation functions satisfy

$$g_{00} = -1, \quad g_{0\alpha} = 0, \quad g_{\alpha\beta} = g_{\alpha\beta}(t),$$

$$\gamma^c{}_{ab} = \gamma^c{}_{ab}(t). \tag{1.86}$$

Since the congruence $\mathbf{n}$ is hypersurface–orthogonal, the vector fields $\mathbf{e}_\alpha$ generate a Lie algebra with structure constants† $\gamma^\mu{}_{\alpha\beta}$. This Lie algebra is in fact equivalent to the Lie algebra of KVFs (MacCallum 1973, page 89). One can thus classify the Bianchi cosmologies using either the structure constants $C^\mu{}_{\alpha\beta}$ ($\hat{n}^{\alpha\beta}$ and $\hat{a}_\alpha$) or the spatial commutation functions $\gamma^\mu{}_{\alpha\beta}$ ($n^{\alpha\beta}$ and a_α).

The remaining freedom in the choice of the frame is a time–dependent linear transformation

$$\mathbf{e}'_\alpha = A_\alpha{}^\beta(t)\mathbf{e}_\beta. \tag{1.87}$$

One can use this freedom to introduce a set of three *orthonormal* spatial vectors, so that

$$g_{\alpha\beta} = \delta_{\alpha\beta},$$

in which case the frame $\{\mathbf{n}, \mathbf{e}_\alpha\}$ is orthonormal, and the commutation functions $\gamma^c{}_{ab}(t)$ are the basic variables. The line–element is given by

$$ds^2 = -dt^2 + \delta_{\alpha\beta}\boldsymbol{\omega}^\alpha(t)\boldsymbol{\omega}^\beta(t),$$

where the $\boldsymbol{\omega}^\alpha$ are the 1–forms dual to the vector fields $\mathbf{e}_\alpha$. We refer to this as the *orthonormal frame approach*.

On the other hand, one can use the freedom (1.87) to introduce a set of *time–independent* spatial vectors $\mathbf{E}_\alpha$, i.e.

$$[\mathbf{n}, \mathbf{E}_\alpha] = 0.$$

This implies that the commutation functions are constants, so that one can use a constant linear transformation to make them equal to the structure constants of the Lie algebra,

$$[\mathbf{E}_\alpha, \mathbf{E}_\beta] = C^\mu{}_{\alpha\beta}\mathbf{E}_\mu. \tag{1.88}$$

---

† Constant on each hypersurface $t = $ constant.

In this approach, which we shall refer to as *the metric approach*, the basic variables are the frame components of the metric $g_{\alpha\beta}(t)$. The 1–forms $\mathbf{W}^\alpha$ which are dual to the frame vectors $\mathbf{E}_\alpha$ are also group–invariant and time–independent and satisfy

$$d\mathbf{W}^\alpha = -\tfrac{1}{2}C^\alpha{}_{\mu\nu}\mathbf{W}^\mu \wedge \mathbf{W}^\nu,$$

where d denotes the exterior derivative and $\wedge$ the wedge product (e.g. Mac-Callum 1994, page 185). They will be referred to as *canonical* 1–forms. Their components, in terms of local coordinates, are given in MacCallum (1979, page 548). In the metric approach the line–element is given by

$$ds^2 = -dt^2 + g_{\alpha\beta}(t)\mathbf{W}^\alpha\mathbf{W}^\beta$$

or, in terms of an arbitrary time variable, by

$$ds^2 = -N(\tilde{t})^2 d\tilde{t}^2 + g_{\alpha\beta}(\tilde{t})\mathbf{W}^\alpha\mathbf{W}^\beta. \tag{1.89}$$

The metric form (1.89) permits time–dependent transformations (1.87) of the spatial frame vectors which preserve (1.88). These transformations form a group, the *automorphism group*, and are used to simplify the spatial metric $g_{\alpha\beta}$ by introducing new automorphism variables.

1.5.3 Evolution and constraint equations

We now give the evolution equations, constraints and Weyl tensor for the Bianchi models relative to a group–invariant orthonormal frame $\{\mathbf{n}, \mathbf{e}_\alpha\}$, where $\mathbf{n}$ is the unit normal to the group orbits. Since $\mathbf{n}$ is tangent to a hypersurface–orthogonal congruence of geodesics, the equations are obtained by setting

$$\partial_\alpha(\ ) = 0, \quad \dot{u}_\alpha = 0, \quad \omega_\alpha = 0,$$

in the general equations in Section 1.4. This means that the spatial differential operators ∂_α do not appear in the equations, and that all variables depend only on time t. The differential operator ∂_0 is differentiation with respect to t, which we denote by an overdot.

Einstein field equations:

$$\dot{H} = -H^2 - \tfrac{2}{3}\sigma^2 - \tfrac{1}{6}(\mu + 3p), \tag{1.90}$$

$$\dot{\sigma}_{\alpha\beta} = -3H\sigma_{\alpha\beta} + 2\varepsilon^{\mu\nu}{}_{(\alpha}\sigma_{\beta)\mu}\Omega_\nu - {}^3S_{\alpha\beta} + \pi_{\alpha\beta}, \tag{1.91}$$

$$\mu = 3H^2 - \sigma^2 + \tfrac{1}{2}{}^3R, \tag{1.92}$$

$$q_\alpha = 3\sigma_\alpha{}^\mu a_\mu - \varepsilon_\alpha{}^{\mu\nu}\sigma_\mu{}^\beta n_{\beta\nu}. \tag{1.93}$$

Spatial curvature:

$$^3S_{\alpha\beta} = b_{\alpha\beta} - \tfrac{1}{3}(b_\mu{}^\mu)\delta_{\alpha\beta} - 2\varepsilon^{\mu\nu}{}_{(\alpha}n_{\beta)\mu}a_\nu, \tag{1.94}$$
$$^3R = -\tfrac{1}{2}b_\mu{}^\mu - 6a_\mu a^\mu, \tag{1.95}$$

where

$$b_{\alpha\beta} = 2n_\alpha{}^\mu n_{\mu\beta} - (n_\mu{}^\mu)n_{\alpha\beta}.$$

Jacobi identities:

$$\dot{n}_{\alpha\beta} = -Hn_{\alpha\beta} + 2\sigma_{(\alpha}{}^\mu n_{\beta)\mu} + 2\varepsilon^{\mu\nu}{}_{(\alpha}n_{\beta)\mu}\Omega_\nu, \tag{1.96}$$
$$\dot{a}_\alpha = -Ha_\alpha - \sigma_\alpha{}^\beta a_\beta + \varepsilon_\alpha{}^{\mu\nu}a_\mu\Omega_\nu, \tag{1.97}$$
$$0 = n_\alpha{}^\beta a_\beta. \tag{1.98}$$

Contracted Bianchi identities:

$$\dot{\mu} = -3H(\mu+p) - \sigma_\alpha{}^\beta\pi_\beta{}^\alpha + 2a_\alpha q^\alpha, \tag{1.99}$$
$$\dot{q}_\alpha = -4Hq_\alpha - \sigma_\alpha{}^\beta q_\beta - \varepsilon_\alpha{}^{\mu\nu}\Omega_\mu q_\nu + 3a^\beta\pi_{\alpha\beta} + \varepsilon_\alpha{}^{\mu\nu}n_\mu{}^\beta\pi_{\beta\nu}. \tag{1.100}$$

Weyl tensor:

$$E_{\alpha\beta} = H\sigma_{\alpha\beta} - (\sigma_\alpha{}^\mu\sigma_{\mu\beta} - \tfrac{2}{3}\sigma^2\delta_{\alpha\beta}) + {}^3S_{\alpha\beta} - \tfrac{1}{2}\pi_{\alpha\beta}, \tag{1.101}$$

$$H_{\alpha\beta} = -3\sigma^\mu{}_{(\alpha}n_{\beta)\mu} + n_{\mu\nu}\sigma^{\mu\nu}\delta_{\alpha\beta} + \tfrac{1}{2}n_\mu{}^\mu\sigma_{\alpha\beta} - a_\mu\sigma_{\nu(\alpha}\varepsilon_{\beta)}{}^{\mu\nu}. \tag{1.102}$$

The remaining freedom in choosing the orthonormal frame is a time–dependent rotation of the spatial frame vectors. This freedom can be used to diagonalize $n_{\alpha\beta}$:

$$n_{\alpha\beta} = \mathrm{diag}\,(n_1, n_2, n_3). \tag{1.103}$$

Since the third Jacobi identity implies that a_α is an eigenvector of $n_{\alpha\beta}$, one can also obtain

$$a_\alpha = (a_1, 0, 0), \tag{1.104}$$

with

$$a_1 n_1 = 0. \tag{1.105}$$

Comment: This canonical form for $n_{\alpha\beta}$ and a_α is the same as the canonical form for the structure constants $\hat{n}^{\alpha\beta}$ and $\hat{a}_\alpha$ in Table 1.1. These variables thus determine the Bianchi type of the isometry group.

1.5.4 Bianchi models with non–tilted perfect fluid

For a non–tilted perfect fluid (1.20), the source terms (1.47) satisfy

$$q_\alpha = 0, \quad \pi_{\alpha\beta} = 0. \tag{1.106}$$

For class A models ($a_\alpha = 0$) these restrictions, in conjunction with the field equations (1.91) and (1.93) and the Jacobi identities (1.96), imply that

$$\sigma_{\alpha\beta} = \mathrm{diag}\,(\sigma_{11}, \sigma_{22}, \sigma_{33}), \quad \Omega_\alpha = 0, \tag{1.107}$$

(see Ellis & MacCallum 1969, page 118, for more details), and the evolution and constraint equations simplify accordingly. This class of models is analyzed in Chapter 6.

For class B models ($a_\alpha \neq 0$), the situation is more complicated. First, (1.105) implies

$$n_1 = 0. \tag{1.108}$$

Second, the Jacobi identities (1.96) and (1.97) in conjunction with (1.103) and (1.104) imply that if $n_2 n_3 \neq 0$, then

$$a_1^2 = h n_2 n_3, \tag{1.109}$$

where h is a constant which may be identified with the group parameter h. In addition, the Jacobi identity for $\dot{a}_\alpha$, with $\alpha = 2, 3$, implies that

$$\Omega_2 = -\sigma_{13}, \quad \Omega_3 = \sigma_{12}. \tag{1.110}$$

Finally, (1.93) implies that $\sigma_{12} = 0 = \sigma_{13}$ unless $h = -\frac{1}{9}$.

These results divide the Bianchi models of class B with non–tilted perfect fluid[†] into two subclasses, the *non–exceptional* class in which the group parameter h is arbitrary, and

$$\sigma_{12} = \sigma_{13} = \Omega_2 = \Omega_3 = 0, \tag{1.111}$$

and the *exceptional* class for which $h = -\frac{1}{9}$ and $\sigma_{12}^2 + \sigma_{13}^2 \neq 0$. Note that the exceptional class only contains models of Bianchi type VI_h with $h = -\frac{1}{9}$.

These subclasses can also be distinguished by using the action of the isometry group. For all Bianchi group types except VIII and IX (and thus for all class B models), the isometry group G_3 admits an Abelian subgroup G_2 (a quick glance at Table 6.1 on page 110 in Ryan & Shepley (1975) confirms this fact). It follows from Section 1.6.2 (see the statement preceding (1.128), and (1.119)) that *in the non–exceptional class, this Abelian G_2 acts orthogonally transitively*,[‡] with the frame vectors $\mathbf{e}_2$ and $\mathbf{e}_3$ tangent to the

---

[†] These results do not require $\mu \neq 0$, and are thus valid for vacuum solutions.
[‡] See (1.113) for the definition of this term.

group orbits. The non–exceptional class B models are discussed in Chapter 7. The exceptional class, which we shall denote by $VI^*_{-1/9}$, has received little attention in the literature even though (see Table 9.5) it is dynamically more general than the non–exceptional class. A brief description of what is known about the exceptional class is given in Section 8.1.

Finally we note that another important subclass is defined by the condition

$$n_\alpha{}^\alpha = 0,$$

(see Ellis & MacCallum 1969, page 122), and will play an important role in Chapters 6, 7 and 9.

1.5.5 Bianchi models with tilted perfect fluid

In this case, one can choose a frame aligned with the fluid, or one aligned with the normals to the surfaces of homogeneity (King & Ellis 1973). The latter frame is simpler to use in many ways, even though the fluid will not be comoving with it (see equations (1.21), (1.22)), because it is adapted to the spatial symmetry. However the surfaces of homogeneity may change from being spacelike to timelike (Ellis & King 1974); in that case a normal–based frame will break down at this change of symmetry, which is associated with a non–scalar curvature singularity. A fluid–based frame (or null frame) is needed to follow the global evolution of the fluid (Collins & Ellis 1979, Siklos 1978). Tilted models are discussed further in Section 8.4.

1.6 G_2 cosmologies

By a G_2 cosmology, we mean a cosmological model $(\mathcal{M}, \mathbf{g}, \mathbf{u})$ which admits an Abelian group G_2 of isometries whose orbits are spacelike 2–surfaces. A G_2 cosmology can thus be used to model spatial inhomogeneities (e.g. density fluctuations, gravitational waves, inhomogeneous big–bang nucleosynthesis, inhomogeneous inflation) with one degree of freedom. The group orbits can be homeomorphic to R^2 (planar universe), $S^1 \times R^1$ (cylindrical universe), or $S^1 \times S^1$ (2–torus universe).

In recent years this class of models has been studied using a variety of techniques, namely numerical methods (for example, Anninos, Centrella & Matzner 1991, Berger & Moncrief 1993, Shinkai & Maeda 1994), qualitative methods (Isenberg & Moncrief 1990, Hewitt & Wainwright 1990), approximation methods (for example, Liang 1976, Carmeli et al. 1981, Adams et al. 1982), and exact solutions (for example, Carmeli et al. 1981, Carr &

Verdaguer 1983). To date however, because the EFE for this class form a system of partial differential equations (one time and one space variable in this case), little progress has been made in understanding the evolution of G_2 cosmologies from a qualitative point of view.

In many of the models with higher symmetry, discussed in Section 1.2.2, the isometry group G_r ($r \geq 3$) has an Abelian subgroup G_2; hence these models are included in the general class of G_2 cosmologies. These relationships will be discussed in more detail in Section 12.4.

1.6.1 Classification of G_2 cosmologies

There are four special classes of G_2 cosmologies depending on how the group acts on space–time:

(1) *Hypersurface–orthogonal*

The G_2 admits a hypersurface–orthogonal KVF $\boldsymbol{\xi}$,

$$\xi_{[a;b}\xi_{c]} = 0. \tag{1.112}$$

(2) *Orthogonally transitive*

The G_2 acts orthogonally transitively, that is, the 2–spaces orthogonal to the orbits of the G_2 are surface–forming. This is equivalent to the condition

$$\xi_{[a;b}\xi_c\eta_{c]} = 0, \qquad \eta_{[a;b}\eta_c\xi_{d]} = 0, \tag{1.113}$$

where $\boldsymbol{\xi}$ and $\boldsymbol{\eta}$ are two linearly independent KVFs (Carter 1973, page 160).

If we impose (1) and (2), it follows (Wainwright 1981, see Theorem 2.1 on page 1133) that there exists a second hypersurface–orthogonal KVF, which is orthogonal to the first one, leading to class (3).

(3) *Diagonal*

The G_2 admits two hypersurface–orthogonal KVFs, in which case the line–element can be written in diagonal form.

Finally in the diagonal case we can further assume there exists in addition a one–parameter isotropy group, giving the class of inhomogeneous LRS models with plane symmetry (see Section 1.2.2(c)).

(4) *Plane symmetric (LRS)*

The four subclasses and their inter–relationships are shown schematically

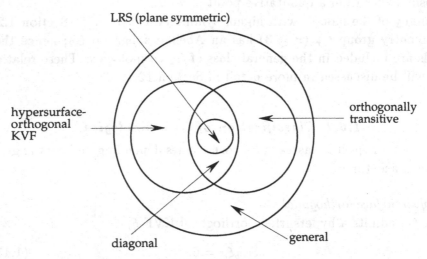

Fig. 1.1. Classification of G_2 cosmologies.

in Figure 1.1.

Condition (1.112) leads to the following restriction on the Ricci tensor,

$$\xi^c R_{c[a}\xi_{b]} = 0, \tag{1.114}$$

while (1.113) gives rise to

$$\eta^d R_{d[a}\eta_b\xi_{c]} = 0, \qquad \xi^d R_{d[a}\xi_b\eta_{c]} = 0, \tag{1.115}$$

(Carter 1973, pages 152–3 and 161). These equations in turn lead to restrictions on the source terms in the field equations.

1.6.2 Choice of orthonormal frame

When applying the orthonormal frame formalism to G_2 cosmologies it is desirable to choose a *group-invariant orbit-aligned* frame $\{e_a\}$, that is an orthonormal frame which is invariant under the G_2:

$$[\boldsymbol{\xi}, \mathbf{e}_a] = 0, \qquad [\boldsymbol{\eta}, \mathbf{e}_a] = 0,$$

where $\boldsymbol{\xi}$ and $\boldsymbol{\eta}$ are independent KVFs, and for which $\mathbf{e}_2$ and $\mathbf{e}_3$ are tangential to the group orbits. The existence of such a frame is established in the proof of Theorem 3.1 in Wainwright (1979).

For a group–invariant orbit–aligned frame it follows (Wainwright 1979) that

$$\partial_A(\gamma^a{}_{bc}) = 0, \tag{1.116}$$

$$\gamma^0{}_{0A} = \gamma^0{}_{1A} = \gamma^1{}_{0A} = \gamma^1{}_{1A} = 0, \quad \gamma^a{}_{23} = 0, \tag{1.117}$$

where here and in the rest of this section we use the convention that capital indices A, B and C take on the values 2 and 3. It follows from (1.61), (1.62) and (1.117) that the complete set of 24 basic variables (1.63) is reduced to the following set of 12 variables,

$$\{H, \sigma_{AB}, \sigma_{1A}, \dot{u}_1, \Omega_1, n_{AB}, a_1\}, \tag{1.118}$$

with

$$\Omega_A = \varepsilon_A{}^C \sigma_{1C}, \tag{1.119}$$

where ε_{AB} is the two–dimensional permutation symbol, and indices A, B, C are raised and lowered using δ_{AB}. Note that the vorticity $\omega_\alpha = 0$, so that e_0 is irrotational, as is the case for any group–invariant timelike vector field orthogonal to the orbits of a G_2 (Wainwright 1981, page 1134). It follows from (1.69), (1.116) and (1.118) that the trace–free Ricci tensor of the hypersurfaces orthogonal to e_0 satisfies

$$^3S_{1A} = 0. \tag{1.120}$$

At this stage we regard e_0 as a preferred vector field; we shall subsequently identify it with u in the general decomposition (1.19) of the stress–energy tensor. The remaining freedom (excluding reflections) in the choice of frame is a rotation

$$\begin{aligned} e_2' &= \cos\phi\, e_2 + \sin\phi\, e_3, \\ e_3' &= -\sin\phi\, e_2 + \cos\phi\, e_3, \end{aligned} \tag{1.121}$$

where ϕ is constant on the orbits of the G_2. In order to describe the transformation properties of the basic variables under (1.121) it is useful to introduce the trace–free quantities (Hewitt & Wainwright 1990)

$$\begin{aligned} \tilde{n}_{AB} &= n_{AB} - \tfrac{1}{2} n_C{}^C \delta_{AB}, \\ \tilde{\sigma}_{AB} &= \sigma_{AB} - \tfrac{1}{2} \sigma_C{}^C \delta_{AB}, \end{aligned} \tag{1.122}$$

and denote the traces by

$$n_+ = \tfrac{1}{2} n_C{}^C, \quad \sigma_+ = \tfrac{1}{2} \sigma_C{}^C. \tag{1.123}$$

It follows that under the rotation (1.121), $H, \sigma_+, \dot{u}_1$ and a_1 transform as scalars, σ_{1A} and Ω_A transform as vectors, $\tilde{\sigma}_{AB}$ and $\tilde{n}_{AB}$ transform as rank 2 tensors, while Ω_1 and n_+ are non–tensorial and transform according to

$$\tilde{\Omega}_1 = \Omega_1 + \boldsymbol{\partial}_0\phi, \quad \tilde{n}_+ = n_+ + \boldsymbol{\partial}_1\phi. \tag{1.124}$$

It is sometimes convenient to write

$$\tilde{\sigma}_{AB} = \sqrt{3}\left(\begin{array}{cc} \sigma_- & \sigma_\times \\ \sigma_\times & -\sigma_- \end{array}\right), \quad \tilde{n}_{AB} = \sqrt{3}\left(\begin{array}{cc} n_- & n_\times \\ n_\times & -n_- \end{array}\right). \tag{1.125}$$

In terms of the variables (1.122), (1.123) and (1.125) the shear scalar is given by

$$\sigma^2 = 3\sigma_+^2 + \tfrac{1}{2}\tilde{\sigma}^{AB}\tilde{\sigma}_{AB} + \sigma_{1A}\sigma_1{}^A = 3(\sigma_+^2 + \sigma_-^2 + \sigma_\times^2) + \sigma_{1A}\sigma_1{}^A. \tag{1.126}$$

We note that one way to fix the frame is to choose e_2 or e_3 to be parallel to a KVF. For example, if e_3 is parallel to a KVF, then

$$\Omega_1 = -\sqrt{3}\sigma_\times, \quad n_+ = -\sqrt{3}n_-, \tag{1.127}$$

(see (A5) in Wainwright 1981).

We now consider the effect of restricting the action of the G_2. Firstly, *the group acts orthogonally transitively if and only if*

$$\sigma_{1A} = 0, \tag{1.128}$$

(use (3.1) in Wainwright 1979). In order to describe the remaining cases, we note that e_3 is hypersurface–orthogonal if and only if

$$\Omega_1 = \sqrt{3}\sigma_\times, \quad \Omega_2 = -\sigma_{13}, \quad n_+ = \sqrt{3}n_-. \tag{1.129}$$

(see the appendix in Wainwright 1981). In the remaining special cases, in which there is at least one hypersurface–orthogonal KVF $\boldsymbol{\xi}$, we choose e_3 to be parallel to $\boldsymbol{\xi}$. Then, using (1.119) and (1.127)–(1.129) we obtain the following sets of basic variables.

Hypersurface-orthogonal:

$$(H, \sigma_+, \sigma_-, \sigma_{12}, \dot{u}_1, a_1, n_\times), \tag{1.130}$$

with

$$\Omega_3 = -\sigma_{12}.$$

Diagonal:

$$(H, \sigma_+, \sigma_-, \dot{u}_1, a_1, n_\times). \tag{1.131}$$

Plane–symmetric:

$$(H, \sigma_+, \dot{u}_1, a_1). \tag{1.132}$$

The restrictions $\sigma_- = 0$ and $n_\times = 0$, which give the plane–symmetric (LRS) case, arise since otherwise there would be a preferred direction in the 23–space.

In order to write the evolution equations, we need the restrictions on the source terms in the general stress–energy tensor (1.19) that arise when we make the choice $\mathbf{e}_0 = \mathbf{u}$. We first perform a $1 + 2$ decomposition of q_α and $\pi_{\alpha\beta}$:

$$q_\alpha = (q_1, q_A), \quad \pi_{\alpha\beta} = (\pi_+, \pi_{1A}, \tilde{\pi}_{AB}),$$

where π_+ and $\tilde{\pi}_{AB}$ are defined as in (1.122) and (1.123). Using (1.19) and the EFE (1.1), the restriction (1.114) on the Ricci tensor implies

$$\xi^a q_a = 0, \quad \xi^a \pi_{a[b} \xi_{c]} = 0,$$

while (1.115) gives rise to

$$\xi^a q_a = 0 = \eta^a q_a, \quad \xi^a \pi_{a[b} \xi_c \eta_{d]} = 0 = \eta^a \pi_{a[b} \eta_c \xi_{d]}.$$

One obtains the following non–zero source terms when the group action is restricted. We decompose $\tilde{\pi}_{AB}$ into π_- and $\pi_\times$, in analogy with (1.125).

Orthogonally transitive:

$$(\mu, p, \pi_+, \tilde{\pi}_{AB}, q_1). \tag{1.133}$$

Hypersurface–orthogonal:

$$(\mu, p, \pi_+, \pi_-, \pi_{12}, q_1, q_2). \tag{1.134}$$

Diagonal:

$$(\mu, p, \pi_+, \pi_-, q_1). \tag{1.135}$$

Plane–symmetric:

$$(\mu, p, \pi_+, q_1). \tag{1.136}$$

1.6.3 Evolution and constraint equations

In defining the basic variables we performed a $1+1+2$ decomposition of the commutation functions with respect to the G_2 orbits. The analogous $1+1+2$ decomposition of the EFE, the Jacobi identities and the contracted Bianchi identities, as given in Section 1.4, may now be performed. In deriving the system of equations given below we have separated the $\alpha = 1$ equations from the $\alpha = A = 2,3$ equations, and have split equations involving a pair of indices AB into their trace and trace–free parts. We also need to decompose the trace–free spatial Ricci tensor $^3S_{\alpha\beta}$ in analogy with (1.122), (1.123) and (1.125):

$$^3\tilde{S}_{AB} = {}^3S_{AB} - \tfrac{1}{2}\,{}^3S_C{}^C\delta_{AB}, \qquad {}^3S_+ = \tfrac{1}{2}\,{}^3S_C{}^C,$$

$$^3\tilde{S}_{AB} = \sqrt{3}\begin{pmatrix} S_- & S_\times \\ S_\times & -S_- \end{pmatrix}. \tag{1.137}$$

In the interests of simplicity *we restrict our considerations to the orthogonally transitive case* (which specializes to the diagonal and plane–symmetric cases). The resulting system of equations involves the basic variables

$$(H, \sigma_+, \tilde{\sigma}_{AB}, \dot{u}_1, \Omega_1, n_+, \tilde{n}_{AB}, a_1), \tag{1.138}$$

the source terms

$$(\mu, p, \pi_+, \tilde{\pi}_{AB}, q_1), \tag{1.139}$$

and the differential operators $\boldsymbol{\partial}_0$ (timelike) and $\boldsymbol{\partial}_1$ (spacelike). For convenience, we define the dual tensors

$$^*\tilde{n}_{AB} = \tilde{n}_A{}^C\varepsilon_{BC} \qquad {}^*\tilde{\sigma}_{AB} = \tilde{\sigma}_A{}^C\varepsilon_{BC}. \tag{1.140}$$

and note that

$$^*\tilde{n}_{AB} = \sqrt{3}\begin{pmatrix} n_\times & -n_- \\ -n_- & -n_\times \end{pmatrix}, \qquad {}^*\tilde{\sigma}_{AB} = \sqrt{3}\begin{pmatrix} \sigma_\times & -\sigma_- \\ -\sigma_- & -\sigma_\times \end{pmatrix},$$

in terms of the notation in (1.125).

Einstein field equations:

$$\boldsymbol{\partial}_0 H = -H^2 - \tfrac{2}{3}\sigma^2 + \tfrac{1}{3}(\partial_1 + \dot{u}_1 - 2a_1)\dot{u}_1 - \tfrac{1}{6}(\mu + 3p), \tag{1.141}$$

$$\boldsymbol{\partial}_0\sigma_+ = -3H\sigma_+ - \tfrac{1}{3}(\partial_1 + \dot{u}_1 + a_1)\dot{u}_1 - {}^3S_+ + \pi_+, \tag{1.142}$$

$$\boldsymbol{\partial}_0\tilde{\sigma}_{AB} = -3H\tilde{\sigma}_{AB} + 2\Omega_1\,{}^*\tilde{\sigma}_{AB} - \dot{u}_1\,{}^*\tilde{n}_{AB} - {}^3\tilde{S}_{AB} + \tilde{\pi}_{AB}, \tag{1.143}$$

$$\mu \;=\; 3H^2 - \sigma^2 + \tfrac{1}{2}\,{}^3R, \tag{1.144}$$

$$q_1 \;=\; 2\partial_1(H + \sigma_+) - 6\sigma_+ a_1 + {}^*\!\tilde{\sigma}^{AB}\tilde{n}_{AB}. \tag{1.145}$$

Spatial curvature:

$$\,{}^3S_+ \;=\; -\tfrac{1}{3}\partial_1 a_1 + \tfrac{1}{3}\tilde{n}^{AB}\tilde{n}_{AB}, \tag{1.146}$$

$$\,{}^3\tilde{S}_{AB} \;=\; (\partial_1 - 2a_1)\,{}^*\!\tilde{n}_{AB} + 2n_+\tilde{n}_{AB}, \tag{1.147}$$

$$\,{}^3R \;=\; 4\partial_1 a_1 - 6a_1^2 - \tilde{n}^{AB}\tilde{n}_{AB}. \tag{1.148}$$

Jacobi identities:

$$\partial_0 n_+ \;=\; (-H + 2\sigma_+)n_+ + \tilde{\sigma}^{AB}\tilde{n}_{AB} + (\partial_1 + \dot{u}_1)\Omega_1, \tag{1.149}$$

$$\begin{aligned}
\partial_0 \tilde{n}_{AB} \;=\;\; & (-H + 2\sigma_+)\tilde{n}_{AB} + 2\Omega_1\,{}^*\!\tilde{n}_{AB} + 2n_+\tilde{\sigma}_{AB} \\
& + (\partial_1 + \dot{u}_1)\,{}^*\!\tilde{\sigma}_{AB},
\end{aligned} \tag{1.150}$$

$$\partial_0 a_1 \;=\; (-H + 2\sigma_+)a_1 - (\partial_1 + \dot{u}_1)(H + \sigma_+). \tag{1.151}$$

Contracted Bianchi identities:

$$\begin{aligned}
\partial_0 \mu \;=\;\; & -3H(\mu + p) - 6\sigma_+\pi_+ - \tilde{\sigma}^{AB}\tilde{\pi}_{AB} \\
& -(\partial_1 + 2\dot{u}_1 - 2a_1)q_1,
\end{aligned} \tag{1.152}$$

$$\begin{aligned}
\partial_0 q_1 \;=\;\; & 2(-2H + \sigma_+)q_1 - \partial_1 p - (\mu + p)\dot{u}_1 \\
& + 2(\partial_1 + \dot{u}_1 - 3a_1)\pi_+ - {}^*\!\tilde{n}^{AB}\tilde{\pi}_{AB}.
\end{aligned} \tag{1.153}$$

Commutator:

$$[\partial_0, \partial_1] = \dot{u}_1\partial_0 + (-H + 2\sigma_+)\partial_1. \tag{1.154}$$

Weyl tensor:

We decompose $E_{\alpha\beta}$ and $H_{\alpha\beta}$, as given by (1.101)–(1.102), in analogy with

(1.137), and note that in the orthogonally transitive case, $E_{1A} = 0 = H_{1A}$.

$$
\begin{aligned}
E_+ &= (H + \sigma_+)\sigma_+ - \tfrac{1}{6}\tilde{\sigma}_{AB}\tilde{\sigma}^{AB} + {}^3S_+ - \tfrac{1}{2}\pi_+, \\[2mm]
\tilde{E}_{AB} &= (H - 2\sigma_+)\tilde{\sigma}_{AB} + {}^3\tilde{S}_{AB} - \tfrac{1}{2}\pi_{AB}, \\[2mm]
H_+ &= \tfrac{1}{2}\tilde{n}_{AB}\tilde{\sigma}^{AB}, \\[2mm]
\tilde{H}_{AB} &= (\boldsymbol{\partial}_1 - a_1)\overset{*}{\tilde{\sigma}}_{AB} + 3\sigma_+\tilde{n}_{AB} + 2n_+\tilde{\sigma}_{AB},
\end{aligned}
\tag{1.155}
$$

where ${}^3S_+$ and ${}^3\tilde{S}_{AB}$ are given by (1.146) and (1.147). It follows from (1.35) that

$$
\begin{aligned}
C_{abcd}C^{abcd} &= 8[(\tilde{E}_{AB}\tilde{E}^{AB} + 6E_+^2) - (\tilde{H}_{AB}\tilde{H}^{AB} + 6H_+^2)], \\[2mm]
C_{abcd}{}^*C^{abcd} &= 16(\tilde{E}_{AB}\tilde{H}^{AB} + 6E_+H_+).
\end{aligned}
\tag{1.156}
$$

2

Friedmann–Lemaître universes

G. F. R. ELLIS

University of Cape Town,

J. WAINWRIGHT

University of Waterloo

In this chapter, we give a brief introduction to the FL universes, which are the standard models of present cosmology. In Section 2.1 we give the Robertson–Walker (RW) metric, the field equations, and some special solutions which play an important role in the sequel, and in Section 2.2 we introduce the observational parameters. In Section 2.3 we formulate the field equations in a dimensionless form and give a qualitative analysis of the evolution. This analysis serves as an introduction to the more general qualitative analysis of the Bianchi models, given in Chapters 6 and 7. In interpreting observational data (see Chapter 3) one is interested in models which are 'close to FL' during some epoch. In Section 2.4 we thus give a formal definition of this class of models.

2.1 Basic properties and special solutions

The FL models are based on the assumption that the universe is isotropic about every point, which necessarily implies homogeneity. The resulting RW metric depends on a parameter $k = 0$ or ± 1, and in comoving coordinates has the form

$$ds^2 = -dt^2 + \ell^2(t)d\Omega^2, \tag{2.1}$$

where the space sections are of constant curvature:

$$d\Omega^2 = dr^2 + f^2(r)(d\theta^2 + \sin^2\theta d\phi^2), \tag{2.2}$$

with

$$f(r) = \sin r, \ r, \ \sinh r, \tag{2.3}$$

depending on whether $k = +1, 0$ or -1, respectively. The spatial curvature ^3R is given by

$$^3R = \frac{6k}{\ell^2}. \tag{2.4}$$

The coordinate t measures proper time along the world lines with tangent vector

$$\mathbf{u} = \frac{\partial}{\partial t}. \tag{2.5}$$

The function $\ell(t)$ in the line–element is called the *length scale function*. In the cases $k = \pm 1$ it is uniquely determined by the line–element, while if $k = 0$, $\ell(t)$ is determined up to a constant multiplicative factor since the r coordinate can be rescaled.

The symmetry of the space–time forces the stress–energy tensor to have the algebraic form of a perfect fluid, with energy density μ and pressure p:

$$T_{ab} = \mu u_a u_b + p(g_{ab} + u_a u_b). \tag{2.6}$$

The evolution of the models is governed† by the conservation equation

$$\dot{\mu} = -3\frac{\dot{\ell}}{\ell}(\mu + p), \tag{2.7}$$

the Raychaudhuri equation

$$\frac{\ddot{\ell}}{\ell} = -\tfrac{1}{6}(\mu + 3p), \tag{2.8}$$

and the Friedmann equation

$$3\frac{\dot{\ell}^2}{\ell^2} = \mu - \frac{3k}{\ell^2}, \tag{2.9}$$

which is a first integral of the other two whenever $\dot{\ell} \neq 0$ (e.g. Weinberg 1972, page 472). We note that (2.7) and (2.8) lead directly to an elementary singularity theorem:

if $\mu > 0$ and $\mu + 3p > 0$ for all t, and $\dot{\ell}(t_0) > 0$, then there exists a time $t_b < t_0$ such that $\ell(t_b) = 0$ and $\lim\limits_{t \to t_b^+} \mu = +\infty$.

The *Hubble scalar* H, which measures the rate of expansion of the universe, is given by

$$H = \frac{\dot{\ell}}{\ell}, \tag{2.10}$$

† Equations (2.7)–(2.9) follow from (1.50), (1.42) and (1.57) respectively, as specialized to a perfect fluid (i.e. $q_a = 0, \pi_{ab} = 0$), on using (2.4), (2.10) and (2.55).

where ˙ denotes differentiation with respect to t. This formula follows when the definition (1.31) of H in a general model is applied to the line–element (2.1) and vector field (2.5). The *deceleration parameter* q, which measures whether the expansion is speeding up or slowing down, is defined by

$$q = -\frac{\ddot{\ell}\ell}{(\dot{\ell})^2}. \qquad (2.11)$$

The *density parameter* Ω, which describes the dynamical effect of the matter density, is defined by

$$\Omega = \frac{\mu}{3H^2}. \qquad (2.12)$$

Both q and Ω are dimensionless, while H has dimensions of $(time)^{-1}$. These variables play a fundamental role in the FL models and also in more general models (see Sections 5.2 and Chapters 12 and 13).

The Friedmann equation (2.9) can be expressed in terms of Ω, giving

$$\Omega - 1 = \frac{k}{H^2\ell^2}. \qquad (2.13)$$

This equation shows that the value of Ω determines the geometry of the 3–spaces $t = $ constant, and vice versa:

$$
\begin{aligned}
&\Omega > 1 \quad \text{corresponds to closed 3–spaces} \quad (k = +1), \\
&\Omega = 1 \quad \text{corresponds to flat 3–spaces} \quad\;\; (k = 0), \qquad (2.14) \\
&\Omega < 1 \quad \text{corresponds to open 3–spaces} \quad\; (k = -1).
\end{aligned}
$$

In the galactic epoch one can idealize the matter–energy content as *dust* $(p = 0, \mu > 0)$, while in the pre–decoupling epoch, the dominant contribution will be *radiation* $(p = \frac{1}{3}\mu, \mu > 0)$. These two cases can be treated simultaneously by writing

$$p = (\gamma - 1)\mu, \qquad (2.15)$$

where $\gamma = 1$ (dust) or $\gamma = \frac{4}{3}$ (radiation). An extreme case is $\gamma = 0$, in which case (2.7) implies $\mu = -p = \Lambda$, a constant, in which case the source is interpreted as a *cosmological constant*. Note that (2.7), in conjunction with (2.15) implies that

$$\mu = \frac{C}{\ell^{3\gamma}}, \qquad (2.16)$$

where C is a constant. Then (2.9) determines the length scale function $\ell(t)$, giving a complete description of the dynamics.

More realistic models can be constructed by using a two–fluid description of the source. Provided that both fluids have the same 4–velocity (2.5), the stress–energy tensor will still be of the form (2.6), and hence compatible with the RW metric (2.1), with

$$\mu = \mu_1 + \mu_2, \quad p = p_1 + p_2. \tag{2.17}$$

We assume that the two fluids are non–interacting, so that each component separately satisfies the conservation equation (2.7) and hence (2.16). Two particular cases are of interest:

(1) a combination of radiation and dust,

$$p_1 = \tfrac{1}{3}\mu_1 > 0, \quad p_2 = 0, \quad \mu_2 > 0, \tag{2.18}$$

so that the model will describe the transition from a radiation–dominated epoch to a matter–dominated epoch,

(2) dust with a cosmological constant,

$$p_1 = 0, \quad \mu_1 > 0, \quad p_2 = -\mu_2, \quad \mu_2 = \Lambda, \tag{2.19}$$

so that the model will describe the effects of a cosmological constant on the evolution.

We can treat both cases simultaneously by writing

$$p_1 = (\gamma_1 - 1)\mu_1, \quad p_2 = (\gamma_2 - 1)\mu_2, \tag{2.20}$$

where $\gamma_1 = \tfrac{4}{3}$, $\gamma_2 = 1$ in case (i), and $\gamma_1 = 1$, $\gamma_2 = 0$ in case (ii).

The most important single–fluid solutions are listed below.

(1) *Flat* FL *universe* $(k = 0)$.
The line–element is

$$ds^2 = -dt^2 + t^{\frac{4}{3\gamma}}[dr^2 + r^2(d\theta^2 + \sin^2\theta d\phi^2)], \tag{2.21}$$

and the energy density and pressure are

$$\mu = \frac{4}{3\gamma^2 t^2}, \quad p = (\gamma - 1)\mu, \quad 0 < \gamma \le 2.$$

In the case of dust $(\gamma = 1)$, this solution is the *Einstein–de Sitter universe* (Einstein & de Sitter 1932).

(2) *de Sitter universe* $(k = 0)$.
The line–element is

$$ds^2 = -dt^2 + e^{\sqrt{\frac{\Lambda}{3}}t}[dr^2 + r^2(d\theta^2 + \sin^2\theta d\phi^2)], \tag{2.22}$$

and the energy density and pressure are

$$\mu = \Lambda, \quad p = -\Lambda,$$

where Λ is the cosmological constant.

(3) *Milne universe* ($k = -1$).
The line–element is

$$ds^2 = -dt^2 + t^2[dr^2 + \sinh^2 r(d\theta^2 + \sin^2\theta d\phi^2)], \qquad (2.23)$$

and $\mu = 0$, $p = 0$.

(4) *Einstein static universe* ($k = +1$).
The line–element is

$$ds^2 = -dt^2 + \ell^2[dr^2 + \sin^2 r(d\theta^2 + \sin^2\theta d\phi^2)], \qquad (2.24)$$

where $\ell > 0$ is a constant. The energy density and pressure are

$$\mu = 3\ell^{-2}, \quad p = -\ell^{-2},$$

(i.e. $\gamma = \frac{2}{3}$). One can also interpret the source as consisting of two fluids, with

$$\mu_1 = \left(\frac{2-3\gamma_2}{\gamma_1-\gamma_2}\right)\ell^{-2}, \quad \mu_2 = \left(\frac{3\gamma_1-2}{\gamma_1-\gamma_2}\right)\ell^{-2},$$

and $p_1 = (\gamma_1 - 1)\mu_1$, $p_2 = (\gamma_2 - 1)\mu_2$, where $0 \le \gamma_2 < \frac{2}{3} < \gamma_1 \le 2$. Einstein's choice was $\gamma_1 = 1$, $\gamma_2 = 0$, so that $\mu_2 = \ell^{-2} = \Lambda$, where Λ is the cosmological constant.

Case (1) is the simplest non–empty, expanding cosmological model, and is widely used by astrophysicists. It may, however, be incompatible with observations (see Section 3.5.1). Cases (2), (3) and (4) are not of direct interest as cosmological models, but we shall see that (2) and (3) are important as asymptotic states of matter–filled models, while (4) is a saddle point in some phase planes.

2.2 Observational parameters

In this section we introduce the observational parameters that characterize a given FL model at a particular time, and that enable one to compare the model with observational data. We restrict our considerations to a single–fluid model with equation of state (2.15).

The initial singularity in a FL model occurs when $\ell(t) = 0$. We choose the origin of t so that $\ell(0) = 0$. Then the present–day value of t, denoted by t_0, represents the *age of the universe*. The present–day values of H, q and Ω are denoted by

$$H_0 = H(t_0), \quad q_0 = q(t_0), \quad \Omega_0 = \Omega(t_0),$$

and H_0 is called the *Hubble constant*. The four constants

$$\{t_0, H_0, q_0, \Omega_0\}, \tag{2.25}$$

are referred to as the *observational parameters*.

The observational parameters (2.25) are not independent since the EFE lead to two restrictions. The first is an immediate consequence of (2.8), (2.11) and (2.12):

$$q_0 = \tfrac{1}{2}(3\gamma - 2)\Omega_0. \tag{2.26}$$

To derive the second, we define a dimensionless scale function

$$y = \frac{\ell(t)}{\ell_0}, \tag{2.27}$$

where $\ell_0 = \ell(t_0)$. Then (2.16) can be written

$$\mu = \frac{\mu_0}{y^{3\gamma}}, \tag{2.28}$$

where $\mu_0 = \mu(t_0)$. Evaluating (2.13) at t_0 gives

$$\Omega_0 - 1 = \frac{k}{H_0^2 \ell_0^2}. \tag{2.29}$$

It now follows from (2.27), (2.28), (2.29) and (2.12) that (2.9) can be written in the form

$$\left(\frac{dy}{dt}\right)^2 = H_0^2 F(y, \Omega_0), \tag{2.30}$$

where

$$F(y, \Omega_0) = 1 + \Omega_0(y^{2-3\gamma} - 1). \tag{2.31}$$

It follows from (2.30) that

$$t_0 H_0 = \int_0^1 \frac{1}{\sqrt{F(y, \Omega_0)}} \, dy.$$

In other words, $t_0 H_0$ is a function of Ω_0,

$$t_0 H_0 = \alpha(\Omega_0), \tag{2.32}$$

where

$$\alpha(\Omega_0) = \int_0^1 \frac{1}{\sqrt{F(y, \Omega_0)}} \, dy. \tag{2.33}$$

We shall refer to $t_0 H_0$, which is dimensionless, as the *age parameter*. For $\gamma = 1$, the integral can be evaluated, giving

$$\alpha(\Omega_0) = \begin{cases} (\Omega_0 - 1)^{-3/2}[-(\Omega_0 - 1)^{1/2} + \frac{1}{2}\Omega_0 \cos^{-1}(\frac{2}{\Omega_0} - 1)], & \Omega_0 > 1 \\[2mm] \frac{2}{3}, & \Omega_0 = 1 \\[2mm] (1 - \Omega_0)^{-3/2}[(1 - \Omega_0)^{1/2} - \frac{1}{2}\Omega_0 \cosh^{-1}(\frac{2}{\Omega_0} - 1)]. & \Omega_0 < 1 \end{cases} \tag{2.34}$$

An important property is that $\alpha(\Omega_0)$ is a decreasing function, which satisfies $\alpha(0) = 1$, and $\alpha(\Omega_0) \to 0$ as $\Omega_0 \to +\infty$. It follows that

$$\alpha(\Omega_0) < 1 \quad \text{for all} \quad \Omega_0 > 0. \tag{2.35}$$

The restrictions (2.26) and (2.32)–(2.33) can be generalized to an n–fluid model: Ω_0 is replaced by $\Omega_{A,0}, A = 1, \ldots, n$, the density parameters for the n fluids, and $F(y, \Omega_0)$ in (2.31) is replaced by

$$F(y, \Omega_{A,0}) = 1 + \sum_{A=1}^{n} \Omega_{A,0}(y^{2-3\gamma_A} - 1), \tag{2.36}$$

where the γ_A are the equation of state parameters. Equation (2.26) becomes

$$q_0 = \frac{1}{2} \sum_{A=1}^{n} (3\gamma_A - 2)\Omega_{A,0}, \tag{2.37}$$

(see Wainwright 1996). For the case of dust and a cosmological constant, i.e. $\gamma_1 = 1$, and $\gamma_2 = 0$, we obtain

$$F(y, \Omega_m, \Omega_\Lambda) = 1 + \Omega_m(y^{-1} - 1) + \Omega_\Lambda(y^2 - 1), \tag{2.38}$$

and

$$q_0 = \frac{1}{2}\Omega_m - \Omega_\Lambda, \tag{2.39}$$

where $\Omega_m \equiv \Omega_{1,0}$ and $\Omega_\Lambda \equiv \Omega_{2,0}$.

The Hubble constant H_0 and deceleration parameter q_0 can, in principle, be determined by measuring the luminosities (or equivalently, apparent magnitudes) of galaxies. The redshift parameter z is defined by

$$z = \frac{\lambda_0 - \lambda_1}{\lambda_1}, \tag{2.40}$$

where λ_0 and λ_1 are the observed and emitted wavelengths, respectively. It is a well–known result (e.g. Wald 1984, pages 102–4) that for the RW metric with length scale function $\ell(t)$,

$$1 + z = \frac{\ell(t_0)}{\ell(t_1)}, \tag{2.41}$$

where t_1 and t_0 are the times of emission and reception. The luminosity distance d_L is defined by

$$d_L = \left(\frac{L}{4\pi\mathcal{F}}\right)^{\frac{1}{2}},$$

where L is the absolute luminosity (assumed known) and $\mathcal{F}$ is the apparent luminosity (i.e. the measured flux). For any RW geometry, it follows that

$$d_L = H_0^{-1}[z + \tfrac{1}{2}(1 - q_0)z^2 + \cdots], \tag{2.42}$$

(e.g. Weinberg 1972, pages 441–2). For a dust FL model, one can derive an exact formula (e.g. Weinberg 1972, page 485, equation (15.3.24), and Terrell 1977). We also refer to Sandage (1995) for a detailed discussion of the classical cosmological tests.

The restrictions imposed on the observational parameters by observational data are discussed in Section 3.5.

2.3 Qualitative analysis

In the simple fluid models, even in the case where there are several matter components one can solve for ℓ at least implicitly (e.g. Harrison 1967, and Cohen 1967) and hence describe the evolution in detail. One can, however, understand the evolution of the whole class in a simple and direct way by using qualitative methods. For this purpose it is helpful to express the field equations (2.8), (2.9) and the conservation equation (2.7) in terms of dimensionless variables.

We begin by using the length scale function ℓ to define a dimensionless time variable τ:

$$\ell = \ell_0 e^{\tau}, \tag{2.43}$$

where ℓ_0 is the value of ℓ at some arbitrary reference time. It is convenient to think of (2.11) with (2.10) as giving an evolution equation for H. It follows from (2.10), (2.11), and (2.43) that

$$\frac{dt}{d\tau} = \frac{1}{H}, \tag{2.44}$$

and

$$\frac{dH}{d\tau} = -(1 + q)H. \tag{2.45}$$

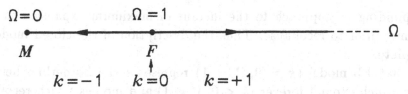

Fig. 2.1. State space for the single–fluid FL models with $\gamma > \frac{2}{3}$. The equilibrium points F and M describe the flat FL and Milne universes, respectively.

Equation (2.8) serves to relate q, as defined by (2.11), to the source terms. For a perfect fluid with linear equation of state (2.15), it follows that

$$q = \tfrac{1}{2}(3\gamma - 2)\Omega. \tag{2.46}$$

Furthermore, (2.7), (2.44) and (2.45) lead to an evolution equation for the density parameter Ω as defined by (2.12):

$$\frac{d\Omega}{d\tau} = [2q - (3\gamma - 2)]\Omega. \tag{2.47}$$

On substituting from (2.46) we obtain

$$\frac{d\Omega}{d\tau} = -(3\gamma - 2)(1 - \Omega)\Omega. \tag{2.48}$$

This dimensionless DE describes the evolution of the single–fluid FL models with linear equation of state (2.15).

The state space of the DE (2.48) is the non–negative Ω–axis. There are two equilibrium points, namely $\Omega = 1$, corresponding to the flat FL universe (2.21), and $\Omega = 0$, corresponding to the Milne universe (2.23). The one–dimensional state space is shown in Figure 2.1 for the case $\gamma > \frac{2}{3}$, which includes dust or radiation. We note that

$$\lim_{\tau \to -\infty} \Omega = 1 \tag{2.49}$$

for all models, showing that near the initial singularity the evolution of all models is approximated by the flat FL universe. For open models (i.e. $k = -1, \Omega < 1$),

$$\lim_{\tau \to +\infty} \Omega = 0, \tag{2.50}$$

showing that at late times, the evolution of open models is approximated by the Milne universe. For closed models (i.e. $k = +1$, $\Omega > 1$),

$$\lim_{\tau \to \tau_{\max}} \Omega = +\infty,$$

corresponding to approach to the instant of maximum expansion in a finite time, prior to recollapse. Thus the description of the closed models is incomplete.

The flat FL model ($k = 0$, $\Omega = 1$) represents the borderline between models which expand forever ($k < 0$, $\Omega < 1$) and models which recollapse ($k = 0$, $\Omega > 1$). We thus refer to the matter density in the flat model as the critical density, denoted μ_{crit}. It follows from (2.12) that the density parameter can be written

$$\Omega = \frac{\mu}{\mu_{\mathrm{crit}}},$$

indicating that if at a particular time the density is sufficiently high, i.e. $\mu > \mu_{\mathrm{crit}}$, then the model will recollapse.

Figure 2.1, using Ω as the state variable, gives a complete description of the asymptotic behaviour of the single–fluid FL models, but does not contain information about the Hubble scalar H (i.e. the rate of expansion of the universe), which is governed by (2.45) with (2.46). In order to incorporate information about the expansion, one can regard a point $(\Omega, H) \in \mathbf{R}^2$ as defining the state of the universe at an instant of time. Then (2.48) and (2.45), with (2.46), form an autonomous DE for (Ω, H), and one can portray the evolution of the universe by drawing the orbits of this DE (see Section 4.1.2 for this terminology) in the ΩH–plane, as in Figure 2.2.

The age constraint (2.32) was derived for an arbitrary time t_0. We can drop the subscript and rewrite the constraint as

$$H = \frac{1}{t}\alpha(\Omega),$$

where $\alpha(\Omega)$ is given by (2.34) in the case of dust ($\gamma = 1$). This equation describes the curves of constant time ($t = $ constant) in the ΩH–plane, as shown in Figure 2.2. We shall make use of this diagram in Section 3.5.4 when using observational data to restrict the observational parameters (2.25).

One can extend the analysis of the Ω–space to multi–fluid models. The first to do this were Stabell & Refsdal (1966), who considered the two–fluid case, specifically dust with a cosmological constant ($\gamma_1 = 1, \gamma_2 = 0$), and plotted the 'evolution curves', i.e. orbits in the $q\Omega$–plane† (see their Figure 4, page 384). Madsen *et al.* (1992, page 1409) give the phase portraits for the

† These authors actually used $\frac{1}{2}\Omega$ as one of the variables, which they denoted by σ (not to be

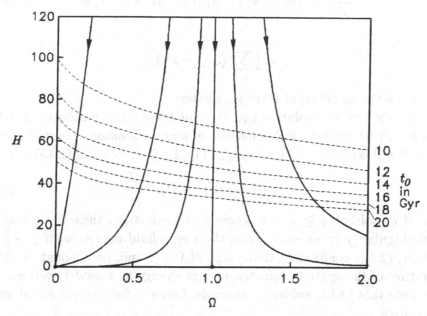

Fig. 2.2. The ΩH–plane for the dust ($\gamma = 1$) FL models. The solid curves are typical orbits of the DE (2.45) and (2.48), and the dotted curves are curves of constant time. The range of variables $0 \leq \Omega \leq 2$, $0 \leq H \leq 120$ accommodates the present epoch (the units for H are km s^{-1} Mpc^{-1}; see the Appendix to Chapter 3).

cases $\gamma_1 = 1$, $\gamma_2 = 0$ and $\gamma_1 = \frac{4}{3}$, $\gamma_2 = 0$. Coley & Wainwright (1992), using different variables, give phase portraits for the cases $\gamma_1 = \frac{4}{3}$, $\gamma_2 = 1$ and $\gamma_1 = 1$, $\gamma_2 = 0$ (see their Figures 1 and 3). We refer to Section 8.2 for more details. Ehlers & Rindler (1989) consider three–fluid models, containing radiation, dust and a cosmological constant. They use the three density parameters $\Omega_1 = \Omega_{rad}, \Omega_2 = \Omega_{dust}$ and $\Omega_3 = \Omega_\Lambda$ as independent variables. The phase portraits for the two–fluid cases $\gamma_1 = \frac{4}{3}$, $\gamma_2 = 1$ and $\gamma_1 = 1$, $\gamma_2 = 0$ are shown in their Figures 2(ii) and 2(iv), respectively.‡

We note that the evolution equations for an n–fluid model using the density parameters $\Omega_1, \cdots, \Omega_n$ as variables assume a simple form. The deriva-

confused with our shear scalar). This paper was written before the density parameter Ω had assumed its current prominent role in theoretical cosmology.

‡ In all of these formulations the state space is unbounded, since for models that recollapse, Ω diverges at the instant of maximum expansion. This deficiency can be circumvented by introducing a 'compactified density parameter', that enables one to describe the complete evolution of all two–fluid FL models using a bounded subset of the plane (Wainwright 1996, Section 3.3). In this representation the Einstein static model is a saddle point.

tion of (2.46) and (2.47) can easily be generalized to give

$$\frac{d\Omega_A}{d\tau} = [2q - (3\gamma_A - 2)]\Omega_A, \quad A = 1, \cdots, n, \tag{2.51}$$

where

$$q = \tfrac{1}{2} \sum_{A=1}^{n} (3\gamma_A - 2)\Omega_A, \tag{2.52}$$

and γ_A are the equation of state parameters.

We finally give an evolution equation for the density parameter Ω which holds for all FL models, *independently of any assumptions about the source terms*. It follows from (2.7), (2.8), (2.11), (2.12), (2.44) and (2.45) that

$$\frac{d\Omega}{d\tau} = -2q(1 - \Omega). \tag{2.53}$$

Thus, if q satisfies $q \geq b > 0$ where b is a constant, then $\Omega \to 1$ at the initial singularity ($\tau \to -\infty$), as in the single–fluid models with $\gamma > \frac{2}{3}$. In addition, (2.53) confirms that the sign of $\Omega - 1$ and, on account of (2.13), the nature of the spatial geometry, cannot change as a model evolves.

We note that (2.53) holds, for example, for an n–fluid model described by (2.51), with

$$\Omega = \sum_{A=1}^{n} \Omega_A, \tag{2.54}$$

and q given by (2.52). In the language of Chapter 4, the conditions $\Omega < 1$, $\Omega = 1$ and $\Omega > 1$ define invariant sets of the flow determined by the DE (2.51).

2.4 Universes close to Friedmann–Lemaître

The high symmetry of the FL models leads to strong restrictions on the kinematic quantities, namely

$$\sigma_{ab} = 0, \quad \omega_{ab} = 0, \quad \dot{u}_a = 0. \tag{2.55}$$

This in turn implies that the Weyl tensor is zero,

$$E_{ab} = 0, \quad H_{ab} = 0 \tag{2.56}$$

and that the spatial gradients of all invariantly defined scalars are zero, for example

$$\hat{\nabla}_a \mu = 0, \quad \hat{\nabla}_a H = 0, \tag{2.57}$$

where $\hat{\nabla}_a = h_a{}^b \nabla_b$, and h_{ab} is the projection tensor (1.25).

The conditions (2.55) and (2.56) in fact characterize the FL models:

(1) a cosmological model which satisfies the Einstein field equations with perfect fluid source is a FL model if and only if (2.55) holds, (e.g. Krasinski 1996, Section 3.2).

(2) a cosmological model which satisfies the Einstein field equations with perfect fluid source and equation of state $p = p(\mu)$ is a FL model if and only if (2.56) holds, (e.g. Wainwright 1996, Section 4.1).

We wish be able to specify when a cosmological model is 'close to' an expanding FL model in a well–defined and invariant sense. It is not enough to require that the quantities (2.55)–(2.57) are small (i.e. less than some specified constant $\epsilon \ll 1$), since in an anisotropic ever–expanding model the shear tends to zero irrespective of whether the model isotropizes. The appropriate quantities are the dimensionless ratios formed by normalizing the quantities (2.55)–(2.57) with the expansion Θ or the Hubble scalar H. This leads us to consider scalars such as

$$\frac{\sigma}{H}, \quad \frac{|E_{ab}|}{H^2}, \quad \text{and} \quad \frac{|\hat{\nabla}_a \mu|}{H\mu},$$

where

$$|E_{ab}| = (E_{ab} E^{ab})^{\frac{1}{2}}, \quad |\hat{\nabla}_a \mu| = (\hat{\nabla}_a \mu \hat{\nabla}^a \mu)^{\frac{1}{2}}.$$

It is important to note that restricting σ_{ab}, ω_{ab} and $\dot{u}_a$ to be small relative to H does not guarantee that E_{ab}/H^2, for example, will be small, since this quantity is related to the time derivative of σ_{ab}, which may not be small, relative to H^2. It is thus necessary to impose stronger restrictions than may at first sight appear to be appropriate. On the other hand, it is not known what constitutes a minimal set for ensuring that a cosmological model is 'close to FL' in an appropriate way. It is clear, however, that a restriction on $\hat{\nabla}_a \mu$ will imply a corresponding restriction on $\dot{u}_a$, provided that an equation of state $p = p(\mu)$ holds. The definition below, which for simplicity is given for the case of a perfect fluid, should be regarded as tentative.

Definition 2.1. A cosmological model $(\mathcal{M}, \mathbf{g}, \mathbf{u})$ with perfect fluid source and equation of state $p = p(\mu)$ is said to be *close to Friedmann–Lemaître* in some open set $\mathcal{U}$ if and only if for some suitably small constant $\epsilon \ll 1$ the following inequalities hold in $\mathcal{U}$:

$$\frac{\sigma}{H} < \epsilon, \quad \frac{\omega}{H} < \epsilon, \quad \frac{|\hat{\nabla}_a\mu|}{H\mu} < \epsilon, \quad \frac{|\hat{\nabla}_a H|}{H^2} < \epsilon,$$

$$\frac{|E_{ab}|}{H^2} < \epsilon, \quad \frac{|H_{ab}|}{H^2} < \epsilon. \tag{2.58}$$

It is clear that if the inequalities (2.58) hold in some open set $\mathcal{U}$, they may not hold at earlier or later times, or at other points that are spatially distant. We note that if (2.58) holds, the metric in $\mathcal{U}$ can be locally written in almost–RW form (Stoeger *et al.* 1995).

Observations of galaxies can in principle lead to bounds on the above dimensionless quantities which quantify the departure from an exact FL model. The necessary theoretical framework was provided in the 1960s by Kristian & Sachs (1966), who gave a detailed analysis of observations in an arbitrary cosmological model which satisfies the field equations with dust as source. They assumed that all quantities can be expanded in a power series in s, an affine parameter along the past null geodesics, and derived the following fundamental equation:

$$\frac{f_1}{f_0} = 1 - (u_{a;b})_0 e^a e^b r - \tfrac{1}{2}(u_{a;bc})_0 e^a e^b e^c r^2 + \cdots \tag{2.59}$$

(see their page 386, equation (36)), where f_1 and f_0 are the frequencies of emission and reception, r is the luminosity distance (see their page 392, equation (68)), and e^a is a suitably normalized null vector giving the direction of the source. The coefficient of r depends only on the Hubble constant and the shear tensor, while the coefficient of r^2 also depends on the electric part of the Weyl tensor, the vorticity and certain spatial derivatives of the kinematic quantities, called the differential velocity gradient (see their pages 393–4, equations (70)–(74)). In addition, observations of anisotropies in the cosmic microwave background radiation can also lead to bounds on the above dimensionless quantities, provided one introduces a weak Copernican Principle (see Maartens *et al.* 1995a,b, and Section 3.2.5).

3

Cosmological observations

G. F. R. ELLIS

University of Cape Town

J. WAINWRIGHT

University of Waterloo

The FL universes, based on the RW metric, are the standard models of current cosmology. In this chapter we discuss cosmological observations with a view to assessing the evidence for these models.

There are two stages in this process of assessment:

(1) to discuss to what extent observations require the universe to be close to FL during the different epochs in its evolution,

(2) assuming the universe is close to FL, to discuss the observational constraints on the parameters that characterize an FL universe.

We group the observations that pertain to the first stage under three headings, namely, *discrete sources* (Section 3.1), the *cosmic microwave background radiation* (Section 3.2) and the *light–element abundances* arising from nucleosynthesis in the early universe (Section 3.3). In Section 3.4 we assess the extent to which these observations require the universe to be close to FL in different epochs. We do not discuss events at earlier epochs (e.g. baryogenesis; see Kolb & Turner 1990, Chapter 6) since we regard our current knowledge of the physics concerned as too tentative to lead to reliable constraints. Finally, in Section 3.5 we discuss the 'best–fit' FL parameters and the 'age problem'.

3.1 Observations of discrete sources

Observations of discrete sources (primarily galaxies, radio sources, infrared sources and quasars) provide information about the structure of the universe in the galactic epoch (say $z \lesssim 5$). These observations can be grouped into three classes, *number count surveys*, that provide direct evidence concerning isotropy about our position, *redshift surveys*, that provide information about inhomogeneities in the distribution of galaxies, and *peculiar velocity surveys*, that describe deviations from a uniform Hubble flow. We consider each in turn.

3.1.1 Number count surveys

A number count survey for galaxies is determined by binning counts from data on angular positions and apparent magnitudes (or angular diameters) for all galaxies in a specified region of the sky, up to a specified limiting apparent magnitude. A classic example is the Lick galaxy count survey taken in the 1960s, which is represented visually in Peebles (1993, see Figure 3.9, page 41). A modern version of the Lick survey used an automatic plate measuring system (APM) to create a survey of two million galaxies in a field of the sky covering 1.3 steradians near the south galactic pole (Maddox *et al.* 1990). Radio source count surveys have also been done (see Peebles 1993, Figure 3.10, page 43, for an example). Initially it was hoped to use these surveys to determine the sign of the spatial curvature and to test for spatial homogeneity; the attempt failed because of unknown source evolution (e.g. Peebles 1993, pages 22–3).

3.1.2 Redshift surveys

Number count surveys provide two–dimensional maps of regions of the sky, up to a limiting magnitude, but do not give reliable information about the distribution in space because the sources are not accurate standard candles, so apparent magnitude does not directly correlate with distance. A redshift survey is constructed by measuring the redshifts for all galaxies in a number count survey. By converting the redshift to a distance† using Hubble's law, one can create three–dimensional maps of the distribution of galaxies in a region of space. The maps are actually drawn in 'velocity space', using two angles and the 'velocity of recession' cz as coordinates, so as to avoid dependence on the uncertainties in determining the Hubble parameter (see Section 3.5.1). This type of survey, which requires many redshift measurements, became feasible in the late 1970s when measurement efficiencies increased. A visual representation of one of the most influential surveys, the CfA2 (Centre for Astrophysics) survey (de Lapparent *et al.* 1986), is shown in Peebles (1993, see Figure 3.6, pages 36–7). These surveys

---

† This reliance on redshift to represent distance inevitably results in distortion, as it does not allow for galaxy peculiar velocities relative to the local Hubble flow; thus for example a galaxy cluster will appear elongated in the radial direction from the observer (the 'finger of God' effect), because of random velocities within the cluster (e.g. Coles & Lucchin 1995, pages 70 and 404).

have shown structure (i.e. voids, superclusters, walls) on increasingly large scales, a finding that initially took the astronomical community by surprise. Research in this area is continuing. Geller, Huchra and collaborators are extending the CfA2 survey; another extensive survey near completion is the Las Campanas survey (LCRS), which will include 25,000 galaxies with a median redshift of $cz \sim 30,000$ km s^{-1} (i.e. $z \sim 0.1$, corresponding to a distance of $300h^{-1}$ Mpc), which is significantly deeper than previous surveys. We refer to Strauss & Willick (1995) for a comprehensive discussion of the variety, history and findings of redshift surveys.

3.1.3 Peculiar velocity surveys

Redshift surveys are unable to separate the two contributions to a galaxy's observed redshift: the *cosmological* component associated with the expansion of the universe (the large–scale Hubble flow) and the *peculiar* component associated with small–scale motions caused by local mass contributions. A peculiar velocity survey is constructed by making redshift–independent distance measurements d for all galaxies in a redshift survey. It is then assumed that the radial component of the peculiar velocity u can be calculated using $u = cz - H_0 d$, where H_0 represents the Hubble constant of the background RW metric that approximates the real universe.

Starting with the work of Rubin in the 1970s (see Rubin 1977 for a summary) observers have performed surveys on increasingly large scales and have discovered large–scale coherent motions (bulk flows) relative to the average Hubble flow. The bulk flow reported by Lauer & Postman (1994) has the largest scale and has generated considerable controversy. We refer to Strauss & Willick (1995) for a detailed discussion of this topic (galaxian distance indicator relations in Section 6.1, a history of the observations in Section 7.1 and the difficulties in constructing consistent peculiar velocity surveys in Section 7.2).

3.2 The cosmic microwave background radiation

The cosmic microwave† background radiation (CBR) has played an increasingly dominant role in cosmology since its discovery in 1965. It is interpreted as the relic radiation of a hot dense epoch, and is usually assumed to have travelled freely to us since decoupling of matter and radiation took place at

† We note that background radiation in other parts of the spectrum is also observed, for example, the X–ray background (e.g. Coles & Lucchin 1995, pages 418-19 and Peebles 1993, pages 43-4).

a time corresponding to a redshift‡ of about 1000. Much of the recent theoretical work on the CBR has related to using the deviations from isotropy to test theories of structure formation (e.g. White *et al.* 1994, and Scott *et al.* 1995), and to test the predictions of inflationary cosmology (e.g. Smoot & Steinhardt 1993). Our primary interest in the CBR lies in the fact that its high isotropy forms the strongest evidence that the universe is close to FL, at least from last–scattering to the present. Our main goal in this section is to describe and assess the theoretical work that has been used to reach this conclusion.

3.2.1 *Observations*

Much effort, culminating in the measurements by the Cosmic Background Explorer (COBE) satellite, has been expended in making precise measurements of the intensity of the CBR. There are two goals in making these measurements:

(1) to determine to what extent the spectrum is that of a black–body, and to determine the corresponding temperature by measuring the intensity at different frequencies, and

(2) to determine deviations from isotropy by comparing the intensity (and hence the temperature) in different directions, obtaining estimates for the angular fluctuation.

$$\frac{\Delta T}{T}(\theta, \phi) = \frac{T(\theta, \phi) - T_0}{T_0}, \tag{3.1}$$

where T_0 is the present mean temperature over the sky.

The COBE satellite has confirmed that the CBR spectrum is black–body to a high degree of accuracy, with a temperature $T_0 = 2.726 \pm 0.10$ K (e.g. Partridge 1994, page A157). As regards anisotropy, the dominant feature is a dipole variation with $\Delta T/T \approx 10^{-3}$, which is interpreted† as a Doppler effect due to our motion relative to a cosmological restframe defined by the radiation (*the CBR frame*). The conclusion is that the Local Group of galaxies is moving at a velocity of $v_{CBR} = 627 \pm 22$ km s^{-1} relative to the CBR frame in a direction $(\ell, b) = (276° \pm 3°, 30° \pm 3°)$ (White *et al.* 1994, page 319). Here (ℓ, b) are the usual galactic coordinates (e.g. Coles & Lucchin 1995, page 61).

---

‡ It is possible that reheating of the intergalactic matter took place at a late stage, in which case the effective surface of last–scattering is much closer — at a redshift of between 10 and 100.

† If this interpretation is correct, there should be a corresponding anisotropy in the number counts, irrespective of evolution and selection effects (Ellis & Baldwin 1984). At present the number count surveys have not reached a level of sensitivity to allow the detection of this effect.

When the dipole anisotropy is removed by taking this velocity into account there remains an anisotropy on angular scales $\Delta\theta \geq 10°$ of $\Delta T/T \approx 10^{-5}$, as detected by COBE (see Partridge 1994, pages A159–A160 for a summary) and now confirmed by other groups (e.g. Hancock *et al.* 1994). On smaller angular scales, the observations give an upper bound on $\Delta T/T$ of order 10^{-5}; several claims of detection of anisotropy have been made (e.g. Netterfield *et al.* 1995), but to date none has been confirmed by a second group (see also Coles & Lucchin 1995, page 384).

3.2.2 *Origins and analysis of temperature fluctuations*

When analyzing the anisotropies in the CBR it is necessary to distinguish between fluctuations at large angular scales and at small angular scales. By large angular scales we mean $\Delta\theta > \theta_h$, where θ_h (h for horizon) is the angle subtended by the light rays emitted at last–scattering from two points whose linear separation is twice the distance to the particle horizon at that time. In a FL model,

$$\theta_h \approx 3°[1300\Omega_0/z]^{\frac{1}{2}}, \tag{3.2}$$

where $z \gg 1$ is the redshift at last–scattering (e.g. Panek 1986, page 418). If $z = 1300$, $\Omega_0 = 0.1$, then $\theta_h \approx 1°$. The significance of this is that fluctuations on an angular scale $\theta \gtrsim \theta_h$ have not been affected by causal processes prior to last–scattering (provided inflation has not taken place). These fluctuations, which relate to the structure of the universe on the largest scales, are the ones that are most relevant in determining closeness to FL.

Various physical phenomena which can generate small temperature fluctuations in the CBR on large angular scales $\Delta\theta \gtrsim \theta_h$ have been described in the literature:

(1) Density fluctuations at the surface of last–scattering: the Sachs–Wolfe (1967) effect.

(2) Intervening density perturbations such as voids and superclusters of redshifts $0 < z < 4$: the Rees–Sciama effect. The analyses use 'Swiss–Cheese' models (Rees & Sciama 1968, and for example Dyer 1976, Mészáros 1994). Dekel (1994, page 395) links bulk velocities in the Hubble flow ($v \approx$ 300 km s^{-1} across a distance $d \approx 100h^{-1}$ Mpc)†, presumably generated by large–scale density inhomogeneities, with temperature fluctuations of order 10^{-5} on scales of 1°.

(3) Large–scale anisotropy in the Hubble flow. This anisotropy, described

† See Appendix for explanation of h and cosmological units.

by the normalized shear σ/H of the fundamental congruence and viewed as having a primordial origin, contributes to the fluctuations through an integral from last–scattering to the present.

Most theoretical analyses of these effects start with the assumption that *the universe is close to FL back to last–scattering*, which permits the use of a perturbed FL model. The goal of these analyses, which are discussed in Sections 3.2.3 and 3.2.4, is to relate quantities which describe the departure from an exact FL model to the temperature fluctuation $\Delta T/T$. Equally important for our purposes is to assess whether the CBR observations do require the universe to be close to FL. This question is discussed in Section 3.2.5, where we invoke a 'weak Copernican Principle'.

3.2.3 The Sachs–Wolfe analysis

The classic paper by Sachs & Wolfe (1967) considered perturbations of the flat FL universe, and (using a particular gauge) determined the CBR anisotropies that would result from perturbations of the density and velocity fields at the surface of last–scattering. As these could also be related to expected present–day matter velocities and inhomogeneities, this in principle enabled one to relate such present day features to the observed CBR anisotropies.

This paper has been the foundation of most further analyses of the CBR anisotropies, at large angular scales. These analyses have considered background FL universes with both flat and curved spatial sections, and have addressed the question of gauge–invariance. The key point in these derivations is that the observed radiation temperature T_0 is related to the emitted temperature T_1 by the formula

$$T_0 = \frac{T_1}{1+z},\qquad (3.3)$$

where z is the redshift of the matter that emitted the radiation at the surface of last–scattering. This redshift is calculated in the perturbed universe model using the fundamental relation

$$1 + z = \frac{(k^a u_a)_1}{(k^b u_b)_0},\qquad (3.4)$$

where k^a is tangent to the null geodesics joining the emitter (subscript 1) to the observer (subscript 0), and u^a is the fundamental 4–velocity. One evaluates this relation by integrating the null geodesic equations from the surface of last–scattering to the observation event.

The reader should be warned that the so–called *Sachs–Wolfe effect* is described by a variety of (superficially) different formulas in the literature. The problem is partly the lack of a uniform notation, and partly that different authors include different effects. The simplest formula, describing the effect of adiabatic density fluctuations, is

$$\frac{\Delta T}{T} \approx \frac{1}{3} \frac{\delta \rho}{\rho} \left(\frac{\lambda}{ct}\right)^2, \tag{3.5}$$

where ρ is the density, λ is the scale of the perturbation at time t, and ct approximates the distance to the horizon. This formula can be derived heuristically, by thinking in terms of photons climbing out of a potential well (see Coles & Lucchin 1995, page 374); however, it is gauge–dependent. More complicated formulas, derived using (3.3) and (3.4) are given by Sachs & Wolfe (1967, page 83, equation (43)), and, for example, by Panek (1986, page 421, equation (53)), Russ *et al.* (1993, page 4554, equation (10) or (13)) and by White *et al.* (1994, page 363, equation (43)). We note that Panek (1986) uses the Bardeen (1980) gauge–invariant variables, while Russ *et al.* (1993) use the gauge–invariant variables of Ellis & Bruni (1989). Finally we mention that Stoeger *et al.* (1994) have used so–called observational coordinates, which are adapted to null geodesics and hence simplify the calculation.

3.2.4 Bounds on shear and vorticity

The effect of large–scale anisotropy in the Hubble flow on the temperature fluctuations has been analyzed by applying the basic relations (3.3) and (3.4) in spatially homogeneous models. These analyses relate kinematic scalars such as σ/H and ω/H to the temperature fluctuation $\Delta T/T$. In the simplest case, namely LRS Bianchi I models, $\Delta T/T$ depends only on one angle θ, and one can integrate the null geodesics explicitly. The result is

$$\frac{\Delta T}{T}(\theta) = \pm \frac{4}{3\sqrt{3}} \left(\frac{\sigma}{H}\right)_0 (1 + z_{\mathrm{d}})^{\frac{3}{2}} P_2(\cos \theta), \tag{3.6}$$

where σ/H is evaluated at the present time, z_{d} is the redshift at last–scattering and P_2 is a Legendre polynomial (e.g. Barrow *et al.* 1983, page 398). In this case the temperature fluctuation is a quadrupole; since no peculiar velocities are permitted, there is no dipole contribution. The observational result $\Delta T/T \approx 10^{-5}$, with $z_{\mathrm{d}} = 1000$ in (3.6), yields

$$\left(\frac{\sigma}{H}\right)_0 \approx 1.6 \times 10^{-9}. \tag{3.7}$$

In models of more general Bianchi type the geodesic equations cannot be integrated exactly. The approach taken has been to either restrict consideration to perturbed FL models,† or to integrate the geodesic equations numerically. The first systematic treatment using perturbed FL models was that of Collins & Hawking (1973a), which included all Bianchi types compatible with the RW symmetry, i.e. types I, V, VII_0, VII_h and IX (see Table 1.1). As regards numerical work, Bajtlik *et al.* (1986) have given a detailed analysis and have produced temperature maps of the sky for non-tilted Bianchi V models. Overall, these analyses have shown that in Bianchi models there are two main types of anisotropy, a quadrupole, and a 'hot spot', with or without a dipole. In addition, in Bianchi types VII_0 and VII_h the null geodesics exhibit a spiralling behaviour. We refer to Barrow *et al.* (1983) and Barrow *et al.* (1985, in particular, page 922) for more details.

As regards bounds on the shear, Bajtlik *et al.* (1986) give the bound

$$\left(\frac{\sigma}{H}\right)_0 < 3 \times 10^{-8}, \qquad (3.8)$$

for non-tilted Bianchi V models (weaker than the bound (3.7) for Bianchi I). Barrow (1987a, page 270) has updated the older bounds of Collins & Hawking (1973a) for $(\sigma/H)_0$ (see their Table 1, page 309) in types VII_0, VII_h and IX. Barrow *et al.* (1985) have given upper bounds for the vorticity $(\omega/H)_0$ and the associated peculiar velocities (i.e. tilt) in types V, VII_0, VII_h and IX (see their Tables 1–4, pages 932–3, and equations (5.31) and (5.32) for IX).

We finally note that most analyses to date only consider those Bianchi types that are compatible with the RW symmetry (i.e. they exclude types II, IV, VI_h and VIII; see our Table 1.1). The only exceptions of which we are aware are Raine (1987), who investigated the temperature fluctuation in a Bianchi VI_h model with a flat quasi-isotropic epoch (see Section 5.3.3), and the qualitative analysis of Doroshkevich *et al.* (1975). We stress that flat quasi-isotropic epochs can occur in models of all Bianchi types, making them potentially compatible with observations (see Section 15.2.1).

3.2.5 A model-independent analysis

It has been shown by Stoeger, Maartens & Ellis (1995), generalizing a theorem by Ehlers *et al.* (1968), that if the CBR is measured to be almost isotropic by every member of a family of geodesically moving observers in a

---

† We note that Doroshkevich *et al.* (1975) argue against restricting consideration to perturbed FL models, and use a piecewise-approximation analysis (Doroshkevich *et al.* 1973) to analyze the CBR in models of types VII, VIII and IX.

space–time region $\mathcal{U}$, then the geometry is close to FL in $\mathcal{U}$, in the sense of (2.58) . The proof follows from the Einstein–Liouville equations for freely propagating radiation. So we can argue as follows: we observe the CBR to be isotropic to a high degree of accuracy from our space–time position ('here and now'). If we are not in a special space–time position, the same should be true for all observers; but then the theorem applies and shows the universe is close to FL. Note that the conditions for the theorem are not observationally verifiable, for we cannot view the universe from other space–time positions; but the inference is made plausible by explicit adoption of a weak Copernican Principle, namely the assumption that we are not privileged observers, so what we see should be typical of all observers. Application of this argument shows it is plausible the universe is close to FL in a region $\mathcal{U}$ of arbitrary spatial extent,[†] extending in time[‡] back from today to last–scattering.

Analyzing the proof more closely enables one to bound the deviation from a FL model in the region $\mathcal{U}$, by obtaining limits on the kinematic quantities, the density gradient and the Weyl tensor in terms of the assumed limits on the temperature anisotropy (Maartens *et al.* 1995a,b). This analysis shows that if we interpret the CBR dipole as due to our motion relative to the CBR frame, then we obtain the following bounds on the quantities in (2.58):

$$\frac{\sigma}{H} < C_1\epsilon, \quad \frac{\omega}{H} < C_1\epsilon, \quad \frac{|\hat{\nabla}_a\mu|}{H\mu} < C_1\epsilon, \quad \frac{|\hat{\nabla}_aH|}{H^2} < C_1\epsilon, \qquad (3.9)$$

$$\frac{|E_{ab}|}{H^2} < C_2\epsilon, \quad \frac{|H_{ab}|}{H^2} < C_3\epsilon, \qquad (3.10)$$

where C_1 is a constant of order unity, C_2 is of order 10^2, C_3 is of order 10, and ϵ is the order of magnitude of the CBR quadrupole anisotropy (about 10^{-5}).

3.3 Big–bang nucleosynthesis (BBN)

Big–bang nucleosynthesis (conventionally abbreviated to BBN) provides a link between observations and the very early universe, when the temperature was of the order of 10^9K (redshift $z \approx 10^9$). The idea originated in the pioneering work of Gamov, Alpher and Herman in the late 1940s, and has been pursued intensively since the 1960s, when the discovery of the CBR

---

† Or at least of the same order as the present horizon size; caution suggests we should not assume too much about what happens far outside our past light cone (Ellis 1975).

‡ When applying the weak Copernican Principle in time, extending back to last–scattering, one idealizes the notion of observer.

provided evidence of an early hot epoch in the universe. Our goal is to briefly describe the background and the current results, and to address the issue of to what extent the analysis constrains the universe to be close to FL. We refer to Coles & Lucchin (1995, pages 166–75) for an introduction to this topic and to Kolb & Turner (1990, Chapter 4) for a detailed discussion of the physics.

3.3.1 Standard BBN

Observations of various astronomical objects show that the fraction of helium (^{4}He) by mass, denoted by Y_p, is roughly $Y_p \approx 0.25$. Nucleosynthesis in stars, however, is unable to explain such a high value, which provides an observational motivation for investigating nucleosynthesis in the early universe. The detailed study of nuclear reactions in a hot expanding universe shows that this process can produce ^{4}He and the light isotopes ^{3}He, D and ^{7}Li, but no significant amounts of any heavier nuclei. The standard analysis assumes a FL model with radiation equation of state, and calculates the abundances of the light isotopes numerically because of the complex chain of reactions that take place. It is found that *the abundances depend on the baryon–to–photon ratio η*, least sensitively in the case of ^{4}He (see Copi *et al.* 1995, page 193, Figure 1). For a given η, the mass fraction Y_p for ^{4}He, is determined very accurately (to within 0.001), but, due to uncertainties in the detailed physics, the ratios for ^{3}He, D and ^{7}Li are determined less accurately (within 10–25%; see Copi *et al.* 1995, page 194).

3.3.2 Observations of light–element abundances

The next step is to compare these theoretical predictions with the observed abundances. There is a major difficulty. The objects used to observe the light–element abundances are at low redshift,† and thus the observed abundances cannot be compared directly with the calculated primordial abundances: one has to take into account that nuclear reactions in stars and elsewhere during the galactic epoch affect the abundances. Copi *et al.* (1995, page 195) give $0.221 < Y_p < 0.243$ as a reasonable estimate for the primordial abundance of helium as inferred from the observations. A recent re-evaluation of the systematic errors by Sasselov & Goldwirth (1995) suggests that the upper bound could be as large as 0.255. Observations likewise lead

---

† Recent observations of Jakobsen *et al.* (1994), however, provide evidence for primordial helium in the intergalactic gas at redshifts greater than 3 (Longair 1995, page 495).

to abundances for the other light isotopes, expressed as fractions by numbers: $D/H \approx 10^{-5}$, $^3He/H \approx 10^{-5}$, $^7Li/H \approx 10^{-10}$ (Copi *et al.* 1995, pages 194–6). The final conclusion is that if η satisfies the bounds

$$2.5 \times 10^{-10} < \eta < 6 \times 10^{-10}, \tag{3.11}$$

then there is agreement between predicted and measured abundances of all four light elements (Copi *et al.* 1995, page 192). We note, however, a re–analysis by Hata *et al.* (1995) questions whether the observations and predictions are compatible.

3.3.3 Effect of anisotropy and inhomogeneity

Standard BBN is based on the flat FL universe with radiation equation of state. We now discuss what is known about the effect of using an anisotropic or inhomogeneous universe model. A significant factor in determining the helium abundance is the relation $T(t)$ between the radiation temperature and time at nucleosynthesis. For the flat radiation model we have

$$T(t) \quad \alpha \quad t^{-\frac{1}{2}}, \tag{3.12}$$

while for a Bianchi I model in the epoch in which shear is dominant,

$$T(t) \quad \alpha \quad t^{-\frac{1}{3}}, \tag{3.13}$$

as first pointed out by Hawking & Tayler (1966, equations (1) and (3)). This difference simply reflects the difference in the rate of expansion, as described by the Hubble scalar, $H = \frac{1}{2}t^{-1}$ in FL versus $H = \frac{1}{3}t^{-1}$ in a Kasner–like model. These authors gave a heuristic argument to the effect that as the anisotropy increases from zero the helium production will initially increase but then decrease to zero. This conclusion has been reached by other authors: Peebles (1966) rescaled the time–temperature relation with a constant factor in order to mimic the effect of anisotropy while integrating the equations in a FL model (see his equation (7), page 545 and Figures 1 and 2, page 550), while Olson (1978) integrated the full Bianchi I equations (*ibid*, figure 1, page 778). Thorne (1967) was the first to calculate the helium abundance in a Bianchi I model. Rothman & Matzner (1984) incorporated more detailed physics in the analysis. From their paper one can infer a bound on the shear during nucleosynthesis:

$$Y_p < 0.26 \quad \text{requires} \quad \left(\frac{\sigma}{H}\right)_{t=t_n} < 0.2, \tag{3.14}$$

(*ibid*, table 4 and page 1660; the quantity ρ_β/ρ_{rad} corresponds to Σ^2/Ω, where Ω and Σ^2 are defined by our (5.16) and (5.17), with $\Sigma^2 + \Omega = 1$ for Bianchi I models). This places very strong limits on $(\sigma/H)_{t=t_0}$ in these models.

As regards more general Bianchi types, Barrow (1976) considered all types compatible with a FL model, and obtained bounds on $(\sigma/H)_{t=t_n}$. Subsequently Barrow (1984) gave a simplified analysis of helium production in four Bianchi models with anisotropic spatial curvature. These models all satisfy the FL relation (3.12), although the constant of proportionality differs, and for this reason give a helium abundance which does not differ greatly from the FL result. For example, using the Collins–Stewart type II solution Y_p is only increased by 0.01 even though $(\sigma/H)_{t=t_n} \approx 0.43$ (see Barrow 1984, Table 1, page 225), while in the Collins VI_0 solution Y_p increases by 0.05 and $(\sigma/H)_{t=t_n} \approx 0.87$. The models used by Barrow are in fact self–similar, which implies that dimensionless quantities such as σ/H and Ω are constant (see our Table 6.3; the values of Σ and hence of σ/H can be obtained from our Section 6.2.1). These models are thus not potential models of the universe over an extended period of time. They do make the point, however, that significant amounts of shear at $t = t_n$ do not necessarily lead to helium abundances in serious conflict with observations. We further note that Matravers *et al.* (1984) have shown that LRS Bianchi V models with tilt can produce acceptable helium abundances. In a subsequent paper Matzner *et al.* (1986) questioned the validity of this analysis, and used timescale arguments to show that helium production will be raised with respect to that in FL universes for Bianchi universes of types I, V and IX for acceptable parameter values. They also conjectured that this is true for all Bianchi types.

It is of interest that, because the shear in a Bianchi I model is monotone decreasing, the bound (3.14) leads to a bound on $(\sigma/H)_{t=t_0}$ that is stronger than the bound (3.7) deduced from the CBR observations (e.g. Olson 1978). A similar extrapolation can be made in non–tilted Bianchi V models, for which the shear is also decreasing, but not more generally.

The effect of an inhomogeneous background model on BBN has been investigated using the numerical code for G_2 cosmologies developed by Centrella & Wilson (1983, 1984). Kurki–Suonio (1989) has described a series of numerical experiments and reports that the inhomogeneity has only a moderate effect on the abundances (*ibid*, page 337). Further research on this question would be desirable.

3.4 Is the universe close to FL?

In order for a FL universe to be a viable model of the real universe during the galactic epoch, a number of compatibility conditions have to be satisfied.

First, in an ideal FL universe the matter density is uniform, while in the real universe the matter is not distributed homogeneously even on the scale of clusters of galaxies, as revealed by redshift surveys. In order for the real universe to be described by a FL model it is thus necessary that there exists a distance scale d Mpc over which the distribution of matter is homogeneous (e.g. 'cubes of side 300 Mpc are indistinguishable'). As emphasized by Ellis *et al.* (1996) the number d is an important parameter in our attempts to model the universe. At the present time it is not clear on what scale the matter distribution can be regarded as homogeneous, although it appears that the scale must be well in excess of $100h^{-1}$ Mpc (see for example Tully *et al.* 1992). Ellis *et al.* (1996) conclude that homogeneity on a scale of $300h^{-1}$ Mpc is compatible with the data but cannot be inferred from it.

Second, in an ideal FL universe the Hubble flow is uniform (isotropic) while in the real universe the peculiar velocity surveys reveal deviations from uniformity. In order for the real universe to be described by a FL model, it is necessary that there exists a distance scale beyond which the peculiar velocities are small compared with the Hubble velocity, so that sufficiently distant galaxies and clusters will define a frame of reference (the *large–scale Hubble frame*) relative to which the large–scale Hubble flow will be (highly) isotropic. In principle, one can use the peculiar velocity surveys to determine the velocity v_H of the Local Group relative to this frame. Once a large–scale Hubble frame has been established, any small residual (large–scale) anisotropies in the flow would lead to bounds on quantities which characterize departures from a FL model such as σ/H and ω/H. The requirement that the Hubble frame and the CBR frame coincide in an FL model leads to the consistency requirement $v_H = v_{CMB}$ (one speaks of the 'convergence of the local velocities'). At present it is not clear on what scale convergence occurs, although the results of Lauer & Postman (1994) suggest that the scale may be in excess of $100h^{-1}$ Mpc.

It should be kept in mind that at present the redshift and peculiar velocity surveys do not extend beyond a redshift of $z = 0.1$ ($d \approx 300h^{-1}$ Mpc). The number count surveys extend to greater distances as do redshift measurements of individual galaxies. These observations do not reveal any gross departures from isotropy, but have not reached sufficient sensitivity to provide strong upper bounds on possible anisotropies.

The strongest evidence for the universe being close to FL in the galactic epoch, however, is the isotropy of the CBR, based on adoption of the weak Copernican Principle. The analysis in Section 3.2.5 provides bounds on the dimensionless scalars that describe closeness to FL in terms of observational bounds on $\Delta T/T$. These bounds also apply to the post–decoupling period, prior to galaxy formation.

As regards the pre–decoupling epoch, the main observational link is through BBN. It is certainly satisfying that there is a range of η values for which the FL abundances are compatible with the observed abundances. However, the fact that only a limited range of anisotropic and inhomogeneous models have been analyzed, combined with the uncertainties in the detailed physical processes and in the analyses of the observed abundances, precludes making a strong claim that the universe must be close to FL at the time of nucleosynthesis.

3.5 The best–fit FL parameters and the age problem

Despite the uncertainties associated with the hypothesis of a RW geometry in the galactic era (i.e. the averaging scales which give a homogeneous matter distribution and an isotropic Hubble flow have not been unambiguously established), it is reasonable to accept the hypothesis and attempt to use observations to determine the parameters that characterize a FL model.

As discussed in Section 2.2 a single–fluid FL model is characterized by four observational parameters $\{t_0, H_0, q_0, \Omega_0\}$. These parameters satisfy two constraints, namely (2.26) and (2.32), which for dust ($\gamma = 1$) read:

$$q_0 = \tfrac{1}{2}\Omega_0, \qquad t_0 H_0 = \alpha(\Omega_0), \qquad (3.15)$$

where $\alpha(\Omega_0)$ is given by (2.34). Thus, the models can in principle be falsified by observations. In particular, since $\alpha(\Omega_0) \leq 1$ for all Ω_0, we have the *age inequality*

$$t_0 H_0 \leq 1. \qquad (3.16)$$

In this section we give a brief discussion of the restrictions that are imposed on $\{t_0, H_0, q_0, \Omega_0\}$ by observational data.

3.5.1 Hubble constant and age estimates

The Hubble constant H_0 can in principle be determined by measuring the luminosities and redshifts of galaxies (see (2.42)). A major difficulty is that for nearby galaxies ($d \leq 3$ Mpc), whose distances are reliably known, the peculiar velocities can be comparable to the Hubble velocity. On the other hand, for more distant galaxies there are uncertainties in the distance scale. In particular, the distance to the Virgo cluster is not reliably known. Fukugita *et al.* (1993) summarize the two standard methods for calibrating the extra–galactic distance scale, the one leading to a 'low' H_0 (50 ± 10 km s^{-1} Mpc^{-1}) and the other to 'high' H_0 (80 ± 10 km s^{-1} Mpc^{-1}). Since then, observation of Cepheid variable stars in the Virgo cluster by two groups (Pierce *et al.* 1994, Freedman *et al.* 1994) have lent support to the high value, as has recent work using Type Ia and Type II supernovae (summarized in Ellis *et al.* 1996). Ellis *et al.* (1996) use a Bayesian estimation procedure to combine all estimates and find

$$66 < H_0 < 82 \, \text{km} \, \text{s}^{-1} \, \text{Mpc}^{-1}, \tag{3.17}$$

at a 95% confidence interval. However, one should note that a consensus has not been reached: the value of the Hubble parameter is the subject of ongoing controversy.

A lower bound for the age of the universe t_0 can be estimated by determining the age of various astronomical objects. The oldest stars are typically observed in globular clusters, and their ages are estimated to be 15 ± 3 Gyr (Peebles *et al.* 1991, page 773). Fukugita *et al.* (1993) comment that stellar theories cannot accommodate a universe younger than† 13 Gyr i.e.

$$t_0 > 13 \, \text{Gyr}, \tag{3.18}$$

(see page 312). We refer to Ellis *et al.* (1996) for more details of the methods and for a discussion of the uncertainties and errors involved in making these estimates. One can also date lunar and meteoritic rock using long–lived radioactive isotopes such as U^{238}, leading to estimates for the age of the galaxy of 9–16 Gyr (Symbalisty & Schramm 1981). We refer to Sciama (1993, page 59) for further references.

If we combine the Hubble parameter estimate (3.17) with the age estimates we can obtain bounds for the age parameter $\alpha(\Omega_0)$ in (3.15). Using the lower bound (3.18) and taking an upper bound $t_0 < 20$ Gyr (somewhat arbitrarily) we obtain

$$0.87 < \alpha(\Omega_0) < 1.68. \tag{3.19}$$

In deriving these bounds we use the approximate conversion formula (3.25), given in the Appendix. It follows that *the bounds (3.17) and (3.18) are not compatible with the Einstein–de Sitter universe*, for which $\alpha(\Omega_0) = \frac{2}{3}$.

---

† In a recent analysis of the 17 oldest globular clusters by Chaboyer *et al.* (1996), the one–sided 95% confidence limit lower bound was found to occur at an age of 12.07 Gyr and the median age was 14.56 Gyr.

In addition there is the possibility that the age inequality (3.16) will be violated, necessitating the introduction of a cosmological constant. In any case, since $\alpha(\Omega_0)$ is a decreasing function, the lower bound in (3.19) leads to an upper bound for Ω_0. It follows from (2.34) that

$$0.87 < \alpha(\Omega_0) \quad \text{implies} \quad \Omega_0 < 0.2. \tag{3.20}$$

We note that this result assumes a zero cosmological constant, and is the result of a complex chain of argument that takes an optimistic view of quoted error estimates. It should be treated with some caution.

3.5.2 Direct determination of Ω_0

Following Sciama (1993, page 61) we can distinguish three different types of estimate for Ω_0, labeled Ω_{vis}, Ω_b, Ω_{dyn}.

(1) *The observed baryonic density* Ω_{vis}. This quantity represents the visible baryons in galaxies and clusters, and is estimated using mass–to–luminosity ratios. In a recent analysis, Persic & Salucci (1992) find $\Omega_{vis} \approx 0.003$.

(2) *The derived baryonic density* Ω_b. This quantity is derived from the BBN analysis, which leads to the bound (3.11) on the ratio $\eta = n_b/n_\gamma$, where n_b and n_γ are the baryon and photon number densities. The densities n_b, n_γ vary as the universe expands, but η remains a constant. The observed present CBR temperature T_0 determines the present photon energy density n_γ. Knowing η, one can successively calculate the present values of n_b and μ_b. The resulting bounds on Ω_b depend on the Hubble constant:

$$0.009 \leq \Omega_b h^2 \leq 0.02, \tag{3.21}$$

(Copi *et al.* 1995).

(3) *The dynamical matter density.* This quantity is estimated by determining how much mass is needed to cause the observed orbital motions in galaxies and clusters, using galaxy rotation curves at galactic scales, and at cluster scales by applying the virial theorem to the galaxies in a cluster. On even larger scales, one can use the large–scale streaming flows of matter detected by the peculiar velocity surveys. For example, a method devised by Bertschinger & Dekel (1989), called POTENT, uses peculiar velocity surveys to construct a map of the density field in space and hence estimate Ω_0 (e.g. Nusser & Dekel 1993). There seems to be agreement that $\Omega_{dyn} \geq 0.1$ (Sciama 1993), while some dynamical estimates on scales $d \geq 30 \, h^{-1}$ Mpc give values up to 1 (Peebles 1993, page 476 and Strauss & Willick 1995,

Section 7.5.1; however, see Bahcall *et al.* 1995 and White *et al.* 1993). One might hope that Ω_{dyn} would represent the total matter density, giving Ω_0; however a uniform distribution of matter would not be detected by the dynamical tests.

The differences in the previous estimates are important:

$$\Omega_{vis} < \Omega_b < \Omega_{dyn}. \tag{3.22}$$

The fact that $\Omega_{vis} < \Omega_b$ implies that a significant amount of baryonic matter (perhaps as high as 90%) is *dark*. We refer to Sciama (1993, pages 62–3) for a brief discussion of the possibilities, and to Carr (1994) for further details. We also mention that the statistics of gravitational lensing observations have the potential to give estimates of the dark matter density (e.g. Peebles 1993, pages 336–9). Secondly, $\Omega_b < \Omega_{dyn}$ implies the existence of *non-baryonic dark matter*. Since non–baryonic dark matter has not been detected to date, this state of affairs is controversial (see Peebles 1993, page 677, for comments) and might lead one to question the upper bound in (3.21) provided by BBN. Indeed the re–assessment by Sasselov & Goldwirth (1995) led to an increase in the upper bound (3.18) and the possibility that $\Omega_b \approx \Omega_{dyn}$.

In summary, we note that Coles & Ellis (1994) have reviewed the various observational evidence for Ω_0 and suggest that a good present estimate is

$$0.1 \lesssim \Omega_0 \lesssim 0.3, \tag{3.23}$$

indicating that an open rather than a closed FL model fits the universe better (assuming zero cosmological constant). However one must bear in mind the remark above: a uniform density will not be detected by dynamical estimates, so the value could be higher if there is a distribution of non–baryonic dark matter that is uniform on very large scales.

3.5.3 *Direct determination of* q_0

The classical cosmological tests, for example the magnitude–redshift relation, in principle enable one to determine the deceleration parameter q_0, and hence the density parameter Ω_0 via (3.15). This effort has not succeeded, basically owing to observational selection effects and poor statistics at the detection limit, together with lack of standard candles and in particular an inability to adequately understand the evolution of classes of sources as one goes back into the past. However there is now a prospect that supernovae light curves might enable us to determine q_0 directly (Ellis *et al.* 1996).

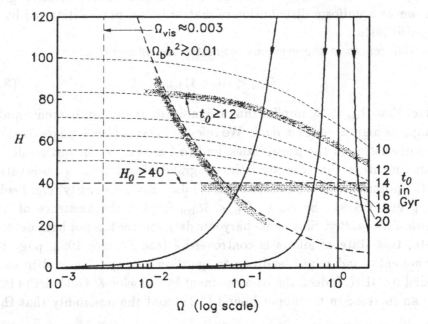

Fig. 3.1. Observational constraints on dust FL models in the ΩH–plane. The units for H are km s^{-1} Mpc^{-1}.

3.5.4 Observational constraints on state space

An advantage of using a dynamical systems approach in analyzing the evolution of cosmological models is that the observational data restrict the region of state space in which the present state of the universe may lie, thereby giving a visual interpretation of the observations. One can do this for the dust FL models by using the ΩH–plane, as in Section 2.3 (see Figure 2.2). It is convenient to use a log scale on the Ω–axis, as in Figure 3.1 (see Gott *et al.* 1974, page 544, Figure 1). We have drawn the age lower limit $t_0 \geq 12$ Gyr, the BBN lower bound $\Omega_b h^2 \geq 0.01$ and the Hubble lower bound $H_0 \geq 40$ as an illustrative example. The figure clearly illustrates the severe constraints that a 'high' value of H_0 imposes. However, it also shows that there is a region of the FL state space that is compatible with all current observational data, in particular (3.17)–(3.20) (see Ellis *et al.* 1996).

Appendix to Chapter 3
Cosmological units

In this appendix we give the units that are used to measure the Hubble constant, and cosmological distances and times. Extragalactic distances are measured in *megaparsec* (Mpc), where 1 Mpc $\approx 3 \times 10^6$ light years. Cosmological times (e.g. the age t_0 of the universe) are measured in gigayears (Gyr) where 1 Gyr $= 10^9$ years. The Hubble constant H_0 is traditionally measured in units of km s^{-1} Mpc^{-1}, since the redshift was interpreted as a classical Doppler shift due to a velocity of recession v, giving $v = H_0 d$. To account for the uncertainty in the value of H_0 (see Section 3.5.1) it is customary to write

$$H_0 = 100h \text{ km s}^{-1} \text{Mpc}^{-1}, \tag{3.24}$$

where h is a dimensionless constant. The constant H_0^{-1} can also be expressed in units of time

$$H_0^{-1} \approx 9.8 h^{-1} \text{ Gyr} \approx 3 \times 10^{17} h^{-1} \text{s}, \tag{3.25}$$

(e.g. Coles & Lucchin 1995, page 73). The quantity cH_0^{-1}, where $c \approx 3 \times 10^5$ km s^{-1} is the speed of light in vacuo, has units of length:

$$cH_0^{-1} \approx 3000 h^{-1} \text{ Mpc}, \tag{3.26}$$

(*ibid*, page 73). The quantity H_0^{-1} is called the *Hubble time* and gives the timescale of the age of the universe in the standard model, while cH_0^{-1} is called the *Hubble distance* and is of the order of magnitude of the presently visible universe (i.e. the distance to the particle horizon). Finally we note that H_0 determines the present value of the critical density $\mu_{\text{crit},0}$:

$$\mu_{\text{crit},0} \approx 1.9 \times 10^{-29} h^2 \text{ g cm}^{-3}. \tag{3.27}$$

4
Introduction to dynamical systems

R. TAVAKOL

Queen Mary & Westfield College

In this chapter we give a brief overview of some aspects of the theory of dynamical systems. We assume that the reader is familiar with the theory of systems of linear differential equations, and with the elementary stability analysis of equilibrium points of systems of non–linear differential equations (e.g. Perko 1991). We emphasize instead the fundamental concept of the flow and various other geometrical concepts such as α– and ω–limit sets, attractors and stable/unstable manifolds, which have proved useful in applications in cosmology. In the interest of readability we have stated some of the definitions and theorems in a simplified form; full details may be found in the references cited. One important aspect of the theory that we do not discuss due to limitations of space is structural stability and bifurcations. We refer to Perko (1991, chapter 4) for an introduction to these matters. We also note that the discussion of chaotic dynamical systems is deferred until Chapter 11.

To date, applications of the theory of dynamical systems in cosmology have been confined to the finite dimensional case, corresponding to systems of ordinary differential equations, although in Chapter 13 we obtain a glimpse of the potential for using infinite dimensional dynamical systems. We restrict our discussion to the finite dimensional case, referring the interested reader to books such as Hale (1988), Temam (1988a,b) and Vishik (1992) for an introduction to the infinite dimensional case.

4.1 Differential equations and flows

4.1.1 Autonomous differential equations and state space

As described in the Introduction the theory of dynamical systems is used to study physical systems whose state at an instant of time t can be described by an element $\mathbf{x}$ of a *state space* X and whose evolution is governed by an autonomous differential equation on X, written symbolically

$$\frac{d\mathbf{x}}{dt} = \mathbf{f}(\mathbf{x}), \tag{4.1}$$

where $\mathbf{f} : X \to X$. In this chapter we deal with the case $X = \mathbb{R}^n$, $\mathbf{x} = (x_1, \ldots, x_n)$, in which case (4.1) represents a *system of n ordinary differential equations*. In this case we shall for brevity refer to (4.1) as *a differential equation (DE) on $\mathbb{R}^n$* and write it in the form

$$\mathbf{x}' = \mathbf{f}(\mathbf{x}), \quad \mathbf{x} \in \mathbb{R}^n. \tag{4.2}$$

The function $\mathbf{f} : \mathbb{R}^n \to \mathbb{R}^n$ can be interpreted as a *vector field* on $\mathbb{R}^n$,

$$\mathbf{f}(\mathbf{x}) = (f_1(\mathbf{x}), \ldots, f_n(\mathbf{x})).$$

We shall assume that $\mathbf{f}$ is (at least) of class C^1 on $\mathbb{R}^n$.

Certain special types of DE are of particular interest in applications.

Linear DE:

If $\mathbf{f}$ is a linear function, i.e. $\mathbf{f}(\mathbf{x}) = A\mathbf{x}$, where A is an $n \times n$ matrix of real numbers, then the DE (4.2) is linear:

$$\mathbf{x}' = A\mathbf{x}. \tag{4.3}$$

Gradient DE:

If $\mathbf{f}$ is the gradient of a C^2 scalar field $Z : \mathbb{R}^n \to \mathbb{R}$, i.e. $\mathbf{f}(\mathbf{x}) = -\nabla Z(\mathbf{x})$, then (4.2) is a gradient DE:

$$\mathbf{x}' = -\nabla Z(\mathbf{x}). \tag{4.4}$$

Hamiltonian DE:

If $\mathbf{x} = (q^1, \ldots, q^n, p_1, \ldots, p_n) \in \mathbb{R}^{2n}$, and $\mathbf{f}(\mathbf{x}) = (\partial\mathcal{H}/\partial p_A, -\partial\mathcal{H}/\partial q^A)$, where $\mathcal{H} : \mathbb{R}^{2n} \to \mathbb{R}$ is a C^1 function called the *Hamiltonian*, then (4.2) is a Hamiltonian DE,

$$\frac{dq^A}{dt} = \frac{\partial\mathcal{H}}{\partial p_A}, \quad \frac{dp_A}{dt} = -\frac{\partial\mathcal{H}}{\partial q^A}, \tag{4.5}$$

where $A = 1, \ldots, n$. The evolution of conservative mechanical systems can be described by a Hamiltonian DE, with $\mathcal{H}$ of the form

$$\mathcal{H}(q^A, p_A) = \tfrac{1}{2}g^{AB}(q^C)p_A p_B + V(q^C).$$

where $g^{AB} = g^{BA}$. The first term is the kinetic energy and $V(q^C)$ is the potential energy.

4.1.2 Basic theorems

A *solution* of the DE $\mathbf{x}' = \mathbf{f}(\mathbf{x})$ on $\mathbb{R}^n$ is a function $\psi : \mathbb{R} \to \mathbb{R}^n$ which satisfies

$$\psi'(t) = \mathbf{f}(\psi(t)),$$

for all $t \in \mathbb{R}$ (or possibly only for t in a finite interval). The image of a solution in $\mathbb{R}^n$ is called an *orbit of the DE*. The evolution in time of the associated physical system is described by the motion of the state vector $\mathbf{x} \in \mathbb{R}^n$ along an orbit of the DE. The DE implies that the vector field $\mathbf{f}(\mathbf{x})$ is tangent to the orbit, and that it can be thought of as the velocity of the moving point in state space (not to be confused with the velocity of a physical particle).

We begin by stating the standard existence–uniqueness theorem for the initial value problem for a DE on $\mathbb{R}^n$.

Theorem 4.1 *(Existence–Uniqueness).* Consider the initial value problem

$$\mathbf{x}' = \mathbf{f}(\mathbf{x}), \qquad \mathbf{x}(0) = \mathbf{a} \in \mathbb{R}^n.$$

If $\mathbf{f} : \mathbb{R}^n \to \mathbb{R}^n$ is of class $C^1(\mathbb{R}^n)$, then for all $\mathbf{a} \in \mathbb{R}^n$, there exists an interval $(-\delta, \delta)$ and a unique function $\psi_{\mathbf{a}} : (-\delta, \delta) \to \mathbb{R}^n$ such that

$$\psi_{\mathbf{a}}'(t) = \mathbf{f}(\psi_{\mathbf{a}}(t)), \qquad \psi_{\mathbf{a}}(0) = \mathbf{a}.$$

Proof See Hirsch & Smale (1974, page 162).   □

The existence–uniqueness theorem is a local result — it guarantees existence of a solution in some interval $(-\delta, \delta)$ centred at $t = 0$. Since we are interested in the long–term behaviour of solutions, we would like the solutions to be defined for all $t \in \mathbb{R}$. We can extend the interval of definition of the solution $\psi_{\mathbf{a}}(t)$ by successively reapplying the theorem, and in this way obtain a *maximal interval of definition* of the solution $\psi_{\mathbf{a}}(t)$, denoted by $(t_{\min}, t_{\max})$. The next theorem shows that solutions can be extended indefinitely unless the solution diverges in a finite time. Here and in the rest of this chapter, $\| \ \ \|$ denotes the standard norm in $\mathbb{R}^n$.

Theorem 4.2 *(Maximality).* Let $\psi_{\mathbf{a}}(t)$ be the unique solution of the DE $\mathbf{x}' = \mathbf{f}(\mathbf{x})$, where $\mathbf{f} \in C^1(\mathbb{R}^n)$, which satisfies $\psi_{\mathbf{a}}(0) = \mathbf{a}$, and let $(t_{\min}, t_{\max})$ denote the maximal interval on which $\psi_{\mathbf{a}}(t)$ is defined. If $t_{\max}$ is finite, then

$$\lim_{t \to t_{\max}^-} \| \psi_{\mathbf{a}}(t) \| = +\infty.$$

Proof See Hirsch & Smale (1974, pages 171–2). ☐

This result is also valid for the left–hand limit, leading to the following corollary.

Corollary 4.1. Consider the DE $\mathbf{x}' = \mathbf{f}(\mathbf{x})$, $\mathbf{f} \in C^1(\mathbb{R}^n)$, and let $D \subset \mathbb{R}^n$ be a compact set. If a maximally extended solution $\psi_{\mathbf{a}}(t)$ lies in D, then the solution is defined for all $t \in \mathbb{R}$.

Comment: In Theorems 4.1 and 4.2, and in Corollary 4.1, one can replace '$\mathbf{f} \in C^1(\mathbb{R}^n)$' by '$\mathbf{f} \in C^1(U)$', where $U \subset \mathbb{R}^n$ is an open set (see the theorems in Hirsch & Smale 1974 referred to above). Corollary 4.1, with $D \subset U$, can thus be applied in cases where the DE is defined only on an open subset of $\mathbb{R}^n$.

The next theorem shows that if $\| \mathbf{f}(\mathbf{x}) \|$ does not increase too rapidly as $\| \mathbf{x} \| \to +\infty$ then all solutions can be extended indefinitely.

Theorem 4.3 (*Extendibility*). If $\mathbf{f} : \mathbb{R}^n \to \mathbb{R}^n$ is continuous, and there exists a constant M such that $\| \mathbf{f}(\mathbf{x}) \| \leq M \| \mathbf{x} \|$ for all $\mathbf{x} \in \mathbb{R}^n$, then any solution of the DE $\mathbf{x}' = \mathbf{f}(\mathbf{x})$ is defined for all $t \in \mathbb{R}$.

Proof See Nemytskii & Stepanov (1960, theorem 1.31, page 9). ☐

Comment: Theorem 4.3 implies that one can modify a given DE $\mathbf{x}' = \mathbf{f}(\mathbf{x})$, $\mathbf{x} \in \mathbb{R}^n$, so that the orbits are unchanged, but such that all solutions are defined for all $t \in \mathbb{R}$. The idea is to re–scale the vector field $\mathbf{f}$ so as to make it bounded,

$$\mathbf{f}(\mathbf{x}) \to \lambda(\mathbf{x})\mathbf{f}(\mathbf{x}),$$

where $\lambda(\mathbf{x}) : \mathbb{R}^n \to \mathbb{R}$ is a C^1–function which is *positive* on $\mathbb{R}^n$, in order to preserve the direction of time (e.g., $\lambda(\mathbf{x}) = [1 + \| \mathbf{f}(\mathbf{x}) \|]^{-1}$ will suffice).

Corollary 4.2. If $\mathbf{f} : \mathbb{R}^n \to \mathbb{R}^n$ is $C^1(\mathbb{R}^n)$, and $\lambda : \mathbb{R}^n \to \mathbb{R}$ is C^1 and positive, then $\mathbf{x}' = \mathbf{f}(\mathbf{x})$ and $\mathbf{x}' = \lambda(\mathbf{x})\mathbf{f}(\mathbf{x})$ have the same orbits, and λ can be chosen so that all solutions of the second DE are defined for all $t \in \mathbb{R}$.

4.1.3 The flow of a DE

If solutions of a DE $\mathbf{x}' = \mathbf{f}(\mathbf{x})$ can be extended for all times, one can define the flow of the DE.

Definition 4.1. Consider a DE $\mathbf{x}' = \mathbf{f}(\mathbf{x})$, $\mathbf{x} \in \mathbb{R}^n$, where $\mathbf{f}$ is of class $C^1(\mathbb{R}^n)$, whose solutions are defined for all $t \in \mathbb{R}$. Let $\psi_{\mathbf{a}}(t)$ be the unique maximal solution which satisfies $\psi_{\mathbf{a}}(0) = \mathbf{a}$. The *flow of the DE* is defined to be the one–parameter family of maps $\{\phi_t\}_{t \in \mathbb{R}}$, of $\mathbb{R}^n$ into itself such that

$$\phi_t(\mathbf{a}) = \psi_{\mathbf{a}}(t) \quad \text{for all} \quad \mathbf{a} \in \mathbb{R}^n. \tag{4.6}$$

Technical Point: It may happen that the solutions of the DE are extendible for $t \to +\infty$ but not for $t \to -\infty$. In this case one can define a *positive semi–flow* ϕ_t^+ of the DE by replacing '$t \in \mathbb{R}$' by '$t \in \mathbb{R}^+$' in the definition of flow. Similarly, if the solutions are extendible for $t \to -\infty$ but not for $t \to +\infty$, one can define a *negative semi–flow* ϕ_t^-.

The next theorem shows that a flow defines a one–parameter *group* of maps.

Theorem 4.4 *(Group property of a flow)*. Let $\{\phi_t\}$ be the flow of a DE $\mathbf{x}' = \mathbf{f}(\mathbf{x})$. Then

$$
\begin{aligned}
\phi_0 &= I & &\text{(the identity map)} \\
\phi_{t_1 + t_2} &= \phi_{t_1} \circ \phi_{t_2}, & &\text{for all } t_1, t_2 \in \mathbb{R} & &\text{(law of composition)} \\
\phi_{-t} &= (\phi_t)^{-1}, & &\text{for all } t \in \mathbb{R} & &\text{(the inverse map)}
\end{aligned}
\tag{4.7}
$$

Proof See Hirsch & Smale (1974, pages 175–6).   ☐

Comment: The flow ϕ_t is defined in terms of the solution functions $\psi_{\mathbf{a}}(t)$ of the DE by (4.6). It is important to note the difference between $\psi_{\mathbf{a}}(t)$ and $\phi_t(\mathbf{a})$ conceptually:

- *For a specified* $\mathbf{a} \in \mathbb{R}^n$, *the solution* $\psi_{\mathbf{a}} : \mathbb{R} \to \mathbb{R}^n$ *gives the state of the system* $\psi_{\mathbf{a}}(t) \in \mathbb{R}^n$ *for all* $t \in \mathbb{R}$, *with* $\psi_{\mathbf{a}}(0) = \mathbf{a}$ *initially.*

- *For a specified* $t \in \mathbb{R}$, *the flow* $\phi_t : \mathbb{R}^n \to \mathbb{R}^n$ *gives the state of the system* $\phi_t(\mathbf{a}) \in \mathbb{R}^n$ *at time* t *for all initial states* $\mathbf{a}$.

The solution $\psi_{\mathbf{a}}(t)$ satisfies $\psi_{\mathbf{a}}'(t) = \mathbf{f}(\psi_{\mathbf{a}}(t))$, $\psi_{\mathbf{a}}(0) = \mathbf{a}$. Hence $\psi_{\mathbf{a}}'(0) = \mathbf{f}(\mathbf{a})$. By definition of the flow, it follows that

$$\frac{d}{dt}\phi_t(\mathbf{a})\Big|_{t=0} = \mathbf{f}(\mathbf{a}), \tag{4.8}$$

which is simply a statement of the fact that the vector field $\mathbf{f}$ is tangent to the orbits of the DE. Equation (4.8) can also be interpreted as showing that a (sufficiently smooth) flow ϕ_t determines a vector field $\mathbf{f}$, and hence a DE $\mathbf{x}' = \mathbf{f}(\mathbf{x})$.

The next theorem relates the smoothness of the flow to the smoothness of the vector field.

Theorem 4.5 *(Smoothness of a flow).* If $\mathbf{f} \in C^1(\mathbb{R}^n)$, then the flow $\{\phi_t\}$ of the DE $\mathbf{x}' = \mathbf{f}(\mathbf{x})$ consists of C^1 maps.

Proof See Hirsch & Smale (1974, pages 298–300). □

Comment: The significance of this theorem is that the solutions of the DE depend smoothly on the initial conditions. We should note, however, that this smooth dependence on initial conditions does not exclude the (possibly exponential) separation of nearby orbits — the so-called *sensitive dependence on initial conditions* — that is an essential ingredient of chaotic dynamical systems (see Section 11.1 for the definition).

Dynamical systems:

It is appropriate at this time to comment on the use of the term 'dynamical systems'. On the one hand this term is used in an informal sense to mean a physical system of some type which evolves in time. On the other hand it is used in a precise mathematical sense in two ways:

(1) a *continuous dynamical system* means the flow $\{\phi_t\}_{t \in \mathbb{R}}$ of a DE on $\mathbb{R}^n$. More generally, a flow is defined to be a one–parameter family of invertible maps $\phi_t : X \rightarrow X$, $t \in \mathbb{R}$, that satisfy conditions (4.7), where X is a metric space.

(2) a *discrete dynamical system* refers to the one–parameter family of maps $\{g^m\}_{m \in Z}$ generated by iterating an invertible map $g : X \rightarrow X$ (i.e. $g^2 = g \circ g$, and g^{-1} is the inverse map, etc.). If g is not invertible, the family of maps is restricted by $m \geq 0$. The set X is usually a compact subset of $\mathbb{R}^n$. Some examples are given in Chapter 11.

Most of the applications of the theory of dynamical systems in this book involve continuous dynamical systems, although discrete dynamical systems play an important role as regards the oscillatory singularity in the Mixmaster models (see Section 6.4 and Chapter 11).

Technical point: There is a close relationship between the two types of dynamical systems. Given a flow, by sampling points at discrete values of t in some appropriate way one can generate a discrete dynamical system, the classic example being the *Poincaré map* (e.g. Perko 1991, Section 3.4). An unusual example of this process of discretization occurs in Section 6.4 and Chapter 11. Conversely, a discrete dynamical system defined by a map g generates a flow, called the suspension of g (e.g. Arrowsmith & Place 1990, pages 35–6).

In this section we have seen that (subject to certain restrictions) a DE on $\mathbb{R}^n$ determines a flow on $\mathbb{R}^n$, and vice versa:

$$\mathbf{x}' = \mathbf{f}(\mathbf{x}) \longleftrightarrow \{\phi_t\}_{t \in \mathbb{R}}.$$

For a linear DE (4.3) one can write down the flow explicitly:

$$\phi_t(\mathbf{x}) = e^{At}\mathbf{x}, \quad \text{for all} \quad t \in \mathbb{R}, \quad \mathbf{x} \in \mathbb{R}^n. \tag{4.9}$$

Here e^{At} denotes the exponential of the matrix tA (defined as a power series). We refer to Perko (1991, section 1.3) for details. In general, however, one does not expect (or attempt) to find the flow explicitly, since this would be equivalent to finding all solutions of the DE. Instead, the goal is to use the flow as a tool to aid in the qualitative analysis of the whole family of solutions of the DE, as originally proposed by Poincaré. Thus one needs to be able to determine properties of the flow by studying the vector field $\mathbf{f}$, which is what one is given in a particular example.

In the remainder of this chapter we introduce some of the basic concepts associated with dynamical systems. We use the language of flows, and note that many of the concepts are defined in an analogous way for discrete dynamical systems.

4.2 Orbits and invariant sets

4.2.1 Classification of orbits

Definition 4.2. Given a DE (4.2) and the associated flow ϕ_t, the *orbit* through $\mathbf{x}_0$, denoted by $\gamma(\mathbf{x}_0)$, is defined by

$$\gamma(\mathbf{x}_0) = \{\mathbf{x} \in \mathbb{R}^n \mid \mathbf{x} = \phi_t(\mathbf{x}_0),\ t \in \mathbb{R}\}. \tag{4.10}$$

Given a flow ϕ_t, the points in the state space can be divided into two different types:

  (i) equilibrium points
  (ii) ordinary points.

Definition 4.3. An *equilibrium point* x_0 of the DE $x' = f(x)$ satisfies $f(x_0) = 0$ or equivalently, $\phi_t(x_0) = x_0$, for all $t \in R$, where ϕ_t is the corresponding flow.

Comment: An equilibrium point x_0 is interpreted as an *equilibrium state of the physical system*, since the state does not change with time: the velocity of the point in state space is zero, $f(x_0) = 0$, or equivalently, x_0 is unchanged by the action of the flow. For this reason an equilibrium point is also called a *fixed point*.

It follows from (4.10) that the orbit through an equilibrium point is the point itself:

$$\gamma(x_0) = \{x_0\}, \tag{4.11}$$

whereas the orbit through an ordinary point, called an *ordinary orbit*, is a smooth curve with the vector field f as tangent. There are two special types of ordinary orbits, namely periodic and recurrent.

Definition 4.4. Let x_0 be an ordinary point. The orbit $\gamma(x_0)$ is *periodic* means that there exists a $T > 0$ such that $\phi_T(x_0) = x_0$.

Definition 4.5. Let x_0 be an ordinary point such that $\gamma(x_0)$ is a non–periodic orbit. The orbit $\gamma(x_0)$ is *recurrent* means that for all neighbourhoods $N(x_0)$ and for all $T \in \mathbb{R}$, there exists a $t > T$ such that $\phi_t(x_0) \in N(x_0)$.

Comment: If a flow ϕ_t has a periodic orbit of period T, the corresponding physical system can exhibit oscillatory behaviour of period T, whereas if ϕ_t has a recurrent orbit, the corresponding physical system can return arbitrarily close to an earlier state.

The following types of orbits are also important in organizing the phase portraits of a flow.

Definition 4.6. An orbit connecting distinct equilibrium points is referred to as a *heteroclinic orbit*. An orbit connecting an equilibrium point to itself is called a *homoclinic orbit*.

4.2.2 Invariant sets

The concept of an invariant set is one of the most important concepts in the theory of dynamical systems, and plays a major role in this book.

Definition 4.7. A set $S \subset \mathbb{R}^n$ is an *invariant set* of the flow ϕ_t on $\mathbb{R}^n$ (or of the corresponding DE $\mathbf{x}' = \mathbf{f}(\mathbf{x})$) means that for all $\mathbf{x} \in S$ and for all $t \in \mathbb{R}$, $\phi_t(\mathbf{x}) \in S$.

Comments:

   (i) If S is an invariant set and $\mathbf{x}_0 \in S$, then the orbit $\gamma(\mathbf{x}_0)$ belongs to S. Thus *an invariant set is a union of orbits*.

  (ii) One can think of a lower dimensional invariant set as describing a restricted class of physical systems, which satisfy a special property or whose evolution is restricted in some way. Lower dimensional invariant sets often describe the asymptotic behaviour of general classes of solutions as $t \to +\infty$.

Technical point: If S has the property that $\phi_t(\mathbf{x}) \in S$ for all $\mathbf{x} \in S$ and for all $t > 0$, then we say that S is *positively invariant*. It would be appropriate to use this terminology when working with a semi–flow ϕ_t^+. Similarly, if $t > 0$ is replaced by $t < 0$, S is said to be *negatively invariant*. In the case of a discrete dynamical system defined by a non–invertible map g on $\mathbb{R}^n$, one defines a positively invariant set by the requirement that $g^m(\mathbf{x}) \in S$ for all $\mathbf{x} \in S$ and for all $m > 0$.

Examples of invariant sets:

   (i) Single orbits, such as equilibrium points and periodic orbits.

  (ii) Stable, unstable and centre manifolds (see Section 4.3.3).

 (iii) α– and ω–limit sets (see Section 4.4.1).

 (iv) Attractors (see Section 4.4.1).

  (v) Heteroclinic sequences and cycles (see Sections 4.4.4 and 4.4.1).

There is a simple way in which invariant sets arise. Given a DE (4.2) on $\mathbb{R}^n$ and a C^1 function $Z : \mathbb{R}^n \to \mathbb{R}$, the derivative of Z along solutions of (4.2) (i.e. along the flow) is given by

$$Z' = \nabla Z \bullet \mathbf{f}(\mathbf{x}), \tag{4.12}$$

where $\bullet$ is the standard inner product in $\mathbb{R}^n$.

Proposition 4.1. Consider a DE $\mathbf{x}' = \mathbf{f}(\mathbf{x})$ on $\mathbb{R}^n$ with flow ϕ_t. Let $Z : \mathbb{R}^n \to \mathbb{R}$ be a C^1 function which satisfies $Z' = \alpha Z$, where $\alpha : \mathbb{R}^n \to \mathbb{R}$ is a continuous function. Then the subsets of $\mathbb{R}^n$ defined by $Z > 0$, $Z = 0$ and $Z < 0$ are invariant sets of the flow ϕ_t.

Example: Consider the DE on $\mathbb{R}^3$ defined by (6.21), which describes the evolution of Bianchi II universes. One can conclude by inspection that $\Sigma_- = 0$ is an invariant set since the proposition applies with $Z = \Sigma_-$. Similarly $N_1 = 0$ is an invariant set.

4.2.3 Monotone functions

Definition 4.8. Let ϕ_t be a flow on $\mathbb{R}^n$, let S be an invariant set of ϕ_t and let $Z : S \to \mathbb{R}$ be a continuous function. Z is a *monotone decreasing (increasing) function for the flow on S* means that for all $\mathbf{x} \in S$, $Z(\phi_t(\mathbf{x}))$ is a monotone decreasing (increasing) function of t.

Consider a DE (4.2) and the corresponding flow ϕ_t, and suppose that Z is C^1. If

$$Z' \equiv \nabla Z \bullet \mathbf{f} < 0, \quad \text{on} \quad S \tag{4.13}$$

then Z is monotone decreasing on S. If, on the other hand, (4.13) is replaced by the weaker condition

$$Z' \equiv \nabla Z \bullet \mathbf{f} \le 0, \quad \text{on} \quad S, \tag{4.14}$$

and the set on which equality holds contains no orbits, then one can also conclude that Z is monotone decreasing on S. This is how one usually proves that a function is monotone.

Example: Let ϕ_t be a flow determined by a gradient DE (4.4). It follows from (4.12) that the potential function Z in (4.4) satisfies $Z' = - \| \nabla Z \|^2$, which shows that Z is a monotone decreasing function of ϕ_t on any invariant set that does not contain an equilibrium point.

The proposition to follow shows that the existence of a monotone function on an invariant set S simplifies the orbits in S significantly.

Proposition 4.2. Let $S \subset \mathbb{R}^n$ be an invariant set of a flow ϕ_t. If there exists a monotone function $Z : S \to \mathbb{R}$ on S, then S contains no equilibrium points, periodic orbits, recurrent orbits or homoclinic orbits.

Proof See Wainwright & Hsu (1989, page 1418). □

4.2.4 Dulac functions

Another method for excluding periodic orbits for a DE in the plane is provided by the classical Dulac's theorem, which entails finding a *Dulac function*
(the function λ in the theorem).

Theorem 4.6 (*Dulac*). Consider a DE $\mathbf{x}' = \mathbf{f}(\mathbf{x})$ in $\mathbb{R}^2$. If there is a C^1
function $\lambda : \mathbb{R}^2 \to \mathbb{R}$ such that $\nabla \bullet (\lambda \mathbf{f}) > 0$ (or < 0) in a simply–connected
open set $S \subset \mathbb{R}^2$, then there are no periodic orbits in S.

Proof See Perko (1991, page 246).   □

4.3 Behaviour near equilibrium points

The first step in obtaining qualitative information about the solutions of
a DE $\mathbf{x}' = \mathbf{f}(\mathbf{x})$ in $\mathbb{R}^n$, is to study the local properties of the flow in the
neighbourhood of the equilibrium points, i.e. to study their stability. The
essential idea is to linearize the DE at each of the equilibrium points and use
the Hartman–Grobman theorem (see below) to relate the behaviour of the
resulting linear DE to the given non–linear DE. We thus begin by reviewing
some basic results concerning linear DEs in $\mathbb{R}^n$.

4.3.1 Linear DEs in $\mathbb{R}^n$

Given a linear DE $\mathbf{x}' = A\mathbf{x}$ on $\mathbb{R}^n$, we consider the eigenvalues of A (complex in general, and not necessarily distinct) and the associated generalized
eigenvectors (see for example, Perko 1991, page 33). We define three subspaces of $\mathbb{R}^n$,

$$\text{the } \textit{stable subspace} \quad E^{\text{s}} = \text{span}(\mathbf{s}_1, \ldots, \mathbf{s}_{n_s}),$$
$$\text{the } \textit{unstable subspace} \quad E^{\text{u}} = \text{span}(\mathbf{u}_1, \ldots, \mathbf{u}_{n_u}),$$
$$\text{the } \textit{centre subspace} \quad E^{\text{c}} = \text{span}(\mathbf{c}_1, \ldots, \mathbf{c}_{n_c}),$$

where $\mathbf{s}_1, \ldots, \mathbf{s}_{n_s}$ are the generalized eigenvectors whose eigenvalues have
negative real parts, $\mathbf{u}_1, \ldots, \mathbf{u}_{n_u}$ are those whose eigenvalues have positive
real parts and $\mathbf{c}_1, \ldots, \mathbf{c}_{n_c}$ are those whose eigenvalues have zero real parts.
It is well known (see Perko 1991, pages 55–8) that

$$E^{\text{s}} \oplus E^{\text{u}} \oplus E^{\text{c}} = \mathbb{R}^n,$$

and that

$$\mathbf{x} \in E^{\text{s}} \quad \text{implies} \quad \lim_{t \to +\infty} e^{At}\mathbf{x} = \mathbf{0},$$
$$\mathbf{x} \in E^{\text{u}} \quad \text{implies} \quad \lim_{t \to -\infty} e^{At}\mathbf{x} = \mathbf{0}.$$

These statements describe the asymptotic behaviour: all initial states in the stable subspace are attracted to the equilibrium point **0**, while all initial states in the unstable subspace are repelled by **0**. In particular, if dim $E^s = n$, all initial states are attracted to **0**, which is referred to as a *linear sink*, while if dim $E^u = n$, all initial states are repelled by **0**, which is referred to as a *linear source*.

4.3.2 Linearization and the Hartman–Grobman theorem

We are considering a DE $\mathbf{x}' = \mathbf{f}(\mathbf{x})$ on $\mathbb{R}^n$, where $\mathbf{f}$ is of class C^1. If $\mathbf{a}$ is an equilibrium point ($\mathbf{f}(\mathbf{a}) = \mathbf{0}$), the linear approximation of $\mathbf{f}$ at $\mathbf{a}$ becomes

$$\mathbf{f}(\mathbf{x}) \approx D\mathbf{f}(\mathbf{a})(\mathbf{x} - \mathbf{a}),$$

where

$$D\mathbf{f}(\mathbf{a}) = \left(\frac{\partial f_i}{\partial x_j}\right)_{\mathbf{x}=\mathbf{a}} \tag{4.15}$$

is the derivative matrix of $\mathbf{f}$. Thus, with the given DE $\mathbf{x}' = \mathbf{f}(\mathbf{x})$, we associate the linear DE

$$\mathbf{u}' = D\mathbf{f}(\mathbf{a})\mathbf{u}, \tag{4.16}$$

where $\mathbf{u} = \mathbf{x} - \mathbf{a}$, called *the linearization of the DE at the equilibrium point* **a**.

The hope is that the solutions of (4.16) will approximate the solutions of the non–linear DE in a neighbourhood of the equilibrium point **a**. This will be true in a qualitative sense (the idea of topological equivalence) provided that the equilibrium point is hyperbolic.

Definition 4.9. Given a DE $\mathbf{x}' = \mathbf{f}(\mathbf{x})$ on $\mathbb{R}^n$, an equilibrium point $\mathbf{x} = \mathbf{a}$ is *hyperbolic* means that all eigenvalues of $D\mathbf{f}(\mathbf{a})$ have non–zero real part.

Definition 4.10. Two flows ϕ_t and $\tilde{\phi}_t$ on $\mathbb{R}^n$ are *topologically equivalent†* means that there exists a homeomorphism $h : \mathbb{R}^n \to \mathbb{R}^n$ which maps orbits of ϕ_t onto orbits of $\tilde{\phi}_t$, preserving their orientation.

Theorem 4.7 (*Hartman–Grobman*). Consider a DE $\mathbf{x}' = \mathbf{f}(\mathbf{x})$ in $\mathbb{R}^n$, where $\mathbf{f}$ is of class C^1, with flow ϕ_t. If $\mathbf{a}$ is a hyperbolic equilibrium point of the

---

† We refer to Arrowsmith & Place (1990, page 29) for the difference between the closely related concepts of topological equivalence and topological conjugacy. For flows, topological equivalence provides more satisfactory equivalence classes. One can replace 'topologically equivalent' by 'topologically conjugate' in the Hartman–Grobman Theorem (*ibid*, page 69).

DE, then there exists a neighbourhood N of **a** on which ϕ_t is topologically equivalent to the flow of the linearization of the DE at **a**.

Proof See Hartman (1982, pages 244–50).   □

Comment: In colloquial terms the theorem means that in the neighbourhood N of the equilibrium point **a**, the orbits of $\mathbf{x}' = \mathbf{f}(\mathbf{x})$ can be deformed continuously into the orbits of the linearization (4.16), i.e. qualitatively the orbits are the same. In particular, if all eigenvalues λ_i of $\mathbf{Df(a)}$ satisfy $\text{Re}(\lambda_i) < 0$, the point $\mathbf{u} = \mathbf{x} - \mathbf{a} = \mathbf{0}$ will be a linear sink for the DE (4.16). Thus, for the DE $\mathbf{x}' = \mathbf{f}(\mathbf{x})$, the theorem implies that any initial state in N will be attracted to the equilibrium point **a** as $t \to +\infty$. Similarly if $\text{Re}(\lambda_i) > 0$ any initial state in N will be attracted to the equilibrium point as $t \to -\infty$. These considerations motivate the definition to follow.

Definition 4.11. An equilibrium point **a** of a DE $\mathbf{x}' = \mathbf{f}(\mathbf{x})$ in $\mathbb{R}^n$ is called a *local sink* (*local source*) whenever the eigenvalues of the derivative matrix $\mathbf{Df(a)}$ satisfy $\text{Re}(\lambda_i) < 0$ ($\text{Re}(\lambda_i) > 0$) for $i = 1, \ldots, n$. A hyperbolic equilibrium point that is neither a local source or sink is called a *saddle point*.

4.3.3 Stable, unstable and centre manifolds

If **a** is an equilibrium point of the DE $\mathbf{x}' = \mathbf{f}(\mathbf{x})$ on $\mathbb{R}^n$ it is useful to know which orbits are attracted to **a** as $t \to +\infty$, and which are repelled. One can generalize the idea of the stable, unstable and centre subspaces defined for a linear DE in Section 4.3.1. Specifically, the *stable manifold W^s of an equilibrium point* **a** is a differentiable manifold that is tangent to the stable subspace E^s at **a** and such that all orbits in W^s are asymptotic to **a** as $t \to +\infty$. Similarly the *unstable manifold W^u of an equilibrium point* **a** is a differentiable manifold that is tangent to the unstable subspace E^u at **a** and such that all orbits in W^u are asymptotic to **a** as $t \to -\infty$. Finally, a *centre manifold W^c of an equilibrium point* **a** is a differentiable manifold that is tangent to the centre subspace E^c at **a**. A centre manifold contains orbits whose asymptotic behaviour is not determined by the linearization. The dimension of a stable, unstable or centre manifold equals the dimension of the corresponding subspace.

The next theorem asserts the existence of these invariant manifolds and describes their properties. It is convenient to assume that the equilibrium point has been translated to the origin.

Theorem 4.8 (*Invariant Manifold*). Let $x = 0$ be an equilibrium point of the DE $x' = f(x)$ on $\mathbb{R}^n$ and let E^s, E^u and E^c denote the stable, unstable and centre subspaces of the linearization at 0. Then there exists

a stable manifold W^s, tangent to E^s at 0,
an unstable manifold W^u, tangent to E^u at 0, and
a centre manifold W^c, tangent to E^c at 0.

Proof See Guckenheimer & Holmes (1983, page 127), for references, and Wiggins (1990, page 21). □

Technical point: W^s and W^u are uniquely determined, but there can be an infinite number of centre manifolds. There are also subtleties involving the smoothness of W^c. We refer to Arrowsmith & Place (1990, pages 94–8). We also refer to Carr (1981) for more information about centre manifolds.

We now give a simple illustrative example. Other examples arise in the phase portraits in Chapters 6 and 7.

Example: Consider the DE on $\mathbb{R}^2$,

$$x' = x, \qquad y' = -y + x^2,$$

which has an equilibrium point $(0,0)$. The linearization at $(0,0)$ is

$$u' = u, \qquad v' = -v.$$

The DE implies

$$\left(y - \tfrac{1}{3}x^2\right)' = -\left(y - \tfrac{1}{3}x^2\right),$$

so that $y = \tfrac{1}{3}x^2$ is an invariant set, as follows from Proposition 4.1, with $Z = y - \tfrac{1}{3}x^2$. Figure 4.1 shows the orbits of the given DE and the linearization, and illustrates the Invariant Manifold theorem and the Hartman–Grobman theorem.

4.3.4 Non–isolated equilibrium points

We are also interested in the case where the DE $x' = f(x)$ on $\mathbb{R}^n$ admits *non–isolated* equilibrium points, for example, a curve C of equilibrium points, called an *equilibrium set*. In the case of a curve of equilibrium points, the $n \times n$ matrix $Df(x_0)$ necessarily has one zero eigenvalue at each point x_0

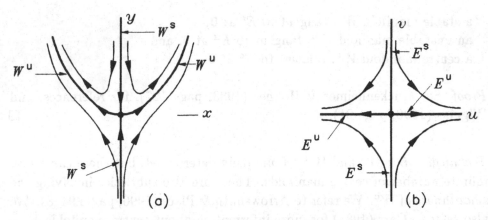

Fig. 4.1. A DE with a saddle point and its linearization

of the equilibrium set (r zero eigenvalues for an r-dimensional equilibrium set (see Bogoyavlensky 1985, page 15)).

Although the equilibrium points in an equilibrium set are necessarily non-hyperbolic, one can apply the Invariant Manifold theorem. Suppose that each point of the equilibrium set C, assumed to be a curve, has a stable manifold of dimension n_s. The union of these manifolds forms an $(n_s + 1)$-dimensional set whose orbits approach a point of C as $t \to +\infty$. We will refer to this set as the *stable set* of the equilibrium set C. Similarly one can define *the unstable set* of C, of dimension $n_u + 1$ say, with $n_u + n_s \le n - 1$. If $n_s = n - 1$ we say that the equilibrium set C is a *local sink* for the DE, and if $n_u = n - 1$, we say that C is a *local source*. In general, if $n_s + n_u = n - 1$, i.e. the equilibrium set has only one zero eigenvalue at each point and all other eigenvalues have $\text{Re}(\lambda) \ne 0$, the equilibrium set is called *normally hyperbolic* (Aulbach 1984, page 36).

We note that equilibrium sets receive little attention in the standard texts on dynamical systems. However, the book by Aulbach (1984) is devoted to this topic.

4.4 Asymptotic evolution and intermediate evolution

One of the main goals in the theory of dynamical systems is to determine the future asymptotic behaviour ($t \to +\infty$), because one is interested in the long-term evolution of the corresponding physical system. In cosmology, one is also interested in the asymptotic behaviour near the initial singularity, i.e. into the past. In applying dynamical systems theory in cosmology one thus seeks to introduce a dimensionless time variable which tends to $-\infty$ at the initial singularity [see (5.9)], and to $+\infty$ at late times, with a view to studying both the past and future asymptotic behaviour ($t \to -\infty$). In this section we introduce some of the concepts and tools that can be used to investigate the asymptotic behaviour of dynamical systems. The most important concept is the ω–limit set, first introduced by G. D. Birkhoff (see Birkhoff 1927, page 197).

4.4.1 Limits sets and attractors

The simplest type of asymptotic behaviour is that the physical system, starting in state $\mathbf{a}$, approaches an equilibrium state as $t \to +\infty$, i.e. the orbit through $\mathbf{a}$ approaches an equilibrium point $\mathbf{x}^*$. We say that the ω–limit set of $\mathbf{a}$ is $\mathbf{x}^*$ and write $\omega(\mathbf{a}) = \{\mathbf{x}^*\}$. The next simplest case is that the behaviour of the physical system becomes periodic asymptotically, i.e. the orbit through $\mathbf{a}$ approaches a periodic orbit γ^*. We write $\omega(\mathbf{a}) = \gamma^*$.

These examples motivate the definition to follow.

Definition 4.12. Let ϕ_t be a flow on $\mathbb{R}^n$, and let $\mathbf{a} \in \mathbb{R}^n$. A point $\mathbf{x} \in \mathbb{R}^n$ is *an ω–limit point of* $\mathbf{a}$ means there exists a sequence $t_n \to +\infty$ such that $\lim_{n\to\infty} \phi_{t_n}(\mathbf{a}) = \mathbf{x}$. The set of all ω–limit points of $\mathbf{a}$ is called *the ω–limit set of* $\mathbf{a}$, denoted $\omega(\mathbf{a})$.

Comment: The α–limit set $\alpha(\mathbf{a})$ is defined similarly, by using a sequence $t_n \to -\infty$.

The theorem to follow describes the elementary properties of ω–limit sets.

Theorem 4.9 *(ω–limit sets)*. Let ϕ_t be a flow on $\mathbb{R}^n$. For all $\mathbf{a}$ the ω–limit set $\omega(\mathbf{a})$ is a closed, invariant set. If the positive orbit through $\mathbf{a}$ is bounded, then $\omega(\mathbf{a})$ is non–empty and connected.

Proof See Nemytskii & Stepanov (1960, pages 338–43). □

Comment: Other examples of ω–limit sets are the *closure of a homoclinic orbit*, or a *heteroclinic cycle*, which is a sequence of equilibrium points $\mathbf{x}_0, \mathbf{x}_1, \ldots, \mathbf{x}_n = \mathbf{x}_0$ joined by heteroclinic orbits, coherently ordered in time (see Figure 4.2).

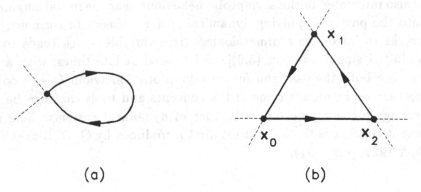

(a)   (b)

Fig. 4.2. (a) Closure of a homoclinic orbit. (b) A heteroclinic cycle.

Technical point: Any ω–limit set is a closed invariant set, but the converse is not true, e.g. the closure of a single heteroclinic orbit cannot be an ω–limit set. There is another important property of ω–limit sets, namely the property of being ϕ_t–connected, first introduced by Dowker & Friedlander (1954), which ensures that the invariant set has a coherent ordering in time, so that other orbits can approach the set asymptotically. See LeBlanc *et al.* (1995) for the definition and an application in cosmology.

The ω–limit set $\omega(\mathbf{a})$ of a point $\mathbf{a}$ describes the future asymptotic behaviour of the physical system when it starts in the initial state $\mathbf{a}$. A single ω–limit set $\omega(\mathbf{a})$ may not be physically relevant since the initial state $\mathbf{a}$ may be special, in the sense that any perturbation of $\mathbf{a}$ changes $\omega(\mathbf{a})$. One thus needs to consider all possible ω–limit sets and determine which ones correspond to randomly chosen initial states. These remarks lead to the concept of attractor: the attractor of a flow is the smallest invariant set that attracts most orbits. Despite the importance of this concept, it was introduced only as recently as 1964 (see Auslander *et al.* 1964). Various definitions have been given. The definition that we use has been found to be suitable for applications in cosmology and is motivated by the approach of Milnor (see Milnor 1985; this paper also gives other references on attractors).

Definition 4.13. Given a flow ϕ_t on $\mathbb{R}^n$, the *future attractor* A^+ is the smallest closed invariant set such that $\omega(\mathbf{a}) \subset A^+$ for all $\mathbf{a} \in \mathbb{R}^n$ apart from a set of measure zero.

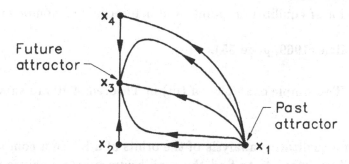

Fig. 4.3. Past and future attractors of a flow in the plane.

Comments:

(i) The past attractor A^- is defined by replacing $\omega(\mathbf{a})$ by the α–limit set $\alpha(\mathbf{a})$.

(ii) The domain of the flow may be a proper subset D of $\mathbb{R}^n$. If D is compact, then each point has a non–empty ω–limit set (see Theorem 4.9), and hence the future attractor is non–empty, $A^+ \neq \emptyset$. If D is not compact, then it is possible that $A^+ = \emptyset$, but if there exists a compact set which all orbits eventually enter (called a trapping set), then as before $A^+ \neq \emptyset$.

Example: For the flow shown in Figure 4.3,

$$A^+ = \{\mathbf{x}_3\}, \qquad A^- = \{\mathbf{x}_1\}.$$

Note that A^+ does not contain all ω–limit sets; e.g. for points $\mathbf{a}$ on the quarter circle, $\omega(\mathbf{a}) = \{\mathbf{x}_4\}$.

4.4.2 Asymptotic behaviour in the plane

Differential equations in the plane are special in that one can give an explicit description of all possible α– and ω–limit sets. This is the content of the theorem to follow.

Theorem 4.10 (*Generalized Poincaré–Bendixson*). Consider a DE $\mathbf{x}' = \mathbf{f}(\mathbf{x})$ on $\mathbb{R}^2$, with $\mathbf{f} \in C^2$, and suppose that there are at most a finite number of equilibrium points. Then any compact limit set is one of the following:

 (i) an equilibrium point,
 (ii) a periodic orbit,
(iii) the union of equilibrium points and heteroclinic or homoclinic orbits.

Proof See Hale (1969, page 55).   □

Comment: Two simple examples of (iii) in Theorem 4.10 are shown in Figure 4.2.

   In giving a qualitative analysis of the orbits of a DE in a compact subset of $\mathbb{R}^2$, one first attempts to find the equilibrium points, analyze their local stability, and determine whether there is a local source and sink (or more than one). If numerical experiments suggest that there are no periodic orbits, one would then attempt to find a Dulac function or a monotone function to prove the non–existence of periodic orbits. One would then try to construct the skeleton consisting of orbits that form the boundary and heteroclinic orbits that join saddle points. With this information, one would then hope to be able to use the generalized Poincaré–Bendixson theorem to find the α– and ω–limit sets and identify the past and future attractor. Of course, one may be unable to implement this process at any stage, in which case one would be forced to rely on numerical experiments.

4.4.3 Asymptotic behaviour in higher dimensions

For a flow in $\mathbb{R}^n$, with $n > 2$, the procedure for giving a qualitative analysis, as described above, breaks down because the generalized Poincaré–Bendixson theorem is not valid. Indeed, as is well known, the dynamical behaviour in higher dimensions can be of great complexity, involving phenomena such as recurrence and chaos (see, for example, Guckenheimer & Holmes 1983, chapters 2, 5, 6, and Wiggins 1990). Thus from an analytic point of view, given an autonomous DE in $\mathbb{R}^n$ with $n > 2$, one cannot expect to be able to do much more than analyze the equilibrium points and any obvious two–dimensional invariant subsets that may exist, before having to resort to numerical experiments. Many of the DEs that arise in cosmology have, however, been found to admit a variety of monotone functions which simplifies the dynamics dramatically (see Proposition 4.2), and enables one to make considerable progress analytically. In particular one can use these functions in conjunction with the LaSalle Invariance Principle and the Monotonicity Principle to obtain information about the α– and ω–limit sets and the past and future attractors.

Theorem 4.11 (*LaSalle Invariance Principle*). Consider a DE $\mathbf{x}' = \mathbf{f}(\mathbf{x})$ on $\mathbb{R}^n$, with flow ϕ_t. Let S be a closed, bounded and positively invariant set of ϕ_t and let Z be a C^1 monotone function. Then for all $\mathbf{x}_0 \in S$,

$$\omega(\mathbf{x}_0) \subseteq \{\mathbf{x} \in S \mid Z' = 0\},$$

where $Z' = \nabla Z \bullet \mathbf{f}$.

Proof See for example Hale (1969, page 296) and Amman (1990, page 234). □

The next theorem can be regarded as a generalization of the LaSalle Invariance Principle.

Theorem 4.12 (*Monotonicity Principle*). Let ϕ_t be a flow on $\mathbb{R}^n$ with S an invariant set. Let $Z : S \to \mathbb{R}$ be a C^1 function whose range is the interval (a, b), where $a \in \mathbb{R} \cup \{-\infty\}$, $b \in \mathbb{R} \cup \{+\infty\}$ and $a < b$. If Z is decreasing on orbits in S, then for all $\mathbf{x} \in S$,

$$\omega(\mathbf{x}) \subseteq \{\mathbf{s} \in \overline{S} \backslash S \mid \lim_{\mathbf{y} \to \mathbf{s}} Z(\mathbf{y}) \neq b\},$$

$$\alpha(\mathbf{x}) \subseteq \{\mathbf{s} \in \overline{S} \backslash S \mid \lim_{\mathbf{y} \to \mathbf{s}} Z(\mathbf{y}) \neq a\}.$$

Proof See LeBlanc *et al.* (1995, page 536). □

In addition, we mention three other results which have proved useful in applications in cosmology, namely Proposition B2 in Hewitt & Wainwright (1993), and Propositions A2 and A3 in LeBlanc *et al.* (1995).

4.4.4 Intermediate behaviour

In addition to the typical asymptotic behaviour as $t \to \pm\infty$, which is described by the future and past attractors, one is also interested in the intermediate behaviour, i.e. how the physical system evolves from early times to late times. In general one cannot expect to describe the intermediate behaviour in as much detail as the asymptotic behaviour. One simple possibility is that there is a finite time interval during which the system is close to an equilibrium state. This situation occurs when the orbit that describes the evolution enters and remains in an ε-neighbourhood of a *saddle equilibrium point*. During this time interval the evolution of the physical system 'slows down' since the velocity $\mathbf{f}(\mathbf{x})$ in state space is close to zero, and the physical state is approximated by the equilibrium state which corresponds to the equilibrium point. We shall refer to such a time interval as a *quasi-equilibrium epoch.*

An orbit that describes a quasi–equilibrium epoch with equilibrium point E_1 will be initially attracted to E_1 by following (shadowing) an orbit in the stable manifold of E_1 and then will subsequently be repelled by following an orbit in the unstable manifold of E_1. If this unstable manifold intersects the stable manifold of another equilibrium point E_2, there is the possibility of a second quasi–equilibrium epoch. In general one can have a finite sequence of such epochs, described by an orbit that shadows a finite heteroclinic sequence, which we now define.

Definition 4.14. A *finite heteroclinic sequence* is a set of equilibrium points $E_0, E_1, \ldots, E_n$, where E_0 is a local source, E_n is a local sink, and the rest are saddles, such that there is a heteroclinic orbit which joins E_{i-1} to E_i, for $i = 1, \ldots, n$.

Comment: For examples of heteroclinic sequences, we refer the reader to Figures 6.5 and 6.8. One can also have infinite (non–terminating) heteroclinic sequences. See Section 6.4.

The proposition to follow justifies the existence of one orbit that shadows another, and is closely related to Theorem 4.5.

Proposition 4.2 (*Approximation property of orbits*). Let ϕ_t be a flow on $\mathbb{R}^n$. For all $\mathbf{x}_1 \in \mathbb{R}^n$, for all $T > 0$ and for all $\epsilon > 0$, there exists a $\delta > 0$ such that for all $\mathbf{x}_2 \in \mathbb{R}^n$ and $t \in \mathbb{R}$,

$$-T \leq t \leq T \text{ and } \| \mathbf{x}_1 - \mathbf{x}_2 \| < \delta \quad \text{implies} \quad \| \phi_t(\mathbf{x}_1) - \phi_t(\mathbf{x}_2) \| < \epsilon.$$

Proof See Sibirsky (1975, page 11).   □

Part two

Spatially homogeneous cosmologies

5

Qualitative analysis of Bianchi cosmologies

G. F. R. ELLIS
University of Capetown

C. UGGLA
University of Stockholm

J. WAINWRIGHT
University of Waterloo

In Section 5.1 we give an overview of the use of qualitative methods in analyzing Bianchi cosmologies, expanding on the brief remarks in the Introduction to the book. Section 5.2 provides an introduction to the use of expansion–normalized variables in conjunction with the orthonormal frame formalism, thereby laying the foundation for the detailed analysis of the Bianchi models with non–tilted perfect fluid source in Chapters 6 and 7. In Section 5.3 we discuss, from a general perspective, the use of dynamical systems methods in analyzing the evolution of Bianchi cosmologies, referring to the background material in Chapter 4.

5.1 Overview

As explained in Section 1.4.2 there are two main approaches to formulating the field equations for Bianchi cosmologies:

(i) the metric approach,
(ii) the orthonormal frame approach.

In the *metric approach* the basic variables are the metric components $g_{\alpha\beta}(t)$ relative to a group–invariant, time–independent frame (see (1.89)). This approach was initiated by Taub (1951) in a major paper. After a number of years researchers became aware that the Bianchi models admitted additional structure, namely the *automorphism group*, which plays an important role in identifying the physically significant variables (also referred to as gauge–invariant variables, or the true degrees of freedom). This group is defined to be the set of time–dependent linear transformations (1.87) of the spatial frame vectors that preserve the structure equations (1.88). Collins & Hawking (1973b) pointed out that one could use this group to replace $g_{\alpha\beta}(t)$ by new automorphism variables, resulting in a considerable simplification of the field equations. Subsequently, the use of automorphism variables was pursued systematically by Siklos, Jantzen and others (see Jantzen 1984, 1987 for details and other references). The significance of the automorphism variables is that in addition to simplifying the field equations, they also provide a complete description of the true degrees of freedom for a given class of Bianchi models. These methods are used to derive reduced evolution equations for Bianchi cosmologies by Rosquist & Jantzen (1988).

In the *orthonormal frame approach*, the basic variables are the commutation functions γ^c_{ab} associated with a group–invariant orthonormal frame (see (1.11) and Section 1.5.2). In the cosmological context this approach was pioneered by Ellis (1967). Generally speaking, the EFE and the Jacobi identities for the orthonormal frame (see Section 1.5.3) provide *first–order* evolution equations for the γ^c_{ab}, provided that one eliminates the remaining freedom in choice of the orthonormal frame (the use of this freedom in particular cases is a non–trivial matter). Another feature of this approach is that since the orthonormal frame is group invariant, and we choose $\mathbf{e}_0 = \mathbf{n}$, the normal to the group orbits, the commutation functions have a direct geometrical and physical interpretation in terms of the kinematical quantities of the timelike vector field $\mathbf{n}$ and the spatial curvature of the group orbits (see (1.61) and Section 1.5.3). This approach is illustrated in detail in Chapters 6 and 7.

As discussed in the Introduction, three methods have been extensively used to analyze the Bianchi field equations from a qualitative point of view:

  (i) piecewise approximation methods,
 (ii) Hamiltonian methods,
(iii) dynamical systems methods.

All three methods have been used in conjunction with the metric approach, historically in the order listed above. On the other hand, only dynamical systems methods have been used with the orthonormal frame approach. However, as will be seen below, piecewise approximations arise in a natural way when using dynamical systems methods.

We now illustrate the nature and role of the three methods by considering the analysis of the well–known 'oscillatory approach to the singularity' in the Bianchi IX models with non–tilted perfect fluid — the so–called Mixmaster models. When using piecewise approximation methods, the field equations are written as second–order ordinary differential equations. The evolution of a cosmological model in time is approximated as a sequence of time intervals during each of which certain terms in the field equations may be neglected. This method led Belinskii *et al.* (1970) to predict that in a generic model the approach to the singularity (into the past) consists of an infinite sequence of time intervals during each of which the evolution is approximated by a Kasner vacuum model, with the transition between different Kasner intervals ('epochs') being approximated by a Taub Bianchi II vacuum model. A similar conclusion was reached independently by Misner (1969a), using Hamiltonian methods. He wrote the field equations as a time–dependent Hamiltonian system for a particle in two dimensions (the $\beta^+\beta^-$–plane). The system was analyzed heuristically by approximating the time–dependent potential in the Hamiltonian by moving 'potential walls', which are assumed to reflect the particle instantaneously. Free motion in the $\beta^+\beta^-$–plane between bounces corresponds to a Kasner epoch of Belinskii *et al.* (1970), and a reflection off a potential wall corresponds to a Bianchi II transition. These methods are discussed in more detail in Chapter 10 (see also Sections 8.4 and 11.3).

The value of these methods is that they lead, without a detailed and lengthy analysis, to a conjecture about the asymptotic behaviour of non–tilted Bianchi IX models near the big–bang. The methods have also been applied to other Bianchi types and to tilted models (see Section 8.4). It is important to note, however, that these methods are heuristic in nature; their predictions should thus be verified by a rigorous analysis. Dynamical systems methods have the potential to fulfill this role.

Dynamical systems methods were first applied in the study of Bianchi cosmologies by Collins (1971), who considered a number of special cases in which the cosmological state space was two–dimensional. Subsequently, also using the metric approach, Bogoyavlensky and Novikov showed that these methods could be used effectively in situations where the dimension of the state space is greater than two (see Bogoyavlensky 1985, for details and other references). In particular, they showed that the oscillatory approach to the singularity could be discussed by means of a past attractor (see Section 4.4.1) of the cosmological dynamical system. The key idea is that, owing to the structure of the orbits in the attractor, a *generic orbit is asymptotically approximated by an infinite sequence of heteroclinic orbits joining equilibrium points* (see Section 4.4.4), which describes the infinite sequence of Kasner epochs and Bianchi II transitions. Finally, the use of expansion–normalized variables in conjunction with the orthonormal frame approach was used to give a simplification and clarification of Bogoyavlensky's ideas (Ma & Wainwright 1992). In terms of these variables, the attractor, referred to as the *Mixmaster attractor* in Section 6.4, has a simple geometric form, and is the union of three hemispheres with a common equatorial circle in the cosmological state space, which is a subset of $\mathbb{R}^5$. In this way, the dynamical systems methods provide a precise description of the oscillatory approach to the singularity. Indeed, the language used in applying the piecewise approximation and Hamiltonian methods can be translated into the language of dynamical systems: the phrases 'neglect terms in the field equations', 'the universe point bounces off a potential wall', and 'undergoes free motion in the $\beta_+\beta_-$–plane' are replaced by the phrase 'the orbit is approximated by (or 'shadows') an orbit of more special Bianchi type'. Up to now, however, a complete proof that the Mixmaster attractor actually is an attractor has not been given. This state of affairs is typical of the results that are obtained when using dynamical systems methods: one can state mathematically well–defined conjectures that can be supported, though not proved, by a combination of analytic arguments and numerical experiments (except in the simpler cases, when complete proofs can be given; see Chapters 6 and 7).

From a qualitative point of view, one can distinguish three aspects of the evolution, namely,

  (i) the behaviour at early times (i.e. near the initial singularity),
 (ii) the behaviour at late times (for models that expand indefinitely),
(iii) the behaviour at intermediate times.

Most of the work using piecewise approximation and Hamiltonian methods has dealt with the evolution near the initial singularity. Some efforts have been made, however, to explore the intermediate and late–time behaviour using piecewise approximations (e.g. Doroshkevich *et al.* 1973, Lukash 1974, and Zel'dovich & Novikov (1983, Section 22.7)). In addition, as described in Chapter 10, Hamiltonian methods have more recently been applied to the late–time behaviour. Unfortunately, neither of these methods seems to be particularly powerful when applied to these regimes, and it is thus necessary to use the more precise dynamical systems methods to obtain reliable results.

5.2 Expansion–normalized variables

In Chapters 6 and 7 we use dynamical systems methods in conjunction with the orthonormal frame formalism to analyze Bianchi cosmologies which have a non–tilted perfect fluid with linear equation of state as source. Our main goal is to give a qualitative description of the evolution of this class of cosmologies, with a view to determining the subset of models which are compatible with observations.

In this section we give some general information about the dimensionless variables and the evolution equations that we use in these chapters.

Although we are here restricting our discussion to Bianchi models with a non–tilted perfect fluid as source, variables of this type can be used in a variety of different situations, as illustrated in Chapters 8, 13 and 14.

5.2.1 The reduced differential equation

The basic idea is to write the EFE in a way that enables one to study the evolution of the various physical and geometrical quantities *relative to the overall rate of expansion of the universe*, as described by the rate of expansion scalar $\Theta = u^a_{;a}$, or equivalently *the Hubble scalar H* (see (1.27) and (1.31)):

$$H = \tfrac{1}{3}\Theta. \tag{5.1}$$

This idea has its origins in the work of Collins (1971), and was subsequently developed more generally by Wainwright (1988), Wainwright & Hsu (1989) and Hewitt & Wainwright (1993). The qualitative analysis of the FL models in Section 2.3 provides a simple introduction to this approach.

For a non–tilted fluid, the 4–velocity **u** is orthogonal to the group orbits. Let t be a global time variable that is constant on the group orbits, and is such that

$$\mathbf{u} = \frac{\partial}{\partial t}.$$

Let $\{\mathbf{e}_a\}$ be a group–invariant orthonormal frame, with

$$\mathbf{e}_0 = \mathbf{u}.$$

We use the commutation functions $\gamma^c{}_{ab}$ associated with the frame $\{\mathbf{e}_a\}$ (see (1.11)):

$$[\mathbf{e}_a, \mathbf{e}_b] = \gamma^c{}_{ab}\mathbf{e}_c,$$

as the basic gravitational field variables (see (1.61) and (1.62)). The $\gamma^c{}_{ab}$ are constant on the group orbits and can thus be regarded as a function of the global time variable t (see (1.86)):

$$\gamma^c{}_{ab} = \gamma^c{}_{ab}(t).$$

Since $\mathbf{e}_0$ is normal to the group orbits, it follows from Section 1.5.3 that the non–zero commutation functions are

$$(\gamma^c{}_{ab}) = (H, \sigma_{\alpha\beta}, \Omega_\alpha, n_{\alpha\beta}, a_\alpha). \tag{5.2}$$

At this stage the remaining freedom in the choice of orthonormal frame needs to be eliminated. This choice effectively specifies the variables Ω_α

implicitly or explicitly (e.g. by specifying them as functions of the $\sigma_{\alpha\beta}$) and also simplifies the other quantities (e.g. choice of a shear eigenframe will result in the tensor $\sigma_{\alpha\beta}$ being represented by two diagonal terms). This leads to a reduced set of variables, consisting of H and the remaining commutation functions, which we denote symbolically by

$$\mathbf{x} = (\gamma^c{}_{ab}|_{reduced}).\tag{5.3}$$

The physical state of the model is thus described by the vector $(H, \mathbf{x})$. The details of this reduction differ for the class A and B models, and in the latter case there is an algebraic constraint of the form

$$g(\mathbf{x}) = 0,\tag{5.4}$$

where g is a homogeneous polynomial. For the class B models, some of the reduced variables are in fact quadratic in the $\gamma^c{}_{ab}$ (see Section 7.1). The preceding remarks thus oversimplify the treatment of the class B models, but nevertheless convey the essential idea.

The idea is now to normalize $\mathbf{x}$ with the Hubble scalar H. We denote the resulting variables by a vector $\mathbf{y} \in \mathbb{R}^n$, and write:

$$\mathbf{y} = \frac{\mathbf{x}}{H}.\tag{5.5}$$

These new variables are *dimensionless*, and will be referred to as *expansion-normalized variables*. It is clear that each dimensionless state $\mathbf{y}$ determines a one–parameter family of physical states $(H, \mathbf{x})$.

The evolution equations for the $\gamma^c{}_{ab}$, given in Section 1.5.3, lead to evolution equations for H and $\mathbf{x}$ (see Sections 6.1.1 and 7.1.1) and hence for $\mathbf{y}$. In deriving the evolution equations for $\mathbf{y}$ from those for $\mathbf{x}$, the *deceleration parameter* q plays an important role. This variable, which initially arises in the FL models, can be defined in any non–tilted Bianchi model, as follows. The Hubble scalar H can be used to define a length scale function ℓ, according to

$$H = \frac{\dot{\ell}}{\ell},\tag{5.6}$$

where $\cdot$ denotes differentiation with respect to t. As in Section 2.1, the deceleration parameter is then defined by

$$q = -\frac{\ddot{\ell}\ell}{\dot{\ell}^2},\tag{5.7}$$

and is a dimensionless quantity. Equivalently, q is related to $\dot{H}$ according to

$$\dot{H} = -(1 + q)H^2.\tag{5.8}$$

In order that the evolution equations define a flow (Section 4.1.3), it is necessary, in conjunction with the rescaling (5.5), to introduce a *dimensionless time variable* τ according to

$$\ell = \ell_0 e^\tau, \tag{5.9}$$

where ℓ_0 is the value of the length scale function at some arbitrary reference time. Since ℓ assumes values $0 < \ell < +\infty$ in an ever–expanding model, τ assumes all real values, with $\tau \to -\infty$ at the initial singularity and $\tau \to +\infty$ at late times. It follows from equations (5.6) and (5.9) that

$$\frac{dt}{d\tau} = \frac{1}{H}, \tag{5.10}$$

and the evolution equation (5.8) for H can be written

$$\frac{dH}{d\tau} = -(1+q)H. \tag{5.11}$$

Since the right–hand sides of the evolution equations for the $\gamma^c{}_{ab}$ in Section 1.5.3 are homogeneous of degree 2 in the $\gamma^c{}_{ab}$ the change (5.10) of the time variable results in H cancelling out of the evolution equation for $\mathbf{y}$, yielding an autonomous DE

$$\frac{d\mathbf{y}}{d\tau} = \mathbf{f}(\mathbf{y}), \quad \mathbf{y} \in \mathbb{R}^n. \tag{5.12}$$

The constraint $g(\mathbf{x}) = 0$ translates into a constraint

$$g(\mathbf{y}) = 0, \tag{5.13}$$

which is preserved by the DE. The functions $\mathbf{f} : \mathbb{R}^n \to \mathbb{R}^n$ and $g : \mathbb{R}^n \to \mathbb{R}$ are polynomial functions in $\mathbf{y}$. An essential feature of this process is that the evolution equation for H, namely (5.11), decouples from the remaining equations (5.12) and (5.13). In other words, the DE (5.12) describes the evolution of the non–tilted Bianchi cosmologies, the transformation (5.5) essentially scaling away the effects of the overall expansion. An important consequence is that the new variables are bounded near the initial singularity (see Chapters 6 and 7, and Section 8.1), and the equilibrium points of the DE represent self–similar solutions (see Section 5.3.3 below).

5.2.2 The density parameter

Two of the field equations have a simple form, independently of how the freedom in the choice of the orthonormal frame is used, namely the *Raychaudhuri equation* and the *generalized Friedmann equation*. Since the cosmological velocity field is geodesic and irrotational, the Raychaudhuri equation

(1.42) reads

$$\dot{H} = -H^2 - \tfrac{2}{3}\sigma^2 - \tfrac{1}{6}(3\gamma - 2)\mu, \tag{5.14}$$

and the generalized Friedmann equation (1.57) reads

$$3H^2 = \sigma^2 - \tfrac{1}{2}\,^3R + \mu. \tag{5.15}$$

In these equations, σ^2 is the shear scalar (1.28) and ^3R is the scalar curvature of the hypersurfaces $t = $ constant, both of which can be expressed in terms of the commutation functions (see (1.61) and Section 1.5.3).

Equation (5.14), in conjunction with (5.8), yields an equation for q that is needed to eliminate q when deriving the dimensionless evolution equation (5.12). In addition, equation (5.15) serves to express the matter density μ in terms of commutation functions.

In keeping with our basic approach, we have to write (5.14) and (5.15) in dimensionless form. First, the matter content is described by the *density parameter* Ω, defined by

$$\Omega = \frac{\mu}{3H^2}, \tag{5.16}$$

where μ is the matter density, in analogy with (2.12) in an FL universe.

Secondly, one obtains a dimensionless measure of the anisotropy by forming the ratio of σ^2 with H^2, i.e. by comparing the shear with the expansion. Thus we define the *shear parameter*† Σ by

$$\Sigma^2 = \frac{\sigma^2}{3H^2}. \tag{5.17}$$

The factor of 3 is included so that in ever–expanding models, Σ^2 has a maximum value of 1.

Thirdly, we define the *curvature parameter* K by

$$K = -\frac{^3R}{6H^2}. \tag{5.18}$$

The factor of $-\tfrac{1}{6}$ is included so that K is positive in ever–expanding models, and has a maximum value of 1.

Equation (5.15) now assumes the form

$$\Sigma^2 + K + \Omega = 1, \tag{5.19}$$

and (5.14), in conjunction with (5.8), yields the following equation for q:

$$q = 2\Sigma^2 + \tfrac{1}{2}(3\gamma - 2)\Omega. \tag{5.20}$$

† The CBR observations lead to an upper bound on Σ. See Section 3.2.4.

In Chapters 6 and 7, we shall show that Σ^2 and K are polynomials in terms of the expansion–normalized variables **y** and thus *(5.19) and (5.20) express Ω and q as polynomials in **y**.*

The density parameter has a simple evolution equation. For a perfect fluid with γ–law equation of state, the contracted Bianchi identities (1.50) imply that

$$\dot{\mu} = -3\gamma H\mu.$$

It follows from (5.8), (5.10) and (5.16) that in this case Ω satisfies the evolution equation

$$\frac{d\Omega}{d\tau} = [2q - (3\gamma - 2)]\Omega. \tag{5.21}$$

Equations (5.19)–(5.21) play a fundamental role in Chapters 6 and 7.

In concluding this section we note that equations (5.9)–(5.11), (5.20) and (5.21) correspond to equations (2.43)–(2.47) in the FL case. The difference is that in the present case the expression for q contains the shear parameter Σ, which is zero in the FL case.

5.3 Cosmological dynamical systems
5.3.1 Invariant sets

In summary, the first step in the analysis is to formulate the field equations, using expansion–normalized variables, as a DE (5.12) in $\mathbb{R}^n$, possibly subject to a constraint (5.13). Since τ assumes all real values (for models that expand indefinitely), the solutions of (5.12) are defined for all τ and hence define a *flow* $\{\phi_\tau\}$ on $\mathbb{R}^n$ (see Section 4.1.3). The evolution of the cosmological models can thus be analyzed by studying the orbits of this flow in the physical region of state space, which is a subset of $\mathbb{R}^n$ defined by the requirement that the energy density be non–negative, i.e.

$$\Omega(\mathbf{y}) \geq 0. \tag{5.22}$$

The *vacuum boundary*, defined by $\Omega(\mathbf{y}) = 0$, describes the evolution of vacuum Bianchi models, and is an invariant set on account of the evolution equation (5.21). This set plays an important role in the qualitative analysis because vacuum models can be asymptotic states, near the big–bang or at late times, for perfect fluid models. There are other invariant sets that are also specified by simple restrictions on **y** and that play a special role: the subsets representing each Bianchi type and the subsets representing higher symmetry models, specifically the FL models and the LRS Bianchi models (the case $s = 3$, $d = 1$ in Section 1.2.2).

It is desirable that the dimensionless state space D in $\mathbb{R}^n$ is a compact set. In this case each orbit will have a non–empty α–limit set and ω–limit set, and hence there will exist a past attractor and a future attractor in state space (see Section 4.4.1). When using expansion–normalized variables, compactness of the state space has a direct physical meaning for ever–expanding models: if the state space is compact then at the big–bang no physical or geometrical quantity diverges more rapidly than the appropriate power of H, and at late times no such quantity tends to zero less rapidly than the appropriate power of H. We find that the state space for Bianchi VII_0 and VIII models is in fact non–compact (see Chapter 6). This lack of compactness manifests itself in the behaviour of the Weyl tensor at late times.

5.3.2 The decoupled equations: clock time and the Hubble parameter

The cosmological DE (5.12) is introduced by scaling away the effects of the overall expansion, and using the logarithm of the scale factor ℓ as the time variable. In order to relate the analysis to observations, however, the decoupled equations (5.10) and (5.11), which determine the Hubble scalar H and the clock time t, have to be brought into play. We shall choose the origin of t so that $t = 0$ corresponds to the initial singularity. The present time t_0, called the *age of the universe*, and the present value of H, denoted H_0 and called *the Hubble constant*, are two quantities for which we have observable bounds. The product

$$\alpha = tH \tag{5.23}$$

is dimensionless. Its present value $\alpha_0 = t_0 H_0$ will be referred to as the *age parameter* (c.f. (2.32) and (3.15)). We shall obtain general theoretical bounds for α, valid for all values of t. Note that on account of (5.10) and (5.11),

$$\frac{d\alpha}{d\tau} = 1 - (1 + q)\alpha. \tag{5.24}$$

We have seen that (5.19) and (5.20) define the deceleration parameter q as a function on state space $q = q(\mathbf{y})$. Given an initial point $\mathbf{y}_0$, which we think of as representing the state of the universe at the present time, let $\mathbf{y} = \phi_\tau(\mathbf{y}_0)$ describe the orbit through $\mathbf{y}_0$, with $\phi_0(\mathbf{y}_0) = \mathbf{y}_0$. Then let $\tilde{q}(\tau)$ denote the deceleration parameter evaluated along this orbit, i.e.

$$\tilde{q}(\tau) = q(\phi_\tau(\mathbf{y}_0)),$$

so that $\tilde{q}(0) = q(\mathbf{y_0})$. Then by (5.11), the Hubble scalar along the orbit through $\mathbf{y_0}$ is

$$H(\tau) = H_0 \exp\left[-\int_0^\tau \{1 + \tilde{q}(u)\}du\right],\qquad(5.25)$$

for all $\tau \in \mathbb{R}$. Here H_0 is freely specifiable. This arbitrariness implies that *each non–singular orbit corresponds to a one–parameter family of physical universes, which are conformally related* by a constant rescaling of the commutation functions, and hence of the orthonormal frame and the metric. By (5.10), the value of t at $\mathbf{y_0}$, denoted $t_0 = t(0)$, is

$$t_0 = \int_{-\infty}^0 \frac{1}{H(\tau)} d\tau,\qquad(5.26)$$

where $H(\tau)$ is given by (5.25). It follows from (5.25) and (5.26) that

$$t_0 H_0 = \int_{-\infty}^0 \exp\left[\int_0^\tau \{1 + \tilde{q}(u)\}du\right] d\tau,\qquad(5.27)$$

i.e. the dimensionless quantity $t_0 H_0$ is uniquely determined by the specified point $\mathbf{y_0}$ in state space. This means that $\alpha = tH$ is a well–defined function on state space, $\alpha = \alpha(\mathbf{y})$. Thus, *observational bounds on $\alpha_0 = t_0 H_0$ (e.g. (3.19)) will restrict the location of the present state $\mathbf{y_0}$ of the universe in state space.*

We can also draw the following conclusion at this stage: given any point $\mathbf{y_0}$ in the dimensionless state space D, and any initial time $t_0 > 0$, there exists a physical universe having dimensionless state $\mathbf{y_0}$ at physical time t_0. Of course the rate of expansion at t_0 will be specified through $H_0 = \alpha_0/t_0$.

By (5.20), the strong energy condition, i.e. $\gamma > \frac{2}{3}$, implies $q \geq 0$, which leads to $t_0 H_0 \leq 1$ by (5.27). Since $\mathbf{y_0}$ is arbitrary, it follows that *in any Bianchi cosmology with non–tilted perfect fluid source and $\gamma > \frac{2}{3}$, we have the upper bound*

$$tH \leq 1,\qquad(5.28)$$

for all $t > 0$. We note that this age inequality is the same as that obtained for the FL models (see (2.32), (2.35) and (3.16)). For ever–expanding models (i.e. $K \geq 0$) we also obtain a lower bound. If $\gamma \leq 2$, it follows from (5.20) and (5.27) that

$$tH \geq \tfrac{1}{3}.$$

It is worth emphasizing that the dimensionless time τ does have a geometrical significance as regards the physical universe, since it is directly related to the length scale by (5.9). For any y_0, let y_1 be the point on the orbit through y_0 corresponding to a change $\Delta\tau$, i.e.

$$y_1 = \phi_{\Delta\tau}(y_0). \tag{5.29}$$

By (5.9) it follows that ℓ_1, the length scale at y_1, is related to ℓ_0 the length scale at y_0, by

$$\frac{\ell_1}{\ell_0} = e^{\Delta\tau}. \tag{5.30}$$

Thus, *for any y_0 and any change $\Delta\tau$, the ratio of the length scales at y_0 and y_1 equals $e^{\Delta\tau}$.*

Any two points y_0 and y_1 on an ordinary orbit in the dimensionless state space determine a unique increment $\Delta\tau$ in the dimensionless time variable according to (5.29), i.e. $\Delta\tau$ is the time taken for the flow to map y_0 into y_1. By (5.10) and (5.25), the corresponding change in the clock time t is

$$t_1 - t_0 = \frac{1}{H_0} \int_0^{\Delta\tau} \exp\left[\int_0^\tau \{1 + \tilde{q}(u)\} du\right] d\tau.$$

Since the integral depends only on y_0 and $\Delta\tau$, we can write this as

$$t_1 - t_0 = \frac{1}{H_0} F(y_0, \Delta\tau). \tag{5.31}$$

Since H_0 is an arbitrary positive constant, *the change in clock time $(t_1 - t_0)$ can assume any positive value.* However since $\alpha_0 = t_0 H_0$ is uniquely determined by y_0 (see (5.27)), it follows that the ratio t_1/t_0 is uniquely determined.

The arbitrariness in $t_1 - t_0$ is related to the fact that an ordinary orbit corresponds to a one–parameter family of universes, all conformally related. To specify a unique universe, one has to choose a reference point y_0 on the orbit, which then determines $\alpha_0 = t_0 H_0$, and then choose a value of H_0. Once this is done, two points y_0 and y_1 do determine a unique clock time interval $t_1 - t_0$ via (5.31).

5.3.3 Equilibrium points and self–similar cosmologies

We have seen that each ordinary orbit in the dimensionless state space corresponds to a one–parameter family of physical universes, which are conformally related by a constant rescaling of the metric. On the other hand, for an equilibrium point y^* of the DE (5.12) (which satisfies $f(y^*) = 0$), q is a constant, i.e. $q(y^*) = q^*$, and (5.25) gives

$$H(\tau) = H_0 e^{(1+q^*)\tau}.$$

In this case, however, the parameter H_0 is no longer essential, since it can be set to unity by a translation of τ, $\tau \to \tau +$ constant. In this case (5.10) implies that

$$Ht = \frac{1}{1+q^*},$$ (5.32)

so that by (5.3) and (5.5) the commutation functions are of the form (constant) $\times t^{-1}$. It follows (Hsu & Wainwright 1986, theorem 3.1; see also Rosquist & Jantzen 1985a, and Jantzen & Rosquist 1986) that the space-time admits a similarity group H_4 whose orbits are four-dimensional. Thus, *to each equilibrium point of the DE (5.12) there corresponds a unique transitively self-similar cosmological model.* In such a model the physical states at different times differ only by an overall change in the length scale (Eardley 1974, Wainwright 1985). Such models are expanding, of course, but in such a way that their dimensionless state does not change. The self-similar models are given explicitly in Chapter 9, and include the flat FL model ($\Omega = 1$) and the Milne model ($\Omega = 0$).

There are a number of other non-vacuum ($\Omega > 0$) equilibrium points $\mathbf{y}^*$ (i.e. self-similar models; see Table 9.2). It is of interest that on account of (5.21), all non-vacuum equilibrium points satisfy

$$q^* = \tfrac{1}{2}(3\gamma - 2),$$ (5.33)

so that (5.32) gives

$$H = \frac{2}{3\gamma}t^{-1},$$ (5.34)

the relation which is valid for the flat FL model. Of course the equilibrium points other than the FL ones correspond to anisotropic models and hence have non-zero shear ($\Sigma^* \equiv \Sigma(\mathbf{y}^*) > 0$).

The equilibrium points determine the asymptotic behaviour of other more general models. If the α-limit set of a point $\mathbf{y}$ is an equilibrium point $\mathbf{y}^*$, then the orbit through $\mathbf{y}$ approaches $\mathbf{y}^*$ as $\tau \to -\infty$. The physical interpretation is that the self-similar model that corresponds to $\mathbf{y}^*$ approximates the dynamics of the model with initial state $\mathbf{y}$, as $\tau \to -\infty$. We will say that this model is *asymptotically self-similar into the past.* A similar interpretation holds if the ω-limit set is an equilibrium point. The term *asymptotically self-similar* without a qualifier will mean that the model has this property into the past and into the future. In this case the orbit that describes the model will be *heteroclinic* (i.e. joins two equilibrium points).

In this situation one can find the asymptotic form of the physical state

$(H, \mathbf{x})$. It follows from (5.23) and (5.24) that

$$\lim_{\tau \to \pm\infty} Ht = \frac{1}{1 + q^*}, \tag{5.35}$$

or equivalently, that

$$H = \frac{1}{1 + q^*} t^{-1} + o(t^{-1}) \tag{5.36}$$

as $t \to +\infty$ or $t \to 0^+$. Equation (5.5) now implies that

$$\mathbf{x} = \frac{\mathbf{y}^*}{1 + q^*} t^{-1} + o(t^{-1}). \tag{5.37}$$

The leading terms in (5.36) and (5.37) give the physical state of the exact self–similar model.

It is important to note, however, that *in general, Bianchi models with non–tilted perfect fluid are not asymptotically self–similar* (see Sections 6.5 and 8.1). Nevertheless, as mentioned in Section 5.1, equilibrium points play an essential role in the more complicated asymptotic behaviour that occurs in Bianchi IX (and VIII) models, by determining the infinite heteroclinic sequences that give rise to the oscillatory approach to the singularity.

Equilibrium points also influence the intermediate evolution by determining finite heteroclinic sequences that join the past attractor to the future attractor (see Section 4.2). The intermediate equilibrium points in the sequence determine quasi–equilibrium epochs, which may be important from an observational point of view.

It is of interest from an observational point of view that of the non–vacuum equilibrium points only one is *universal* in the sense that it exists in the state space of each Bianchi type, namely, the flat FL model (i.e. $\Omega = 1$). This may seem puzzling in view of the fact that the flat FL model is of Bianchi type I. The reason for its universal role is that the state space for a particular Bianchi type is a subset of $\mathbf{R}^n$, whose boundary is the state space for models of more special Bianchi types and hence includes models of type I. Furthermore if the equation of state parameter satisfies $\gamma > \frac{2}{3}$ and the Bianchi type is not type I, the flat FL equilibrium point is a saddle point. We shall refer to a quasi–equilibrium epoch that is determined by the flat FL model as a *flat quasi–isotropic epoch*. Since the flat FL equilibrium point is universal, such epochs will occur in Bianchi universes of all types more general than type I (see also Zel'dovich & Novikov 1983, page 550; see Wainwright & Anderson 1984 for an explicit example of Bianchi type VI_h).

In concluding our discussion of self–similar models we note that the classification of singularities in Section 1.3.4 can be applied to any self–similar solution, and hence, on account of (5.36) and (5.37), to any solution that is asymptotically self–similar into the past. Consider any equilibrium point, $\mathbf{y} = \mathbf{y}^*$, or equivalently, any self–similar solution. By performing a constant rotation if necessary, one can introduce an eigenframe of the expansion tensor, so that

$$(\Theta_{\alpha\beta}) = \text{diag}\,(\Theta_1, \Theta_2, \Theta_3).$$

Since the commutation functions are proportional to t^{-1}, we can define constants p_α by

$$p_\alpha = t\Theta_\alpha, \quad \alpha = 1, 2, 3. \tag{5.38}$$

It follows that the scale factors ℓ_α (see Section 1.3.4) are given by

$$\ell_\alpha = t^{p_\alpha},$$

up to a constant multiple. The exponents p_α thus characterize the different singularity types:

(i)   point, $p_1 > 0,\ p_2 > 0,\ p_3 > 0,$

(ii) cigar, $p_1 > 0,\ p_2 > 0,\ p_3 < 0$ (cycle on $1, 2, 3$),

(iii) pancake, $p_1 > 0,\ p_2 = 0 = p_3$ (cycle on $1, 2, 3$),

(iv) barrel, $p_1 > 0,\ p_2 > 0,\ p_3 = 0$ (cycle on $1, 2, 3$).
$$\tag{5.39}$$

It follows from (5.17), (5.32), (1.28) and (1.30) that

$$\sum_{\alpha=0}^{3} p_\alpha = \frac{3}{1 + q^*}, \quad \sum_{\alpha=1}^{3} p_\alpha^2 = \tfrac{1}{3}(1 + 2\Sigma_*^2)\left(\sum_{\alpha=1}^{3} p_\alpha\right)^2, \tag{5.40}$$

where $\Sigma_* = \Sigma(\mathbf{y}^*)$ is the value of the shear parameter Σ, as defined by (5.17), at the equilibrium point. The results of Section 9.1 show that the singularity types (5.39) are a complete set of possibilities for Bianchi models with non–tilted perfect fluid.

5.3.4 Summary

We now summarize the procedure that we use to analyze the cosmological dynamical system (i.e. flow) that is determined by the DE (5.12), possibly in conjunction with a constraint (5.13).

(1) Determine whether the state space, as defined by the inequality (5.22), is compact.

(2) Identify the lower–dimensional invariant sets, which contain the orbits of models of more special Bianchi type and/or models with additional symmetry.

(3) Find all equilibrium points and analyze their local stability. Where possible identify the stable and unstable manifolds, which may coincide with some of the invariant sets in point (2).

(4) Find monotone functions or Dulac functions in the various invariant sets where possible. In some cases a monotone function is suggested by the form of the DE (5.12) and in some cases by the Hamiltonian formulation of the field equations (see Section 10.3).

(5) Investigate any bifurcations that occur as the equation of state parameter γ (or any other parameter) varies. The bifurcations are associated with changes in the local stability of the equilibrium points.

(6) Having all the information in points (1)–(5) one can hope to formulate precise conjectures about the asymptotic evolution, by identifying the past and future attractors. The past attractor will describe the evolution of a typical universe near the initial singularity while the future attractor will play the same role at late times. The monotone functions in point (4) above, in conjunction with the theorems in Section 4.4.3, may enable some of the conjectures to be proved.

(7) Knowing the stable and unstable manifolds of the equilibrium points it is possible to construct all possible heteroclinic sequences that join the past attractor to the future attractor, thereby gaining insight into the intermediate evolution of the cosmological models.

6

Bianchi cosmologies: non–tilted class A models

In this chapter we use the general procedure described in Sections 5.2 and 5.3 to analyze the evolution of Bianchi models of class A with a non–tilted perfect fluid source. Sections 6.1 and 6.2 follow Wainwright & Hsu (1989), abbreviated WH† in this chapter.

6.1 Evolution equations and state space

6.1.1 The evolution equations

As shown in Section 1.5.4, we can introduce a group–invariant orthonormal frame relative to which the only non–zero commutation functions are the Hubble scalar H and the diagonal components of the shear tensor $\sigma_{\alpha\beta}$ and the matrix‡ $n_{\alpha\beta}$,

$$\sigma_{\alpha\beta} = \mathrm{diag}\,(\sigma_{11}, \sigma_{22}, \sigma_{33}), \quad n_{\alpha\beta} = \mathrm{diag}\,(n_1, n_2, n_3). \qquad (6.1)$$

Since $\sigma_{\alpha\beta}$ is trace–free, it has only two independent components, which it is convenient to label using

$$\sigma_+ = \tfrac{1}{2}(\sigma_{22} + \sigma_{33}), \quad \sigma_- = \tfrac{1}{2\sqrt{3}}(\sigma_{22} - \sigma_{33}). \qquad (6.2)$$

The physical state of the models is thus described by a vector $(H, \mathbf{x})$, where

$$\mathbf{x} = (\sigma_+, \sigma_-, n_1, n_2, n_3). \qquad (6.3)$$

Equation (6.1) implies that the trace–free spatial Ricci tensor $^3S_{\alpha\beta}$, given by (1.94), is also diagonal. We label its components in an analogous manner:

$$^3S_+ = \tfrac{1}{2}(^3S_{22} + {}^3S_{33}), \quad {}^3S_- = \tfrac{1}{2\sqrt{3}}(^3S_{22} - {}^3S_{33}). \qquad (6.4)$$

---

† The equations in this section differ from Wainwright & Hsu (1989) as regards numerical factors since we are here using H instead of the expansion scalar Θ as the normalizing factor.

‡ It is convenient to label the diagonal components of $n_{\alpha\beta}$ with a single index as in Section 1.5.3.

Since the source is a non–tilted perfect fluid the source terms satisfy (1.106). Equations (1.91) and (1.96) now yield the following evolution equations for $\sigma_\pm$ and the n_α:

$$
\begin{aligned}
\dot{\sigma}_\pm &= -3H\sigma_\pm - {}^3S_\pm, \\
\dot{n}_1 &= (-H - 4\sigma_+)n_1, \\
\dot{n}_2 &= (-H + 2\sigma_+ + 2\sqrt{3}\sigma_-)n_2, \\
\dot{n}_3 &= (-H + 2\sigma_+ - 2\sqrt{3}\sigma_-)n_3,
\end{aligned}
\tag{6.5}
$$

where

$$
\begin{aligned}
{}^3S_+ &= \tfrac{1}{6}\left[(n_2 - n_3)^2 - n_1(2n_1 - n_2 - n_3)\right], \\
{}^3S_- &= \tfrac{1}{2\sqrt{3}}(n_3 - n_2)(n_1 - n_2 - n_3).
\end{aligned}
\tag{6.6}
$$

As described in Section 5.2, the dimensionless state is obtained by normalizing the physical state (6.3) with H, giving

$$
\mathbf{y} = (\Sigma_+, \Sigma_-, N_1, N_2, N_3) \in \mathbb{R}^5,
\tag{6.7}
$$

where

$$
\Sigma_\pm = \frac{\sigma_\pm}{H}, \quad N_\alpha = \frac{n_\alpha}{H}.
\tag{6.8}
$$

The evolution equation

$$
\frac{d\mathbf{y}}{d\tau} = \mathbf{f}(\mathbf{y})
$$

for the dimensionless state $\mathbf{y}$ is obtained from (6.8) and (6.5) in conjunction with the decoupled equations (5.10) and (5.11), namely

$$
\frac{dt}{d\tau} = \frac{1}{H}, \qquad \frac{dH}{d\tau} = -(1 + q)H.
$$

Defining

$$
\mathcal{S}_\pm = \frac{{}^3S_\pm}{H^2},
$$

we obtain

$$
\begin{aligned}
\Sigma'_\pm &= -(2 - q)\Sigma_\pm - \mathcal{S}_\pm, \\
N'_1 &= (q - 4\Sigma_+)N_1, \\
N'_2 &= (q + 2\Sigma_+ + 2\sqrt{3}\Sigma_-)N_2, \\
N'_3 &= (q + 2\Sigma_+ - 2\sqrt{3}\Sigma_-)N_3,
\end{aligned}
\tag{6.9}
$$

where by (6.6) and (6.8),

$$S_+ = \tfrac{1}{6}\Big[(N_2 - N_3)^2 - N_1(2N_1 - N_2 - N_3)\Big],$$

$$(6.10)$$

$$S_- = \tfrac{1}{2\sqrt{3}}(N_3 - N_2)(N_1 - N_2 - N_3),$$

and ′ denotes differentiation with respect to τ. It follows from (5.19) and (5.20) that

$$q = \tfrac{1}{2}(3\gamma - 2)(1 - K) + \tfrac{3}{2}(2 - \gamma)\Sigma^2, \qquad (6.11)$$

where

$$K = \tfrac{1}{12}\Big[N_1^2 + N_2^2 + N_3^2 - 2(N_1 N_2 + N_2 N_3 + N_3 N_1)\Big]. \qquad (6.12)$$

and

$$\Sigma^2 = \Sigma_+^2 + \Sigma_-^2. \qquad (6.13)$$

Equations (6.12) and (6.13) follow from (1.28), (1.95), (5.17), (5.18) and (6.8). The density parameter is given by (5.19), namely

$$\Omega = 1 - \Sigma^2 - K. \qquad (6.14)$$

It is helpful to keep in mind the geometric and physical interpretation of the state variables (6.7): $\Sigma_\pm$ *describe the anisotropy in the Hubble flow*, and the N_α *describe the spatial curvature of the group orbits*, via (6.10) and (6.12), and the Bianchi type of the isometry group (see the remark following (1.105)).

6.1.2 *Invariant sets*

The state space for the Bianchi models of class A is the set of all points $(\Sigma_\pm, N_1, N_2, N_3) \in \mathbb{R}^5$ which satisfy the inequality $\Omega \geq 0$, or, by (6.14), $\Sigma^2 + K \leq 1$ where Σ and K are given by (6.12) and (6.13). Since K is not a positive definite quadratic form in the N_α, the state space is unbounded. We shall discuss this matter further, after we have introduced various invariant sets.

First, the form of the DE (6.9) implies that any combination of the conditions

$$N_\alpha > 0, \quad N_\beta = 0, \quad N_\gamma < 0$$

for $\alpha, \beta, \gamma = 1, 2, 3$ defines an invariant set (see Proposition 4.1). Since these conditions determine the Bianchi type of the model these invariant sets are the union of orbits corresponding to models of a specific Bianchi type,

Table 6.1. *The Bianchi invariant sets, with $\alpha = 1, 2, 3$. d is the dimension of the invariant set or, equivalently, the number of parameters in the family of cosmological models, and N is the number of disjoint components.*

Notation	Restrictions on the N_α	d	N
$B(\mathrm{I})$	All zero	2	1
$B_\alpha^\pm(\mathrm{II})$	One non–zero	3	6
$B_\alpha^\pm(\mathrm{VI_0})$	Two non–zero and opposite in sign	4	6
$B_\alpha^\pm(\mathrm{VII_0})$	Two non–zero with the same sign	4	6
$B_\alpha^\pm(\mathrm{VIII})$	All non–zero differing in sign	5	6
$B^\pm(\mathrm{IX})$	All non–zero with the same sign	5	2

and hence will be called the *Bianchi invariant sets*. They are described in Table 6.1. The conventions for labelling the various disjoint components are illustrated below:†

$$B_1^+(\mathrm{VIII}): \quad N_1 < 0, \quad N_2 > 0, \quad N_3 > 0,$$

$$B_1^+(\mathrm{VI_0}): \quad N_1 = 0, \quad N_2 > 0, \quad N_3 < 0,$$

$$B_1^+(\mathrm{II}): \quad N_1 > 0, \quad N_2 = 0, \quad N_3 = 0.$$

The different components for one Bianchi type are mapped into one another under the discrete symmetries of the DE and hence are equivalent (see WH, page 1415 for details). Figure 6.1 (adapted from Collins & Hawking 1973b) shows how the Bianchi invariant sets are related to each other. An arrow pointing from one type to another indicates that the invariant set for the second type lies in the boundary of the invariant set for the first type. Here $B(\mathrm{VIII})$ denotes the union of the six invariant sets $B_\alpha^\pm(\mathrm{VIII})$, etc.

Restrictions on the shear variables also lead to invariant sets, that we shall call *shear invariant sets*. It follows from the DE (6.9) that

$$\Sigma_- = 0 \quad \text{implies} \quad \begin{cases} N_2 = N_3 & \text{for types VII}_0, \text{VIII, IX}, \\ N_2 = -N_3 & \text{for type VI}_0, \end{cases}$$

and that these conditions define invariant sets which are subsets of the corresponding Bianchi invariant sets. For example,

$$S_1^+(\mathrm{VIII}): \quad N_1 < 0, \quad N_2 = N_3 > 0, \quad \Sigma_- = 0,$$
$$S_1^+(\mathrm{VI_0}): \quad N_1 = 0, \quad N_2 = -N_3 > 0, \quad \Sigma_- = 0.$$

† To obtain the other + components, cycle on $(1, 2, 3)$, and to obtain the − components, reverse signs throughout.

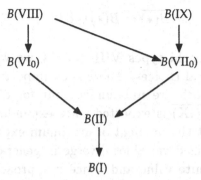

Fig. 6.1. Specialization diagram for the Bianchi invariant sets.

Table 6.2. *The shear invariants sets, with $\alpha = 1, 2, 3$. d is the dimension of the invariant set and N is the number of disjoint components.*

Notation	Class of models	d	N
$S_\alpha(\text{I})$	LRS Bianchi I	1	3
$S_\alpha^\pm(\text{II})$	LRS Bianchi II	2	6
$S_\alpha^\pm(\text{VI}_0)$	Bianchi VI$_0$ $(n_\alpha{}^\alpha = 0)$	2	6
$S_\alpha^\pm(\text{VII}_0)$	LRS Bianchi VII$_0$	2	6
$S_\alpha^\pm(\text{VIII})$	LRS Bianchi VIII	3	6
$S_\alpha^\pm(\text{IX})$	LRS Bianchi IX	3	6

All possibilities are obtained from the basic classification shown in Table 6.2 by cyclically permuting on the labels $1, 2, 3$ and choosing both signs. All the models are locally rotationally symmetric (LRS, see Section 1.2.2) except for the Bianchi VI$_0$ models which are in the subclass with $n_\alpha{}^\alpha = 0$ (see Section 1.5.4).

6.1.3 Properties of the state space

When analyzing models of a particular Bianchi type, described by an invariant set $B(*)$, one has to consider the boundary $\partial B(*)$ of $B(*)$ since an orbit may approach the boundary asymptotically. The boundary consists of Bianchi invariant sets of more specialized Bianchi types and the vacuum subset $\Omega = 0$. Thus one works with the closure

$$\overline{B(*)} = B(*) \cup \partial B(*).$$

For the general Bianchi types VIII and IX we will use $\overline{B_1^+(\text{VIII})}$ and $\overline{B^+(\text{IX})}$ as the canonical choices. There is an important difference between these state spaces: both are unbounded, but for different reasons. The unboundedness of $\overline{B^+(\text{IX})}$ is expected since expanding Bianchi IX models may recollapse, and at the moment of maximum expansion $H = 0$, so that the expansion–normalized variables diverge in general. At this instant the scale factor ℓ has a finite value and hence τ approaches a finite value, i.e. the solutions of the DE (6.9) are in general not extendible into the future ($\tau \to +\infty$). Thus the solutions of the DE define a negative semi–flow but not a flow on $\overline{B^+(\text{IX})}$.

On the other hand, Bianchi VIII and more specialized models expand indefinitely, with ℓ increasing monotonically and assuming all positive values. Thus τ assumes all real values and the DE (6.9) defines a flow ϕ_τ on $\overline{B_1^+(\text{VIII})}$. Nevertheless, $\overline{B_1^+(\text{VIII})}$ is unbounded, although bounds can be placed on some of the variables as follows. One can write the expression (6.12) for the curvature parameter K in the form

$$K = \tfrac{1}{12}\left[N_1{}^2 - 2N_1(N_2 + N_3) + (N_2 - N_3)^2\right]. \tag{6.15}$$

The inequalities $N_1 \leq 0$, $N_2 \geq 0$, $N_3 \geq 0$, valid in $\overline{B_1^+(\text{VIII})}$, imply that $K \geq 0$, and it follows from (6.14) that

$$0 \leq \Sigma \leq 1, \qquad 0 \leq K \leq 1, \qquad 0 \leq \Omega \leq 1.$$

One can now conclude from (6.13) that $-1 \leq \Sigma_\pm \leq 1$ and from (6.15) that $0 \leq N_1{}^2 \leq 12$, $0 \leq (N_2 - N_3)^2 \leq 12$. However, $N_2 + N_3$ can assume arbitrarily large values, although $N_1(N_2 + N_3)$ is bounded.

One might question whether this unboundedness is an artifact of the choice of frame. That it is not follows from a consideration of the Weyl curvature. Equations (6.34)–(6.37) imply that the frame components $\mathcal{E}_-$ and $\mathcal{H}_-$, and hence the dimensionless Weyl invariants $\mathcal{E}_{\alpha\beta}\mathcal{E}^{\alpha\beta} - \mathcal{H}_{\alpha\beta}\mathcal{H}^{\alpha\beta}$ and $\mathcal{E}_{\alpha\beta}\mathcal{H}^{\alpha\beta}$ (see (6.33) and (1.35)), are unbounded. The invariant subset $\overline{B_1^+(\text{VII}_0)}$ ($N_1 = 0$, $N_2 \geq 0$, $N_3 \geq 0$) of $\overline{B_1^+(\text{VIII})}$ is also unbounded, but the remaining invariant subsets, including $\overline{B_1^+(\text{VI}_0)}$ ($N_1 = 0$, $N_2 N_3 \leq 0$) are bounded, since (6.15) now implies that N_2 and N_3 are separately bounded.

6.1.4 Friedmann–Lemaître universes

The one–parameter family of closed FL universes are described by two equivalent ordinary orbits in $B^{\pm}(IX)$, given by

$$\Sigma_{\pm} = 0, \quad N_1 = N_2 = N_3 \neq 0.$$

The flat FL universe is described by an equilibrium point, denoted F in Table 6.3, which lies in the invariant set $B(I)$. This model is also described by an ordinary orbit in $S_1^+(VII_0)$, given by

$$\Sigma_{\pm} = 0, \quad N_1 = 0, \quad N_2 = N_3 > 0,$$

and by the analogous orbits in the other components $S_{\alpha}^{\pm}(VII_0)$. We shall denote these orbits by $F_{\alpha}^{\pm}$. Two distinct representations of the flat FL universe arise because this model admits a group G_3 of type I and a family of groups of type VII_0.

6.1.5 Monotone functions

The DE (6.9) admits a variety of monotone functions which are given in the Appendix. Their existence simplifies the dynamics considerably by excluding periodic and recurrent orbits in all the invariant sets (see Section 4.2.3, Proposition 4.2).

6.2 Stability of the equilibrium points

We begin by listing the equilibrium points of the DE (6.9). Table 6.3 gives the notation for the points, the invariant sets in which they lie, the values of the density parameter, and the corresponding exact solutions, which are self–similar (see section 5.3.3). We also list the subsections of Chapter 9 in which the line–elements are given. The first group of entries in the Table are single points, the second group are one–dimensional sets and $\mathcal{J}$ is a two–dimensional set.

6.2.1 Isolated equilibrium points

We first summarize the results concerning the *isolated* equilibrium points F, $P_{\alpha}^{\pm}(II)$ and $P_{\alpha}^{\pm}(VI_0)$. If $\frac{2}{3} < \gamma < 2$, each of these is *hyperbolic*. We list the eigenvalues, and give a basis for the stable subspace E^s. The standard unit vectors in $\mathbb{R}^5$ are denoted by e_1, e_2, e_3, f_+ and f_-, corresponding to the coordinates N_1, N_2, N_3, Σ_+ and Σ_-. Since these equilibrium points are non–vacuum the deceleration parameter q is given by (5.33).

Table 6.3. *Equilibrium points of the DE (6.9), with $0 < \gamma \leq 2$. The subscript α assumes the values 1, 2 and 3.*

Symbol	Ω	Invariant set	Self–similar solution	Line–element
F	1	$B(\mathrm{I})$	flat FL	9.1.1
$P_\alpha^\pm(\mathrm{II})$	$\frac{3}{16}(6 - \gamma)$	$S_\alpha^\pm(\mathrm{II})$	Collins–Stewart (II)	9.1.2
$P_\alpha^\pm(\mathrm{VI_0})$	$\frac{3}{4}(2 - \gamma)$	$S_\alpha^\pm(\mathrm{VI_0})$	Collins (VI$_0$)	9.1.3
$\mathcal{K}$	0	$B(\mathrm{I})$	Kasner vacuum	9.1.1
$\mathcal{L}_\alpha^\pm$	0	$S_\alpha^\pm(\mathrm{VII_0})$	Taub flat space–time	9.1.6
$\mathcal{F}_\alpha^\pm$	1	$S_\alpha^\pm(\mathrm{VII_0})$	flat FL, $\gamma = \frac{2}{3}$	9.1.1
$\mathcal{F}^\pm(\mathrm{IX})$	$\Omega > 1$	$S^\pm(\mathrm{IX})$	closed FL, $\gamma = \frac{2}{3}$	9.1.7
$\mathcal{J}$	$0 < \Omega < 1$	$B(\mathrm{I})$	Jacobs stiff fluid	9.1.1

Equilibrium point F.

$$\Sigma_+ = \Sigma_- = 0, \qquad N_1 = N_2 = N_3 = 0.$$

Since this equilibrium point corresponds to the flat FL universe we shall refer to it as the *flat FL equilibrium point*.

Eigenvalues.

$$\lambda_1 = \lambda_2 = -\tfrac{3}{2}(2 - \gamma), \qquad \lambda_3 = \lambda_4 = \lambda_5 = \tfrac{1}{2}(3\gamma - 2).$$

Stable subspace. $\quad E^s = \mathrm{span}(f_+, f_-)$.

Conclusions.

C1. If $\frac{2}{3} < \gamma < 2$, orbits which approach F as $\tau \to +\infty$ are of Bianchi type I, and F is a sink in $B(\mathrm{I})$.

C2. If $\frac{2}{3} < \gamma < 2$, orbits which approach F as $\tau \to -\infty$ are of Bianchi types II, VI$_0$, VII$_0$, VIII and IX, and form a three–dimensional unstable manifold.

Note that F is also an equilibrium point for $0 \leq \gamma \leq \frac{2}{3}$. The value $\gamma = 0$ is exceptional, since (6.11) and (5.11) imply that $q = -1$ and hence that $H = \mathrm{constant}$. Thus, unlike all other equilibrium points, the corresponding cosmological model is *not* self–similar. Since the perfect fluid stress–energy tensor with $\gamma = 0$ may be interpreted as a cosmological constant, the corresponding model is locally the de Sitter space–time.

Equilibrium point P_1^+ (II).

$$\Sigma_+ = \tfrac{1}{8}(3\gamma - 2), \quad \Sigma_- = 0, \quad N_1 = \tfrac{1}{4}\big[(2 - \gamma)(3\gamma - 2)\big]^{1/2}, \quad N_2 = N_3 = 0,$$

with $\tfrac{2}{3} < \gamma < 2$. The remaining $P_\alpha^\pm$(II) points are determined by symmetry (see Figure 6.3).

Eigenvalues.

$$\lambda_{1,2} = -\tfrac{3}{4}(2 - \gamma)\big[1 \pm (1 - b^2)^{1/2}\big], \quad \lambda_3 = -\tfrac{3}{2}(2 - \gamma), \quad \lambda_4 = \lambda_5 = \tfrac{3}{4}(3\gamma - 2).$$

where

$$b^2 = (3\gamma - 2)(6 - \gamma)/2(2 - \gamma).$$

Stable subspace. $E^s = \text{span}(e_1, f_+, f_-)$.

Conclusions.

C3. If $\tfrac{2}{3} < \gamma < 2$ orbits which approach $P_\alpha^\pm$(II) as $\tau \to +\infty$ are of Bianchi type II, and $P_\alpha^\pm$(II) is a sink in the invariant set $B_\alpha^\pm$(II).

C4. If $\tfrac{2}{3} < \gamma < 2$ orbits which approach $P_\alpha^\pm$(II) as $\tau \to -\infty$ are of Bianchi types VI$_0$, VII$_0$, VIII and IX, and form a two–dimensional unstable manifold.

Equilibrium point P_1^+ (VI$_0$).

$$\Sigma_+ = -\tfrac{1}{4}(3\gamma - 2), \quad \Sigma_- = 0, \quad N_1 = 0, \quad N_2 = -N_3 = \tfrac{1}{4}\big[(2 - \gamma)(3\gamma - 2)\big]^{1/2},$$

with $\tfrac{2}{3} < \gamma < 2$. The remaining $P_\alpha^\pm$(VI$_0$) points are determined by symmetry (see Figure 6.3).

Eigenvalues.

$$\lambda_{1,2} = -\tfrac{3}{4}(2-\gamma)\big[1\pm(1-r^2)^{1/2}\big], \quad \lambda_{3,4} = -\tfrac{3}{4}(2-\gamma)\big[1\pm(1-s^2)^{1/2}\big], \quad \lambda_5 = \tfrac{3}{2}(3\gamma-2),$$

where

$$r^2 = 2(3\gamma - 2), \qquad s^2 = 4(3\gamma - 2)/(2 - \gamma).$$

Stable subspace. $E^s = \text{span}(e_2, e_3, f_+, f_-)$.

Conclusions.

C5. If $\tfrac{2}{3} < \gamma < 2$, orbits which approach $P_\alpha^\pm$(VI$_0$) as $\tau \to +\infty$ are of Bianchi type VI$_0$, and $P_\alpha^\pm$(VI$_0$) is a sink in $B_\alpha^\pm$(VI$_0$).

C6. If $\tfrac{2}{3} < \gamma < 2$, orbits which approach $P_\alpha^\pm$(VI$_0$) as $\tau \to -\infty$ are of Bianchi type VIII, and form a one–dimensional unstable manifold.

6.2.2 The Kasner equilibrium points

$\mathcal{K}$ is a circle of equilibrium points, given by

$$N_1 = N_2 = N_3 = 0, \quad \Sigma_+^2 + \Sigma_-^2 = 1,$$

where Σ_+ and Σ_- are constant. Since the corresponding models are the Kasner vacuum solutions, we will refer to $\mathcal{K}$ as *the Kasner circle*. It follows from (6.11)–(6.13) that the deceleration parameter q is given by $q = 2$.

The constant values of $\Sigma_\pm$ at these equilibrium points are related to the Kasner exponents of the corresponding Kasner solution (see Section 9.1.1), as follows:

$$p_1 = \tfrac{1}{3}(1 - 2\Sigma_+), \quad p_{2,3} = \tfrac{1}{3}(1 + \Sigma_+ \pm \sqrt{3}\Sigma_-). \tag{6.16}$$

The eigenvalues are

$$\lambda_1 = 6p_1, \quad \lambda_2 = 6p_2, \quad \lambda_3 = 6p_3, \quad \lambda_4 = 3(2 - \gamma), \quad \lambda_5 = 0.$$

The relations

$$p_1 + p_2 + p_3 = 1, \quad p_1^2 + p_2^2 + p_3^2 = 1,$$

hold, implying that exactly one of the eigenvalues is negative except for the points T_α with

$$(p_\alpha) = (1,0,0), \quad (0,1,0) \quad \text{and} \quad (0,0,1),$$

which have three zero eigenvalues. The points T_α correspond to the Taub form of flat space–time (see Section 9.1.6), and will be called *Taub points*.

It is helpful to subdivide $\mathcal{K}$ into three open arcs $\mathcal{K}_\alpha$ separated by the points T_α, as indicated in Figure 6.2. The points Q_α, which are the midpoints of the arcs $\mathcal{K}_\alpha$, correspond to the LRS Kasner solutions (two of the p_α equal but non–zero). Equation (6.16) suggests that we parametrize the Kasner exponents p_α by using a polar angle ψ on the Kasner circle in the $\Sigma_+\Sigma_-$–plane i.e.

$$\Sigma_+ = \cos\psi, \quad \Sigma_- = \sin\psi.$$

It follows that

$$p_1 = \tfrac{1}{3}(1 - 2\cos\psi), \quad p_{2,3} = \tfrac{1}{3}(1 + \cos\psi \pm \sqrt{3}\sin\psi). \tag{6.17}$$

The points T_1, T_2, T_3 are given by $\psi = \pi$, $\frac{\pi}{3}$ and $\frac{5\pi}{3}$, and the points Q_1, Q_2, Q_3 by $\psi = 0$, $\frac{2\pi}{3}$ and $\frac{4\pi}{3}$, respectively.

As is easily verified, the eigenspaces associated with λ_1, λ_2 and λ_3 are parallel to the N_1, N_2 and N_3 axes, respectively. Since one of $\lambda_1, \lambda_2, \lambda_3$ is negative, it follows that only orbits with at least one of N_1, N_2 or N_3 zero can

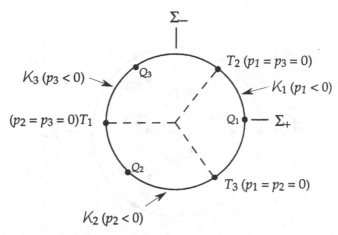

Fig. 6.2. Subsets of the Kasner circle $\mathcal{K} : \Sigma_+^2 + \Sigma_-^2 = 1$, $N_\alpha = 0$.

approach the arcs $\mathcal{K}_\alpha$ as $\tau \to -\infty$ (each arc has a four–dimensional unstable set). By restricting to various Bianchi invariant sets and considering the signs of the remaining eigenvalues, one obtains the following results.

Conclusions.

C7. $\mathcal{K}_1$ is a source in $B_1^\pm(\mathrm{VI_0})$ and in $B_1^\pm(\mathrm{VII_0})$,

C8. $\mathcal{K}_2 \cup T_1 \cup \mathcal{K}_3$ is a source in $B_1^\pm(\mathrm{II})$ and T_1 is a source in $S_1^\pm(\mathrm{II})$,

C9. $\mathcal{K}$ is a source in $B(\mathrm{I})$.

It can be verified that the eigenspaces of λ_1, λ_2 and λ_3 are tangent to the vacuum boundary $\Omega = 0$ at points of $\mathcal{K}$. Since only one of λ_1, λ_2 and λ_3 is negative, it follows that the orbits in $\Omega = 0$ with two of the N_α zero can approach the arcs $\mathcal{K}_\alpha$ as $\tau \to +\infty$ (each arc has a two–dimensional stable set). In particular we have

C10. $\mathcal{K}_1$ is a sink in $B_1^\pm(\mathrm{II})\big|_{\Omega=0}$.

6.2.3 Special equilibrium points

The remaining equilibrium points in Table 6.3 are of a special nature.

$$\mathcal{L}_1^+ : N_1 = 0, \quad N_2 = N_3 = k > 0, \quad \Sigma_+ = -1, \quad \Sigma_- = 0.$$

$\mathcal{L}_1^+$ is a line of equilibrium points emanating from the Taub point T_1 on $\mathcal{K}$. They play an important role in the dynamics in $B_1^+(\mathrm{VII_0})$ (see Section 6.5). We refer to WH (page 1426) for the eigenvalues.

$$\mathcal{F}_1^+(\mathrm{VII_0}) : \Sigma_\pm = 0, \quad N_1 = 0, \quad N_2 = N_3 = b > 0, \quad \gamma = \tfrac{2}{3}$$

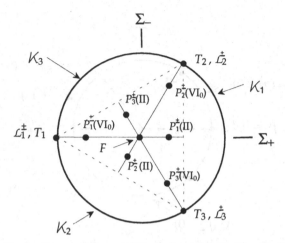

Fig. 6.3. Projections of the equilibrium points in the $\Sigma_+\Sigma_-$-plane.

$$\mathcal{F}^+(\text{IX}): \Sigma_\pm = 0, \quad N_1 = N_2 = N_3 = b > 0, \quad \gamma = \tfrac{2}{3}.$$

These are both a line of equilibrium points, defined only for $\gamma = \tfrac{2}{3}$, emanating from the flat FL equilibrium point F. They mediate the change of stability of F at $\gamma = \tfrac{2}{3}$.

$$\mathcal{J}: \Sigma_+^2 + \Sigma_-^2 < 0, \quad N_1 = N_2 = N_3 = 0, \quad \gamma = 2.$$

$\mathcal{J}$ is a disc of equilibrium points, forming the interior of the Kasner circle. It mediates the change of stability of F at $\gamma = 2$. We refer to WH (page 1426) for the eigenvalues.

6.2.4 Bifurcations

In Figure 6.3, we give an overview of the equilibrium points by drawing their projections in the $\Sigma_+\Sigma_-$-plane. We note that as γ varies, the equilibrium points $P_\alpha^\pm(\text{II})$ and $P_\alpha^\pm(\text{VI}_0)$ move along the radial line segments on which they are located, and coalesce with F when $\gamma = \tfrac{2}{3}$. When $\gamma = 2$, the points $P_\alpha^\pm(\text{VI}_0)$ coalesce with the Taub points T_α on $\mathcal{K}$, and the points $P_\alpha^\pm(\text{II})$ coalesce with three points of the disc $\mathcal{J}$ (on the circle $\Sigma_+^2 + \Sigma_-^2 = \tfrac{1}{4}$). These bifurcations at $\gamma = \tfrac{2}{3}$ and $\gamma = 2$ change the stability of the FL equilibrium point F, as can be seen from the previously given eigenvalues.

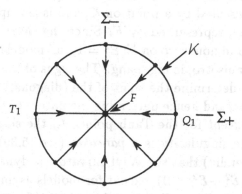

Fig. 6.4. The Bianchi I invariant set $B(I)$

6.3 Asymptotically self–similar models

In this section, we discuss the Bianchi classes which are asymptotically self–similar into the past and into the future (see Section 5.3.3). As regards evolution into the future, we assume that $\frac{2}{3} < \gamma < 2$. The case $0 < \gamma < \frac{2}{3}$ is dealt with for all Bianchi types in Theorem 8.2 in Section 8.3.

6.3.1 Bianchi I models

The Bianchi I invariant subset $B(I)$ is defined by

$$N_1 = N_2 = N_3 = 0, \quad \Omega \geq 0. \tag{6.18}$$

The density parameter (6.14) reduces to

$$\Omega = 1 - \Sigma_+^2 - \Sigma_-^2.$$

The DE (6.9) reduces to

$$\Sigma_+' = -(2-q)\Sigma_+, \quad \Sigma_-' = -(2-q)\Sigma_-, \tag{6.19}$$

where

$$q = \tfrac{1}{2}(3\gamma - 2) + \tfrac{3}{2}(2 - \gamma)(\Sigma_+^2 + \Sigma_-^2).$$

From Table 6.3 we see that the equilibrium points with $\gamma < 2$ which lie in $B(I)$ are F and the Kasner circle $\mathcal{K}$. By (6.19) the orbits are radial lines $\Sigma_+ = C\Sigma_-$ directed towards the origin. Each ordinary orbit thus joins a point on $\mathcal{K}$ to F. The resulting phase portrait is shown in Figure 6.4.

We observe that the flat FL equilibrium point F is the future attractor and the Kasner circle $\mathcal{K}$ is the past attractor. In physical terms, this means that any perfect fluid Bianchi I model is asymptotic at the big–bang to a Kasner model, represented by a point on $\mathcal{K}$, and is asymptotic at late times to the flat FL model, represented by F. Since the shear scalar Σ, given by (6.13), is zero at F and non–zero on $\mathcal{K}(\Sigma^2 = 1)$, all models isotropize into the future but have an anisotropic big–bang. The signs of the Kasner exponents p_α, given by (6.16), determine the signs of the (diagonal) components Θ_α of the expansion tensor, and hence determine the nature of the singularity. For all Kasner points except for the Taub points T_α the singularity is a *cigar*, while for the T_α the singularity is a *pancake*† (see (5.39)). It follows from (6.37) (see the Appendix) that the Weyl curvature is dynamically significant near the big–bang ($\mathcal{E}_+^2 + \mathcal{E}_-^2 \neq 0$), except for models asymptotic to the Taub points, and is dynamically negligible at late times ($\mathcal{E}_\pm \to 0$).

6.3.2 Bianchi II models

The physical region of the Bianchi II invariant set $B_1^+(\mathrm{II})$ is defined by

$$N_1 > 0, \quad N_2 = N_3 = 0, \quad \Omega \geq 0.$$

The density parameter (6.14) reduces to

$$\Omega = 1 - \Sigma_+^2 - \Sigma_-^2 - \tfrac{1}{12}N_1^2. \tag{6.20}$$

The boundary $\partial B_1^+(\mathrm{II})$ is given by the additional condition $N_1\Omega = 0$. It follows that

$$\partial B_1^+(\mathrm{II}) = B(\mathrm{I}) \cup B_1^+(\mathrm{II})\big|_{\Omega=0}$$

The boundary thus consists of Bianchi I orbits and vacuum Bianchi type II orbits. The full state space is the closure

$$\overline{B_1^+(\mathrm{II})} = \partial B_1^+(\mathrm{II}) \cup B_1^+(\mathrm{II})\big|_{\Omega>0},$$

which is a solid half–ellipsoid in $\mathbb{R}^3$.

The DE (6.9) reduces to

$$\begin{aligned} \Sigma_+' &= -(2-q)\Sigma_+ + \tfrac{1}{3}N_1^2, \\ \Sigma_-' &= -(2-q)\Sigma_-, \\ N_1' &= (q - 4\Sigma_+)N_1, \end{aligned} \tag{6.21}$$

where

$$q = \tfrac{1}{2}(3\gamma - 2)\left(1 - \tfrac{1}{12}N_1^2\right) + \tfrac{3}{2}(2 - \gamma)(\Sigma_+^2 + \Sigma_-^2). \tag{6.22}$$

† This remark applies to any model whose orbit is past asymptotic to a point on $\mathcal{K}$.

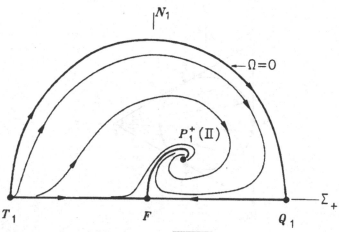

Fig. 6.5. The LRS Bianchi II invariant set $\overline{S_1^+(\mathrm{II})}$, defined by $N_2 = N_3 = 0, \Sigma_- = 0$.

Invariant sets:

Three two–dimensional invariant sets comprise the skeleton for the orbits in $\overline{B_1^+(\mathrm{II})}$, namely $B(\mathrm{I})$, $B_1^+(\mathrm{II})\big|_{\Omega=0}$, and the LRS invariant set $S_1^+(\mathrm{II})$, which is defined by $\Sigma_- = 0$ (see Table 6.2). From Table 6.3 we see that the equilibrium points in $\overline{B_1^+(\mathrm{II})}$, namely F, $P_1^+(\mathrm{II})$ and $\mathcal{K}$, all lie in the skeleton. The condition $\Sigma_- = 0$ defines the points T_1 and Q_1 on $\mathcal{K}$ (see Figure 6.2). We thus obtain the skeleton for the orbits in $\overline{S_1^+(\mathrm{II})}$, shown as the bold curves in Figure 6.5.† The unstable manifold of $P(\mathrm{I})$ is asymptotic to $P_1^+(\mathrm{II})$ since there are no periodic orbits in $S_1^+(\mathrm{II})$ (see Section 6.1.5). The remaining orbits are past asymptotic to T_1 and future asymptotic to $P_1^+(\mathrm{II})$. Some typical orbits are shown as faint curves in Figure 6.5.

The orbits in $B_1^+(\mathrm{II})\big|_{\Omega=0}$ can be found explicitly. It follows from (6.20)–(6.22) that these orbits are given by the intersection of the planes

$$\Sigma_- = k(\Sigma_+ - 2),$$

$k = $ constant, with the vacuum boundary which is the half–ellipsoid

$$\Sigma_+^2 + \Sigma_-^2 + \tfrac{1}{12}N_1^2 = 1, \quad N_1 > 0. \tag{6.23}$$

Each orbit thus joins two points on the Kasner circle $\mathcal{K}$, one point (the α–limit set) lying on the arc $\mathcal{K}_2 \cup T_1 \cup \mathcal{K}_3$, the other (the ω–limit set) lying on the arc $\mathcal{K}_1$ (see Figure 6.2). The projections of these orbits in the $\Sigma_+\Sigma_-$–plane are straight lines which intersect in the point $(2,0)$ when extended outside $\mathcal{K}$ (see Figure 6.6). These orbits in fact correspond to the Taub

† A phase portrait equivalent to Figure 6.5 was first given by Collins (1971).

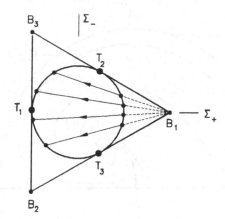

Fig. 6.6. The projections of the Taub orbits with $N_1 > 0$, $N_2 = N_3 = 0$ into the $\Sigma_+\Sigma_-$–plane.

vacuum solutions (see Section 9.2.1) and hence we shall refer to them as the *Taub orbits*. We can now sketch the skeleton of the orbits in the full Bianchi II state space $B_1^+(II)$ as in Figure 6.7.

Asymptotic evolution:

The stability analysis in Section 6.2 shows that the equilibrium point $P_1^+(II)$ is a local sink in $B_1^+(II)$ and that the equilibrium set $\mathcal{K}/\mathcal{K}_1$ is a local source in $B_1^+(II)$. One can use the monotone functions in the Appendix and the Monotonicity Principle (Section 4.4.3), to prove that $P_1^+(II)$ in fact attracts all orbits in $B_1^+(II)$ with $\Omega > 0$.

Proposition 6.1. For all $\mathbf{y} \in B_1^+(II)$ with $\Omega > 0$ and $\frac{2}{3} < \gamma < 2$, the ω–limit set is

$$\omega(\mathbf{y}) = P_1^+(II).$$

Proof See the Appendix.   □

As regards evolution into the past, the situation is more complicated. The analysis in Section 6.2 shows that F has a one–dimensional unstable manifold in $B_1^+(II)$, which in fact lies in the invariant set $S_1^+(II)$ (see Figure 6.5). In other words, there is one special orbit which is past asymptotic to F. One can, in fact, prove that all other ordinary orbits in $B_1^+(II)$ are past asymptotic to $\mathcal{K}\backslash\mathcal{K}_1$.

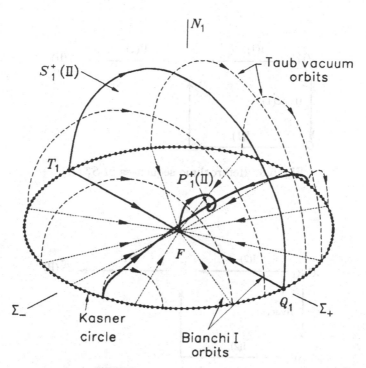

Fig. 6.7. The invariant set $\overline{B_1^+(\mathrm{II})}$, defined by $N_2 = N_3 = 0$, showing the invariant subsets $\overline{S_1^+(II)}$ (Figure 6.5) and $B(I)$ (Figure 6.4).

Proposition 6.2. For all $\mathbf{y} \in B_1^+(\mathrm{II}) \backslash P_1^+(\mathrm{II})$, the α–limit set is

$$\alpha(\mathbf{y}) \subset \mathcal{K} \backslash \mathcal{K}_1 \qquad \text{(except for a set of measure zero),}$$

or

$$\alpha(\mathbf{y}) = F \qquad \text{(a one–dimensional set).}$$

Proof See the Appendix. □

In summary, we have proved that *the equilibrium point P_1^+ (II) is the future attractor* (if $\frac{2}{3} < \gamma < 2$) and *the Kasner arc $\mathcal{K} \backslash \mathcal{K}_1$ is the past attractor in $B_1^+(\mathrm{II})|_{\Omega > 0}$*. In physical terms, these results mean that any perfect fluid Bianchi II model, with $\frac{2}{3} < \gamma < 2$, is asymptotic at late times to the Collins–Stewart model and that all models, except for a set of measure zero, are asymptotic near the big–bang, to a Kasner model. For other analyses of the Bianchi II models using dynamical systems methods, we refer to Bogoyavlensky (1985, page 53), and Uggla (1989).

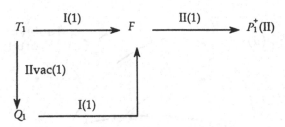

Fig. 6.8. Heteroclinic sequences in $S_1^+(\mathrm{II})$.

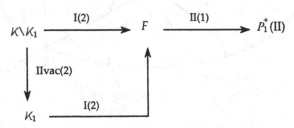

Fig. 6.9. Heteroclinic sequences in $B_1^+(\mathrm{II})$.

The intermediate evolution:

As explained in Section 4.4.4, one can obtain information about the intermediate evolution of a physical system by finding the heteroclinic sequences of the associated dynamical system. For the invariant subset $S_1^+(\mathrm{II})$ (see Figure 6.5), there are two heteroclinic sequences:

$$T_1 \to Q_1 \to F \to P_1^+(\mathrm{II}),$$

$$T_1 \to F \to P_1^+(\mathrm{II}),$$

determined by the saddle points Q_1 and F.

With a view to describing the heteroclinic sequences in the three–dimensional state space we represent the sequences for $S_1^+(\mathrm{II})$ in Figure 6.8, thereby giving a schematic version of Figure 6.5. In the full state space the equilibrium point T_1 replaced by the past attractor $\mathcal{K}\backslash\mathcal{K}_1$ and the saddle point Q_1 is replaced by the saddle set $\mathcal{K}_1$, giving Figure 6.9.

In this diagram, 'I(2)' represents a two–dimensional family of Bianchi I orbits, 'IIvac(2)' represents the two–dimensional family of vacuum Taub orbits, while 'II(1)' represents the one–dimensional unstable manifold of F, which is a unique orbit in $B_1^+(\mathrm{II})$. The diagram thus represents infinitely many heteroclinic sequences which join the past attractor $\mathcal{K}\backslash\mathcal{K}_1$ to the future attractor $P_1^+(\mathrm{II})$. These heteroclinic sequences approximate the evolution of an open set of Bianchi II universes.

6.3.3 Bianchi VI_0 models

The Bianchi VI_0 invariant set $B_1^+(\mathrm{VI}_0)$ is defined by

$$N_1 = 0, \quad N_2 > 0, \quad N_3 < 0, \quad \Omega \geq 0. \tag{6.24}$$

The density parameter (6.14) reduces to

$$\Omega = 1 - \Sigma_+^2 - \Sigma_-^2 - \tfrac{1}{12}(N_2 - N_3)^2. \tag{6.25}$$

The boundary $\partial B_1^+(\mathrm{VI}_0)$ is given by

$$\partial B_1^+(\mathrm{VI}_0) = B(\mathrm{I}) \cup B_2^+(\mathrm{II}) \cup B_3^-(\mathrm{II}) \cup B_1^+(\mathrm{VI}_0)\big|_{\Omega=0}. \tag{6.26}$$

It follows from (6.24) and (6.25) that $\overline{B_1^+(\mathrm{VI}_0)}$ is bounded and hence compact. The DE in $\overline{B_1^+(\mathrm{VI}_0)}$ can be obtained by specializing (6.9). We do not need its specific form.

Invariant sets:

The skeleton for the orbits in $\overline{B_1^+(\mathrm{VI}_0)}$ consists of the invariant sets in the boundary, and the two–dimensional invariant set $S_1^+(\mathrm{VI}_0)$ (see Table 6.2). From Table 6.3 we see that the equilibrium points in $\overline{B_1^+(\mathrm{VI}_0)}$, namely F, $P_2^+(\mathrm{II})$, $P_3^-(\mathrm{II})$, $P_1^+(\mathrm{VI}_0)$ and $\mathcal{K}$ all lie in the skeleton. The orbits in $\overline{S_1^+(\mathrm{VI}_0)}$ for $\frac{2}{3} < \gamma < 2$ are shown in Figure 6.10†, with its skeleton shown as bold curves.

For the vacuum subset $B_1^+(\mathrm{VI}_0)\big|_{\Omega=0}$ we have the following result.

Proposition 6.3. For all $\mathbf{y} \in B_1^+(\mathrm{VI}_0)$ with $\Omega = 0$,

$$\alpha(\mathbf{y}) \subset \mathcal{K}_1 \quad \text{and} \quad \omega(\mathbf{y}) = \{T_1\}.$$

Proof See the Appendix. □

Asymptotic evolution:

The local stability analysis in Section 6.2 shows that the equilibrium point

† A phase portrait equivalent to Figure 6.10 was first given by Collins (1971).

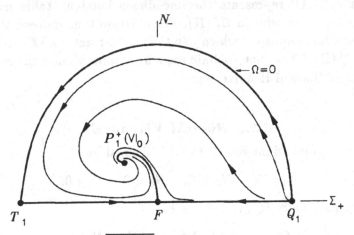

Fig. 6.10. The invariant set $\overline{S_1^+(\mathrm{VI}_0)}$, defined by $N_1 = 0$, $N_2 + N_3 = 0$, $\Sigma_- = 0$. The variable N_- is defined by (6.35).

$P_1^+(\mathrm{VI}_0)$ is a local sink and the equilibrium set $\mathcal{K}_1$ is a local source in $B_1^+(\mathrm{VI}_0)$. One can prove that $P_1^+(\mathrm{VI}_0)$ in fact attracts all orbits in $B_1^+(\mathrm{VI}_0)$ with $\Omega > 0$.

Proposition 6.4. For all $\mathbf{y} \in B_1^+(\mathrm{VI}_0)$ with $\Omega > 0$, and $\frac{2}{3} < \gamma < 2$,

$$\omega(\mathbf{y}) = P_1^+(\mathrm{VI}_0).$$

Proof See the Appendix. □

As regards evolution into the past the situation is considerably more complicated. The local analysis in Section 6.2 shows that $P_2^+(\mathrm{II})$ and $P_3^-(\mathrm{II})$ have a one–dimensional unstable manifold and that F has a two–dimensional unstable manifold. We conjecture that apart from the orbits in these unstable manifolds, all other ordinary orbits in $B_1^+(\mathrm{VI}_0)$ are past asymptotic to $\mathcal{K}_1$. This conjecture is supported by arguments based on the monotone functions, on the known structure of the orbits in the boundary (6.26) and on numerical experiments.

Conjecture: For all $\mathbf{y} \in B_1^+(\mathrm{VI}_0) \backslash P_1^+(\mathrm{VI}_0)$ with $\Omega > 0$ and $\frac{2}{3} < \gamma < 2$,

$$\alpha(\mathbf{y}) \subset \mathcal{K}_1 \qquad \text{(apart from a set of measure zero),}$$
or
$$\alpha(\mathbf{y}) = P_2^+(\mathrm{II}), P_3^-(\mathrm{II}) \quad \text{(a one–dimensional set),}$$
or
$$\alpha(\mathbf{y}) = F \qquad \text{(a two–dimensional set).}$$

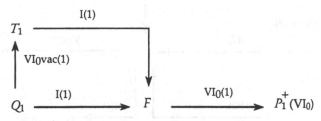

Fig. 6.11. Heteroclinic sequences in $S_1^+(VI_0)$.

In summary, we have proved that P_1^+ *(VI₀) is the future attractor* (if $\frac{2}{3} < \gamma < 2$), and conjecture that $\overline{K}_1$ *is the past attractor*, in $B_1^+(VI)_0|_{\Omega>0}$.

For another analysis of the Bianchi VI₀ models using dynamical systems methods, we refer to Bogoyavlensky (1985, pages 84–5).

The intermediate evolution:

For the invariant subset $S_1^+(VI_0)$, there are two heteroclinic sequences (see Figure 6.10):

$$Q_1 \to T_1 \to F \to P_1^+(VI_0),$$

$$Q_1 \to F \to P_1^+(VI_0),$$

which we represent in Figure 6.11. In the four–dimensional space $\overline{B_1^+(VI_0)}$ we obtain Figure 6.12, which is determined by knowing the past and future attractors, the orbits in the skeleton, and the dimensions of the stable and unstable manifolds. The diagram represents infinitely many heteroclinic sequences which join the past attractor $\overline{K}_1$ to the future attractor $P_1^+(VI_0)$. These heteroclinic sequences approximate the evolution of an open set of Bianchi VI₀ universes. The sequences which contain the flat FL equilibrium point F are of physical interest, since a Bianchi VI₀ model whose orbit is approximated by such a sequence will have a flat quasi–isotropic epoch (see Section 5.3.3).

6.4 The Kasner map and the Mixmaster attractor

The analysis of Section 6.2 shows that there is no local source in the Bianchi VIII or IX state space. In this section we construct a compact invariant set

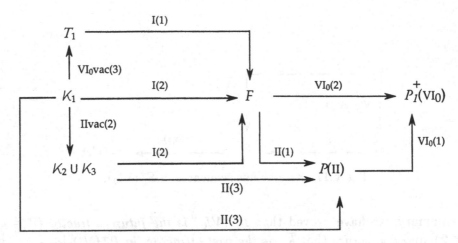

Fig. 6.12. Heteroclinic sequences in $B_1^+(\mathrm{VI}_0)$.

in $\overline{B_1^+(\mathrm{VIII})}$ and in $\overline{B^+(\mathrm{IX})}$ that we conjecture is a past attractor. These invariant sets are the union of the Kasner circle and some of the Taub orbits. The discussion in this section is based on Ma & Wainwright (1992).

There are six families of Taub orbits. Each family lies on a half–ellipsoid, the one for the family with $N_1 > 0$ being given by (6.23). We denote the closure of this half–ellipsoid by E_1^+:

$$E_1^+ : \Sigma_+^2 + \Sigma_-^2 + \tfrac{1}{12}N_1^2 = 1, \quad N_1 \geq 0, \quad N_2 = N_3 = 0. \qquad (6.27)$$

The family with $N_1 < 0$ lies on the half–ellipsoid E_1^-:

$$E_1^- : \Sigma_+^2 + \Sigma_-^2 + \tfrac{1}{12}N_1^2 = 1, \quad N_1 \leq 0, \quad N_2 = N_3 = 0. \qquad (6.28)$$

These two families project onto the same family of lines in the $\Sigma_+\Sigma_-$–plane, shown in Figure 6.6. The Taub orbits with $N_2 \neq 0$ or $N_3 \neq 0$ lie on similar half–ellipsoids, which we denote by $E_2^\pm$ and $E_3^\pm$, e.g.

$$E_2^+ : \Sigma_+^2 + \Sigma_-^2 + \tfrac{1}{12}N_2^2 = 0, \quad N_2 > 0, \quad N_1 = N_3 = 0. \qquad (6.29)$$

The projections of the orbits of these families in the $\Sigma_+\Sigma_-$–plane are obtained by rotating Figure 6.6 clockwise through $\frac{2\pi}{3}$ and $\frac{4\pi}{3}$ radians. The key point is that the Taub orbits and the equilibrium points of the Kasner circle $\mathcal{K}$ lead to the existence of *infinite heteroclinic sequences* which approximate the past asymptotic behaviour of generic orbits. These heteroclinic sequences are defined by a map of $\mathcal{K}$ onto itself.

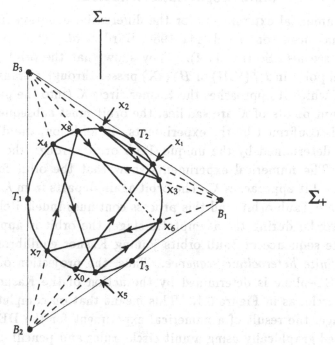

Fig. 6.13. The action of the Kasner map on the Kasner circle, defining a sequence of Kasner equilibrium points.

6.4.1 The Kasner map

We have seen that each non–exceptional Kasner equilibrium point P has a one–dimensional unstable manifold (into the past), formed by the Taub orbit which emanates from P (into the past). This orbit joins P to another point $\tilde{P}$ on the Kasner circle $\mathcal{K}$, which can thus be thought of as the image of P under a mapping. In this way, *the Taub orbits define a map F of $\mathcal{K}$ onto itself.* The exceptional points T_1, T_2 and T_3 are fixed points of F. Thus, given a point P_0 on $\mathcal{K}$, the successive action of this map determines a sequence of Kasner equilibrium points, defined by

$$P_{i+1} = F(P_i), \qquad i = 0, 1, 2, \ldots$$

as illustrated in Figure 6.13. We shall refer to the map F as the *Kasner map,* since its domain is the Kasner circle. This map appears, in terms of different variables, in Bogoyavlensky (1985, pages 57 and 65). Additional properties of the Kasner map, regarded as defining a discrete dynamical system, are discussed in Section 11.4.

6.4.2 Asymptotic behaviour

A series of numerical experiments for the differential equation (6.9) in the case $\gamma < 2$ has been conducted (Ma 1988, Burd *et al.* 1990, Creighton & Hobill 1994; see also Section 11.4). They show that the orbit through a generic initial point in $B_1^+(\text{VIII})$ or $B^+(\text{IX})$ passes through a transient stage at the end of which it approaches the Kasner circle $\mathcal{K}$ (into the past). Since the equilibrium points of $\mathcal{K}$ are saddles, the orbit must subsequently leave $\mathcal{K}$, and this is confirmed by the experiments. In addition, the direction of departure is determined by the unique Taub orbit through the particular point of $\mathcal{K}$. The numerical experiments show that the orbit follows this Taub orbit until it approaches $\mathcal{K}$. The orbit again departs from $\mathcal{K}$, following the appropriate Taub orbit, and this process continues indefinitely.

In other words, during the asymptotic stage, the orbit is approximated by an infinite sequence of Taub orbits joining Kasner equilibrium points, called an *infinite heteroclinic sequence*. Thus, the projection of the orbit into the $\Sigma_+\Sigma_-$–plane is determined by the action of the Kasner map on the Kasner circle, as in Figure 6.13. This means that on completion of the transient stage, the result of a numerical experiment for the DE (6.9) can be constructed graphically using a unit circle, ruler and pencil!

6.4.3 The Mixmaster attractor

Our discussion of the past asymptotic behaviour suggests that in the case $\gamma < 2$ the union of the Kasner circle $\mathcal{K}$, and the family of Taub orbits is the (past) attractor of the DE (6.9), as $\tau \to -\infty$. This set, denoted M_1^+ for the invariant set $B_1^+(\text{VIII})$, is the union of the three ellipsoids E_1^-, E_2^+ and E_3^+

$$M_1^+ = E_1^- \cup E_2^+ \cup E_3^+. \tag{6.30}$$

Their intersection is the Kasner circle:

$$E_1^- \cap E_2^+ \cap E_3^+ = \mathcal{K}.$$

For the Bianchi IX invariant set $B^+(\text{IX})$, the attractor is denoted M^+ and is defined by

$$M^+ = E_1^+ \cup E_2^+ \cup E_3^+. \tag{6.31}$$

Since this attractor gives rise to the so–called Mixmaster oscillatory singularity, we shall refer to it as the *Mixmaster attractor*.

Some analytical evidence that the set M_1^+ (or M^+) is indeed an attractor can be obtained by considering the density parameter Ω and the scalar Δ, which is defined by

$$\Delta = (N_1 N_2)^2 + (N_2 N_3)^2 + (N_3 N_1)^2.$$

The attractor is then given by

$$\Delta = 0, \quad \Omega = 0,$$

since $\Delta = 0$ implies that at least two of the N_α are zero. Thus in order to prove that M_1^+ (or M^+) is an attractor, it is necessary to prove that

$$\lim_{\tau \to -\infty} \Delta = 0, \quad \lim_{\tau \to -\infty} \Omega = 0 \tag{6.32}$$

for almost all orbits.

The functions Δ and Ω are not monotone along orbits, even in a neighbourhood of M_1^+ (or M^+). However, since $q = 2$ on $\mathcal{K}$, it follows from (5.21) that if $\gamma < 2$, then there is a neighbourhood U of $\mathcal{K}$ such that $\Omega' > 0$ along all orbits in U for which $\Omega > 0$. This means that Ω *decreases* exponentially into the past along all orbits in U. In addition, it can be shown using the DE (6.9) that each point of $\mathcal{K}$, except for the points T_α, has a neighbourhood $\tilde{U}$ such that $\Delta' > 0$ along all non–singular orbits in $\tilde{U}$ for which $\Delta > 0$. This means that Δ *decreases* exponentially into the past along all orbits in $\tilde{U}$. Since the universe point, while following a heteroclinic sequence, spends most of the time near the Kasner circle it is plausible that Ω and Δ will overall decrease exponentially, and hence tend to zero as $\tau \to -\infty$. Thus the challenge remains to prove the results (6.32), for almost all orbits. The result of a numerical experiment is shown in Figure 11.1.

6.5 Summary

In this chapter we have given a qualitative description of the evolution of models of Bianchi types I, II and VI_0, by specifying the past and future attractors and a complete set of heteroclinic sequences. All of these models are asymptotically self–similar. We have also given a description of the past asymptotic behaviour of models of Bianchi types VIII and IX in terms of a past attractor, the Mixmaster attractor. As explained in Section 6.1.3, the description of the Bianchi IX models is necessarily incomplete. This deficiency is remedied in Section 8.5.2. As discussed in Section 6.1.3, the invariant sets for the Bianchi VIII and VII_0 models are unbounded. The monotone functions Z_1 and Z_2 in the Appendix imply (unpublished) that in general $N_2 + N_3$ diverges as $\tau \to +\infty$ along orbits in these sets, so that there is no future attractor. The exception is the Bianchi VII_0 *vacuum* models, for which it appears that $N_2 + N_3$ has a finite limit as $\tau \to +\infty$ corresponding to approach to an equilibrium point on the line $\mathcal{L}_1^+$. Into the past the behaviour of Bianchi VII_0 models appears to be analogous to Bianchi VI_0 models.

The Bianchi VIII and VII_0 models clearly require further investigation, which will involve a compactification of the variable $N_2 + N_3$. Whatever the outcome, however, we can say that *Bianchi VIII models and non–vacuum Bianchi VII_0 models are not asymptotically self–similar into the future*: there are no equilibrium points which the orbits can approach.

Finally we comment on the occurrence of matter–dominated singularities. It is part of the folk–lore of Bianchi cosmology that in general 'matter does not matter' near the singularity (see Lifshitz & Khalatnikov 1963, page 200, Belinskii *et al.* 1970, pages 532 and 538). We have provided evidence to support this assertion, in the sense that

$$\lim_{\tau \to -\infty} \Omega = 0$$

for almost all orbits (see the discussion following (6.32)). Since F, $P_\alpha^\pm(\text{II})$ and $P_\alpha^\pm(\text{VI}_0)$ are saddle points in general, however, there are families of special orbits (the unstable manifolds of these equilibrium points) that are past asymptotic to these equilibrium points (see conclusions C2, C4 and C6 in Section 6.2.1). These special orbits describe all universes that have a *matter–dominated big–bang*.† The possible limiting values of Ω, corresponding to F, $P_\alpha^\pm(\text{II})$ and $P_\alpha^\pm(\text{VI}_0)$, respectively, are (see Table 6.3):

$$\lim_{\tau \to -\infty} \Omega = \begin{cases} 1, & \text{occurs for all Bianchi types except I} \\ \frac{3}{16}(6 - \gamma), & \text{occurs for all Bianchi types except I and II} \\ \frac{3}{4}(2 - \gamma), & \text{occurs only for Bianchi VIII.} \end{cases}$$

In the first case, which corresponds to approach to the flat FL model, the initial singularity is an isotropic singularity (Goode & Wainwright 1985). Since the unstable manifold of F is three–dimensional in general, there is a three–parameter family of Bianchi VIII/IX universes with an isotropic singularity.

Appendix to Chapter 6
Weyl curvature

The basic variables $(\Sigma_\pm, N_\alpha)$ determine the Weyl conformal curvature tensor, as follows. Let $E_{\alpha\beta}$, $H_{\alpha\beta}$ denote the electric and magnetic components

† The existence of these special solutions is pointed out by Zel'dovich & Novikov (1983, page 540).

of the Weyl tensor relative to the group invariant frame with $e_0 = u$, defined by (1.34). We define the dimensionless components in the usual way:

$$\mathcal{E}_{\alpha\beta} = \frac{E_{\alpha\beta}}{H^2}, \quad \mathcal{H}_{\alpha\beta} = \frac{H_{\alpha\beta}}{H^2}. \tag{6.33}$$

For class A perfect fluid models, $\mathcal{E}_{\alpha\beta}$ and $\mathcal{H}_{\alpha\beta}$ are diagonal in the standard frame and hence have two independent components. We define

$$\mathcal{E}_+ = \tfrac{1}{2}(\mathcal{E}_{22} + \mathcal{E}_{33}), \quad \mathcal{E}_- = \tfrac{1}{2\sqrt{3}}(\mathcal{E}_{22} - \mathcal{E}_{33}), \tag{6.34}$$

with $\mathcal{H}_\pm$ defined similarly. It is convenient to let

$$N_+ = \tfrac{1}{2}(N_2 + N_3), \quad N_- = \tfrac{1}{2\sqrt{3}}(N_2 - N_3). \tag{6.35}$$

Using equations (1.101) and (1.102) we obtain the following expressions for $\mathcal{E}_\pm$ and $\mathcal{H}_\pm$. For convenience we also include the expressions for Ω, and for the spatial curvature $\mathcal{S}_\pm$ and K in terms of $N_\pm$.

Spatial curvature:

$$\begin{aligned}
K &= N_-^2 + \tfrac{1}{12}N_1(N_1 - 4N_+), \\
\mathcal{S}_+ &= 2N_-^2 - \tfrac{1}{3}N_1(N_1 - N_+), \\
\mathcal{S}_- &= N_-(2N_+ - N_1).
\end{aligned} \tag{6.36}$$

Ricci and Weyl curvature:

$$\begin{aligned}
\Omega &= 1 - (\Sigma_+^2 + \Sigma_-^2) - K, \\
\mathcal{E}_+ &= \Sigma_+ + (\Sigma_+^2 - \Sigma_-^2) + \mathcal{S}_+, \\
\mathcal{E}_- &= \Sigma_- - 2\Sigma_+\Sigma_- + \mathcal{S}_-, \\
\mathcal{H}_+ &= -\tfrac{3}{2}N_1\Sigma_+ - 3N_-\Sigma_-, \\
\mathcal{H}_- &= -3N_-\Sigma_+ + \tfrac{1}{2}(N_1 - 4N_+)\Sigma_-.
\end{aligned} \tag{6.37}$$

Monotone functions

The DE (6.9) admits a variety of monotone functions. We list the functions, the differential equations that they satisfy and the invariant sets S on which they are monotone (see Tables 6.1–6.3 for the notation).

(1) *Bianchi VIII and IX*

$$Z_1 = (N_1 N_2 N_3)^2, \quad Z_1' = 6q Z_1, \tag{6.38}$$

$$S = B_\alpha^\pm(\text{VIII}) \quad \text{or} \quad B^\pm(\text{IX}).$$

(2) Bianchi VI$_0$ and VII$_0$

(i) Let

$$P = N_-^2 + \Sigma_-^2, \quad Q = N_+^2 + 3\Sigma_-^2,$$

where $N_\pm$ are defined by (6.35). Then

$$Z_2 = \frac{P}{Q}, \quad Z_2' = -\frac{4N_2N_3\Sigma_-^2(1+\Sigma_+)}{PQ}Z_2, \tag{6.39}$$

$$S_2 = B_1^\pm(\text{VI}_0)\backslash S_1^\pm(\text{VI}_0) \quad \text{or} \quad B_1^\pm(\text{VII})_0\backslash S_1^\pm(\text{VII}_0).$$

(ii)

$$Z_3 = 1 + \Sigma_+, \quad Z_3' = -2(1-\Sigma)Z_3, \tag{6.40}$$

$$S_3 = \overline{B_1^+(\text{VI}_0)}\Big|_{\Omega=0}\backslash\mathcal{K} \quad \text{or} \quad \overline{B_1^+(\text{VII}_0)}\Big|_{\Omega=0}\backslash(\mathcal{K}\cup\mathcal{L}_1^+)$$

(iii) Let

$$v = -\tfrac{1}{4}(3\gamma - 2), \quad m = -\frac{3v(2-\gamma)}{4(1-v^2)}, \quad \text{with} \quad \tfrac{2}{3} < \gamma < 2. \tag{6.41}$$

Then

$$Z_4 = \frac{|N_2N_3|^m \, \Omega^{1-m}}{(1-v\Sigma_+)^2}, \quad Z_4' = \alpha Z_4, \tag{6.42}$$

where

$$\alpha = \frac{-3(2-\gamma)}{(1-v\Sigma_+)}\left[\frac{(\Sigma_+ - v)^2}{1-v^2} + \Sigma_-^2\right], \tag{6.43}$$

$$S_4 = B_1^\pm(\text{VI}_0)\backslash\{P_1^\pm((\text{VI}_0)\} \quad \text{or} \quad B_1^\pm(\text{VII}_0), \quad \text{with } \Omega > 0.$$

(3) Bianchi II

(i)

$$Z_5 = \Sigma_-^2, \quad Z_5' = -2(2-q)Z_5, \tag{6.44}$$

$$S_5 = B_1^\pm(\text{II})\backslash S^\pm(\text{II}).$$

(ii)

$$Z_6 = \frac{\Sigma_-}{2-\Sigma_+}, \quad Z_6' = -\frac{3(2-\gamma)\Omega}{2-\Sigma_+}Z_6, \tag{6.45}$$

$$S_6 = B_1^+(\text{II})\Big|_{\Omega>0}.$$

(iii) Let

$$v = \tfrac{1}{8}(3\gamma - 2), \quad m = \frac{3v(2 - \gamma)}{8(1 - v^2)}, \quad \text{with } \tfrac{2}{3} < \gamma < 2. \qquad (6.46)$$

Then

$$Z_7 = \frac{N_1^{2m}\Omega^{1-m}}{(1 - v\Sigma_+)^2}, \quad Z_7' = \alpha Z_7, \qquad (6.47)$$

where α is given by (6.43) and (6.46),

$$S_7 = B_1^{\pm}(\mathrm{II}) \backslash \{P_1^{\pm}(\mathrm{II})\}, \quad \text{with} \quad \Omega > 0.$$

The functions Z_4 and Z_7 were discovered by Uggla using the Hamiltonian formulation (see the Appendix to Chapter 10).

(4) *The density parameter*

It follows from (5.20) and (5.21) that for all class A Bianchi types except I and IX, Ω is monotone increasing if $\gamma \leq \tfrac{2}{3}$ and monotone decreasing if $\gamma \geq 2$. For Bianchi I models, Ω is monotone increasing if $\gamma < 2$ and monotone decreasing if $\gamma > 2$.

Proofs

Propositions 6.1–6.4 are proved by applying the Monotonicity Principle (see Section 4.4) to one of the monotone functions Z_A given by (6.38)–(6.47), and the associated invariant set S_A.

Proof of Proposition 6.1

Consider Z_7, noting that $\overline{S}_7 \backslash S_7 = \partial B_1^+(\mathrm{II}) \cup \{P_1^+(\mathrm{II})\}$. Since $Z_7 \to 0$ as $\mathbf{y} \to \partial B_1^+(\mathrm{II})$, $\omega(\mathbf{y}) = P_1^+(\mathrm{II})$. $\square$

Proof of Proposition 6.2

Consider Z_5, noting that $\overline{S}_5 \backslash S_5 = \overline{S_1^+(\mathrm{II})} \cup \mathcal{K}$. Since $Z_5 \to 0$ as $\mathbf{y} \to \overline{S_1^+(\mathrm{II})}$, $\alpha(\mathbf{y}) \subset \mathcal{K}\backslash\{T_1, Q_1\}$. Since $N_2 = 0 = N_3$, only orbits in B(I) are past asymptotic to $\mathcal{K}_1$. Thus for all $\mathbf{y} \in S_5$, $\alpha(\mathbf{y}) \subset \mathcal{K}\backslash\mathcal{K}_1$. But for all $\mathbf{y} \in S_1^+(\mathrm{II})$, $\alpha(\mathbf{y}) = F$ or T_1 (see Figure 6.5) and the result follows. $\square$

Proof of Proposition 6.3

Consider Z_3, noting that $\overline{S}_3 \backslash S_3 = \mathcal{K}$. It follows that $\alpha(\mathbf{y}) \subset \mathcal{K}\backslash T_1$ and $\omega(\mathbf{y}) \subset \mathcal{K}\backslash Q_1$ for all $\mathbf{y} \in S_3$. The local stability properties of $\mathcal{K}$ (Section 6.2.2) imply that $\alpha(\mathbf{y}) \subset \mathcal{K}_1$ and $\omega(\mathbf{y}) = T_1$. $\square$

Proof of Proposition 6.4

Consider Z_4, noting that $\overline{S}_4 \backslash S_4 = \partial B_1^+(\mathrm{VI}_0) \backslash \{P_1^+(\mathrm{II})\}$. Since $Z_4 \to 0$ as $\mathbf{y} \to \partial B_1^+(\mathrm{VI}_0)$, $\omega(\mathbf{y}) = P_1^+(\mathrm{VI}_0)$.   □

7

Bianchi cosmologies: non–tilted class B models

C. G. HEWITT

University of Waterloo

J. WAINWRIGHT

University of Waterloo

In this chapter we use the general procedure described in Sections 5.2 and 5.3 to analyze the evolution of Bianchi models of class B (non–exceptional; see Section 1.5.4) with non–tilted perfect fluid. We give a briefer description than that given for the class A models in Chapter 6, and refer the reader to Hewitt & Wainwright (1993) for many of the details.

7.1 The evolution equation and state space

7.1.1 The evolution equations

It was shown in Section 1.5.4 that the isometry group G_3 for a Bianchi cosmology of class B (non–exceptional) admits an Abelian G_2 which acts orthogonally transitively. We can thus obtain the evolution equations for these models by setting

$$\partial_1(\ ) = 0, \quad \dot{u}_1 = 0,$$

$$q_1 = 0, \quad \pi_+ = 0, \quad \tilde{\pi}_{AB} = 0,$$

in the G_2 evolution equations (1.141)–(1.151) in Section 1.6.3. The differential operator ∂_0 becomes differentiation with respect to clock time t, and will be denoted by an overdot. The resulting set of equations for the variables

$$\{\sigma_+, \tilde{\sigma}_{AB}, \Omega_1, n_+, \tilde{n}_{AB}, a_1\}$$

is under–determined, however, since there is no evolution equation for Ω_1. This property is associated with the non–tensorial character of Ω_1 under the remaining freedom in the choice of frame, described by (1.121) with ϕ a function of t (see (1.124)).

We avoid the problem of how to choose a unique frame by replacing the tensorial variables $\tilde{\sigma}_{AB}, \tilde{n}_{AB}$ by the following scalars:

$$\tilde{\sigma} = \tfrac{1}{6}\tilde{\sigma}^{AB}\tilde{\sigma}_{AB}, \quad \tilde{n} = \tfrac{1}{6}\tilde{n}^{AB}\tilde{n}_{AB},$$

$$\delta = \tfrac{1}{6}\tilde{\sigma}^{AB}\tilde{n}_{AB}, \quad {}^*\delta = \tfrac{1}{6}{}^*\tilde{\sigma}^{AB}\tilde{n}_{AB}. \tag{7.1}$$

These scalars are not independent, and satisfy the identity

$$\tilde{\sigma}\tilde{n} = \delta^2 + {}^*\delta^2, \tag{7.2}$$

(see Hewitt & Wainwright 1993, page 102).

In terms of these variables, the class B constraint (1.145) and the class B first integral (1.109) assume the form

$$ {}^*\delta = \sigma_+ a_1, \tag{7.3}$$

$$\tilde{n} = \tfrac{1}{3}(n_+^2 - \tilde{h}a_1^2), \tag{7.4}$$

respectively, where $\tilde{h} = h^{-1}$, and h is the class B group parameter. We use (7.3) and (7.4) to eliminate ${}^*\delta$ and $\tilde{n}$ as basic variables, but will continue to use $\tilde{n}$ as a shorthand notation. It is also convenient to replace a_1 by the variable $\tilde{a}$ defined by

$$\tilde{a} = a_1^2. \tag{7.5}$$

After making these changes, the basic variables which describe the physical state of a class B Bianchi cosmology are $(H, \mathbf{x})$, where

$$\mathbf{x} = (\sigma_+, \tilde{\sigma}, \delta, \tilde{a}, n_+), \tag{7.6}$$

subject to the constraint

$$\tilde{\sigma}\tilde{n} - \delta^2 - \tilde{a}\sigma_+^2 = 0, \tag{7.7}$$

which arises from (7.2) and (7.3). We shall see below this constraint is preserved by the evolution equations.

The evolution equations for these variables may be obtained by performing the previously described changes on equations (1.141)–(1.151), yielding

$$\begin{aligned}
\dot{\sigma}_+ &= -3H\sigma_+ - 2\tilde{n}, \\
\dot{\tilde{\sigma}} &= -6H\tilde{\sigma} - 4n_+\delta - 4\sigma_+\tilde{a}, \\
\dot{\delta} &= 2(-2H + \sigma_+)\delta + 2(\tilde{\sigma} - \tilde{n})n_+, \\
\dot{\tilde{a}} &= 2(-H + 2\sigma_+)\tilde{a}, \\
\dot{n}_+ &= (-H + 2\sigma_+)n_+ + 6\delta.
\end{aligned} \tag{7.8}$$

As described in Section 5.2, the dimensionless state is obtained by normalizing with H, giving

$$\mathbf{y} = \left(\Sigma_+, \tilde{\Sigma}, \Delta, \tilde{A}, N_+\right) \in \mathbf{R}^5, \tag{7.9}$$

where

$$\Sigma_+ = \frac{\sigma_+}{H}, \quad \tilde{\Sigma} = \frac{\tilde{\sigma}}{H^2}, \quad \Delta = \frac{\delta}{H^2}, \quad \tilde{A} = \frac{\tilde{a}}{H^2}, \quad N_+ = \frac{n_+}{H}. \tag{7.10}$$

The evolution equation

$$\frac{d\mathbf{y}}{d\tau} = \mathbf{f}(\mathbf{y})$$

for the dimensionless state $\mathbf{y}$ is derived from (7.8) and (7.10) in conjunction with the decoupled equations (5.10) and (5.11), namely

$$\frac{dt}{d\tau} = \frac{1}{H}, \quad \frac{dH}{d\tau} = -(1+q)H.$$

We obtain

$$
\begin{aligned}
\Sigma_+' &= -(2-q)\Sigma_+ - 2\tilde{N}, \\
\tilde{\Sigma}' &= -2(2-q)\tilde{\Sigma} - 4N_+\Delta - 4\Sigma_+\tilde{A}, \\
\Delta' &= 2(q + \Sigma_+ - 1)\Delta + 2(\tilde{\Sigma} - \tilde{N})N_+, \\
\tilde{A}' &= 2(q + 2\Sigma_+)\tilde{A}, \\
N_+' &= (q + 2\Sigma_+)N_+ + 6\Delta,
\end{aligned}
\tag{7.11}
$$

where

$$\tilde{N} = \tfrac{1}{3}(N_+^2 - \tilde{h}\tilde{A}), \tag{7.12}$$

and $'$ denotes differentiation with respect to τ. It follows from (5.19) and (5.20) that

$$q = \tfrac{1}{2}(3\gamma - 2)(1 - K) + \tfrac{3}{2}(2 - \gamma)\Sigma^2, \tag{7.13}$$

where

$$K = \tilde{N} + \tilde{A}, \tag{7.14}$$

and

$$\Sigma^2 = \Sigma_+^2 + \tilde{\Sigma}. \tag{7.15}$$

The expressions for K and Σ follow from (1.126), (1.148), (5.17) and (5.18). The constraint equation (7.7) becomes

$$g(\mathbf{y}) = \tilde{\Sigma}\tilde{N} - \Delta^2 - \tilde{A}\Sigma_+^2 = 0, \tag{7.16}$$

with

$$\tilde{A} \geq 0, \quad \tilde{\Sigma} \geq 0, \quad \tilde{N} \geq 0. \tag{7.17}$$

The function g, which defines the constraint, satisfies the DE

$$g' = 4(q + \Sigma_+ - 1)g, \tag{7.18}$$

as follows from (7.11), confirming that $g(\mathbf{y}) = 0$ is an invariant set. We note that the density parameter is given by (5.19), namely

$$\Omega = 1 - \Sigma^2 - K \qquad\qquad (7.19)$$

In summary the evolution of the non–tilted Bianchi models of class B is governed by the DE

$$\frac{d\mathbf{y}}{d\tau} = \mathbf{f}(\mathbf{y}), \qquad \mathbf{y} = (\Sigma_+, \tilde{\Sigma}, \Delta, \tilde{A}, N_+),$$

given by equation (7.11). The function $\mathbf{f} : \mathbb{R}^5 \to \mathbb{R}^5$ is a polynomial function, and so the DE is analytic. The function $\mathbf{f}$ depends on two parameters, the equation of state parameter γ and the parameter $\tilde{h}$ which determines the Bianchi type. For given $\tilde{h}$ and γ, the state space is the subset of $\mathbb{R}^5$ defined by (7.16), (7.17) and the inequality $\Omega \geq 0$. It follows from (7.12), (7.14)–(7.17) and (7.19) that *the state space is compact.* The solutions of the DE (7.11) thus define a flow ϕ_τ on the state space (see Section 4.1.3).

It is helpful to keep in mind the geometric and physical interpretation of the dimensionless state variables (7.9):

  (i) Σ_+ and $\tilde{\Sigma}$ describe the anisotropy in the Hubble flow,

 (ii) $\tilde{A}$ and N_+ describe the spatial curvature (see (7.30)),

(iii) Δ describes the relative orientation of the shear and spatial curvature eigenframes.

7.1.2 Invariant sets

As in Chapter 6, one can identify various Bianchi invariant sets, which we list in Table 7.1, referring to Hewitt & Wainwright (1993, pages 105–6) for further details. The first group are the class B invariant sets, and the second group are the class A invariant sets which satisfy $\tilde{A} = 0$ and hence lie in the boundary of the class B sets. The Bianchi sets also contain various invariant subsets which describe models with higher symmetry. We list these in Table 7.2.

We note that for the invariant sets in Table 7.2, the constraint (7.16) is trivially satisfied, and the description of the evolution is much simplified. For example, for $S^\pm(\text{III})$, (7.11) reduces to a DE for $(\Sigma_+, \tilde{A}, N_+)$ and for $S(\text{VI}_h)$, (7.11) reduces to a DE for $(\Sigma_+, \tilde{A})$.

Table 7.1. *Bianchi invariant sets, with d the dimension of the invariant set, and N the number of components.*

Notation	Restrictions	d	N
B(VI$_h$)	$\tilde{h} = \frac{1}{h} < 0$, $\quad \tilde{A} > 0$	4	1
B$^\pm$(VII$_h$)	$\tilde{h} = \frac{1}{h} > 0$, $\quad \tilde{A} > 0; N_+ > 0$ or $N_+ < 0$	4	2
B$^\pm$(IV)	$\tilde{h} = 0$, $\quad \tilde{A} > 0; N_+ > 0$, or $N_+ < 0$	4	2
B(V)	$\tilde{h} = 0$, $\quad \tilde{A} > 0, \Sigma_+ = \Delta = N_+ = 0$	2	1
B$^\pm$(II)	$\tilde{A} = 0; N_+ > 0$ or $N_+ < 0$	3	2
B(I)	$\tilde{A} = \Delta = N_+ = 0$	2	1

Table 7.2. *Bianchi invariant sets with higher symmetry.*

Notation	Class of Models	Restrictions	d
$S^\pm$(III)	LRS Bianchi III $(h = -1)$	$\tilde{A} > 0$, $\tilde{\Sigma} = 3\Sigma_+^2$, $\Delta = \Sigma_+ N_+$	3
S(VI$_h$)	Bianchi VI$_h$ $(n_\alpha{}^\alpha = 0)$	$N_+ = \Delta = 0$, $3\Sigma_+^2 + \tilde{h}\tilde{\Sigma} = 0$, $\tilde{A} > 0$	2
S(V)	Open FL model	$\tilde{\Sigma} = \Sigma_+ = \Delta = N_+ = 0$, $\tilde{h} = 0$	1
$S^\pm$(VII$_h$)	Open FL model	$\tilde{\Sigma} = \Sigma_+ = \Delta = 0$, $N_+^2 = \tilde{h}\tilde{A} > 0$	1
$S^\pm$(II)	LRS Bianchi II	$\tilde{A} = 0$, $\tilde{\Sigma} = 3\Sigma_+^2$, $\Delta = \Sigma_+ N_+$	2

7.1.3 Properties of state space.

The parameter $\tilde{h} = \frac{1}{h}$ enables one to unify the Bianchi sets $B(\text{VI}_h)$, $B^\pm(\text{IV})$ and $B^\pm(\text{VII}_h)$ into a one–parameter family of compact subsets of $\mathbf{R}^5$, which we denote by $B(\tilde{h})$:

$$B(\tilde{h}) = \begin{cases} \overline{B(\text{VI}_h)} & \text{if} \quad \tilde{h} < 0 \\ \overline{B^\pm(\text{IV})} & \text{if} \quad \tilde{h} = 0 \\ \overline{B^\pm(\text{VII}_h)} & \text{if} \quad \tilde{h} > 0. \end{cases} \tag{7.20}$$

As in Chapter 6, we work with the closure of the Bianchi sets, since the asymptotic behaviour at early and late times is often described by the boundary of the set, which contains models of more specialized Bianchi types. Within this framework, $B(\text{V}) \subset B(0)$, since $B(\text{V})$ is contained in $\overline{B^\pm(\text{IV})}$. In addition

$$B(\text{I}), B^\pm(\text{II}) \subset B(\tilde{h}) \quad \text{for all} \quad \tilde{h} \in \mathbf{R}.$$

We note that $B(\tilde{h})$ is bounded for each $\tilde{h} \in \mathbf{R}$, but the size of the set increases without bound as $\tilde{h} \to +\infty$ (i.e. as $h \to 0^+$). This limiting behaviour is to be expected in view of the fact that the invariant set $B_1^+(\text{VII}_0)$ is unbounded (see Table 6.1 and Section 6.3.1).

One can also unify the lower–dimensional invariant sets of class B (i.e. $\tilde{A} > 0$) by writing

$$S(\tilde{h}) = \begin{cases} \overline{S(\text{VI}_h)} & \text{if } \tilde{h} < 0, \\ \overline{B(\text{V})} & \text{if } \tilde{h} = 0, \\ \overline{S^\pm(\text{VII}_h)} & \text{if } \tilde{h} > 0. \end{cases} \qquad (7.21)$$

There is, however, a lack of continuity in this family at $\tilde{h} = 0$. As $\tilde{h} \to 0^-$, $\overline{S(\text{VI}_h)}$ approaches $\overline{B(\text{V})}$, but as $\tilde{h} \to 0^+$, $\overline{S^\pm(\text{VII}_h)}$ approaches $\overline{S(\text{V})}$, a one–dimensional subset of $\overline{B(\text{V})}$ (see Tables 7.1 and 7.2). These sets, together with the plane wave curve (7.26), are characterized by $\Delta = 0$, $\tilde{A} > 0$. Their phase portraits are shown in Figure 7.3 in the Appendix, for the case $\frac{2}{3} < \gamma < 2$.

Technical point: It is necessary to determine at which points the constraint surface (7.16) is singular (i.e. the points **y** which satisfy $g(\mathbf{y}) = 0$ and $\nabla g(\mathbf{y}) = \mathbf{0}$), since at these points we cannot use the implicit function theorem to eliminate, locally, one of the variables. The gradient of g is

$$\nabla g = (-2\tilde{A}\Sigma_+, \tfrac{1}{3}(N_+^2 - \tilde{h}\tilde{A}), -2\Delta, -(\tfrac{1}{3}\tilde{h}\tilde{\Sigma} + \Sigma_+^2), \tfrac{2}{3}N_+\tilde{\Sigma}).$$

It follows that if $\tilde{h} \geq 0$ the surface is singular at and only at points of the invariant set $S(\tilde{h})$ given by (7.21), and if $\tilde{h} < 0$, at and only at points of $S(\tilde{h}) \cap B(\text{I})$.

7.1.4 Friedmann–Lemaître universes.

The one–parameter family of open FL universes are described by the single ordinary orbit in the invariant sets $S(\text{V})$ and $S^\pm(\text{VII}_h)$, which are described in Table 7.2 and Figure 7.3. Two distinct representations arise because the open FL universe admits groups G_3 of type V and of type VII_h. The flat FL universe is described by an equilibrium point, denoted F in Table 7.3, which lies in the boundary of the state space $B(\tilde{h})$, for all $\tilde{h} \in \mathbf{R}$. The Milne universe is also described by equilibrium points, denoted by M and $M^\pm$ in Table 7.3, which lie in the boundary of $B(\tilde{h})$ for $\tilde{h} = 0$ and $\tilde{h} > 0$, respectively.

Table 7.3. *Equilibrium points of the DE (7.11) and constraint (7.16), with* $\frac{2}{3} < \gamma < 2$.

Symbol	Invariant set	Self–similar solution	Line–element
F	$B(\mathrm{I})$	flat FL	9.1.1
$P^{\pm}(\mathrm{II})$	$S^{\pm}(\mathrm{II})$	Collins–Stewart (II)	9.1.2
$P(\mathrm{VI}_h)$	$S(\mathrm{VI}_h)$	Collins VI_h	9.1.3
M	$S(\mathrm{V})$	Milne	9.1.6
$M^{\pm}$	$S^{\pm}(\mathrm{VII}_h)$	Milne	9.1.6
$\mathcal{K}$	$B(\mathrm{I})$	Kasner vacuum	9.1.1
$\mathcal{L}(\mathrm{VI}_h)$	$B(\mathrm{VI}_h)$	Bianchi VI_h plane wave	9.1.4
$\mathcal{L}^{\pm}(\mathrm{IV})$	$B^{\pm}(\mathrm{IV})$	Bianchi IV plane wave	9.1.4
$\mathcal{L}^{\pm}(\mathrm{VII}_h)$	$B^{\pm}(\mathrm{VII}_h)$	Bianchi VII_h plane wave	9.1.4

7.1.5 Monotone functions.

The DE (7.11) and its specializations admit a number of monotone functions in the invariant sets of dimension less than 4, which are given in the Appendix. However, we have been unable to find a monotone function in the full state space $B(\tilde{h})$, for $\frac{2}{3} < \gamma < 2$.

7.2 Stability of the equilibrium points

The equilibrium points of the dynamical system defined by the DE (7.11) and the constraint (7.16), are the solutions of the system of algebraic equations $\mathbf{f}(\mathbf{y}) = \mathbf{0}$ and $g(\mathbf{y}) = 0$. Table 7.3 gives the notation for the points, the invariant sets in which they lie and the corresponding exact solutions, which are self–similar (see Section 5.3.3). We also list the subsections of Chapter 9 in which the line–elements are given. The first two groups of entries in the Table are single points and the second group are one–dimensional sets. We discuss the most important aspects of the stability of the equilibrium points, referring to Hewitt & Wainwright (1993, pages 107–14) for full details.

7.2.1 Flat FL equilibrium point.

The equilibrium point F is given by

$$\Sigma_+ = \tilde{\Sigma} = \Delta = \tilde{A} = N_+ = 0.$$

It follows from (7.14), (7.15) and (7.19) that the density parameter is given by $\Omega = 1$. The deceleration parameter q is given by (5.33).

Since the constraint surface (7.16) is singular at F it is not possible to eliminate one of the variables using the constraint. The eigenvalues of F in the extended five–dimensional state space, neglecting the constraint, are

$$-\tfrac{3}{2}(2-\gamma), \quad -3(2-\gamma), \quad (3\gamma-4), \quad \tfrac{1}{2}(3\gamma-2), \quad \tfrac{1}{2}(3\gamma-2).$$

If $0 < \gamma < \tfrac{2}{3}$, all eigenvalues are negative, and F is a *local sink in the extended state space*, and hence *in the physical state space* $B(\tilde{h})$.

If $\tfrac{2}{3} < \gamma < 2$, two of the eigenvalues are always negative and two of the eigenvalues are always positive. It follows that F is a saddle in the extended state space and, hence that F *behaves like a saddle point in the state space* $B(\tilde{h})$. Since the constraint surface is singular at F, we are unable to describe how the stable and unstable manifolds of F in the extended state space intersect the constraint surface. However, we know that all orbits in $B(I)$, a two–dimensional set, are future asymptotic to F (see Figure 7.1), and we know of two particular orbits that are past asymptotic to F, namely the orbit of the open FL models in $S(\tilde{h})$ (see Figure 7.3) and the unstable manifold of F in $S^+(II)$ given in Chapter 6 (see Figure 6.5). We thus anticipate that there is a two–dimensional family of orbits that are past asymptotic to F.

7.2.2 Collins VI_h equilibrium point

The equilibrium point $P(VI_h)$ is given by

$$\Sigma_+ = -\tfrac{1}{4}(3\gamma-2), \qquad \tilde{\Sigma} = -3h\Sigma_+^2, \qquad \Delta = 0$$

$$\tilde{A} = -\tfrac{9}{16}h(3\gamma-2)(2-\gamma), \qquad N_+ = 0, \tag{7.22}$$

with $\tfrac{2}{3} < \gamma < 2$. The density parameter is given by

$$\Omega = \tfrac{3}{4}[(2-\gamma) + h(3\gamma-2)],$$

and the deceleration parameter q is given by (5.33). The equilibrium point $P(VI_h)$ lies in the interior ($\Omega > 0$) of the state space $B(\tilde{h})$ if and only if

$$\tilde{h} < \tilde{h}_c(\gamma) \equiv -\frac{(3\gamma-2)}{2-\gamma}, \tag{7.23}$$

where $\tilde{h} = 1/h$.

The constraint is non–singular at $P(VI_h)$, and the eigenvalues may be

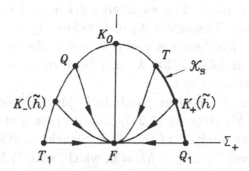

Fig. 7.1. The Kasner parabola $\mathcal{K}(\Sigma_+^2 + \tilde{\Sigma} = 1)$, showing the source $\mathcal{K}_s(\Sigma_+ > \frac{1}{2})$ and the orbits in the invariant set $B(I)$.

determined in the usual way by first using the constraint to eliminate $\tilde{\Sigma}$. The eigenvalues are:

$$-\tfrac{3}{4}(2 - \gamma)(1 \pm \sqrt{1 - r^2}), \qquad -\tfrac{3}{4}(2 - \gamma)(1 \pm \sqrt{1 - s^2}),$$

where

$$r^2 = \frac{2(3\gamma - 2)}{2 - \gamma}[(2 - \gamma) + h(3\gamma - 2)], \qquad s^2 = \frac{2r^2}{2 - \gamma}.$$

Thus, for all values of $\tilde{h}$ for which $P(\text{VI}_h)$ lies in the interior of $B(\tilde{h})$, $P(\text{VI}_h)$ *is a local sink in the state space* $B(\tilde{h})$. The stability of this equilibrium point has also been studied by Bogoyavlensky (1985, page 22), and by Uggla & Rosquist (1988, page 777).

7.2.3 Kasner equilibrium points

$\mathcal{K}$ is a parabolic arc of equilibrium points given by

$$\tilde{\Sigma} + \Sigma_+^2 = 1, \qquad \Delta = \tilde{A} = N_+ = 0. \tag{7.24}$$

Since the corresponding cosmological models are the Kasner vacuum solutions, we will refer to $\mathcal{K}$ as the *Kasner parabola*. This is shown in Figure 7.1.

The eigenvalues of $\mathcal{K}$ in the extended state space are

$$0, \quad 2\left[1 + \Sigma_+ \pm \sqrt{3(1 - \Sigma_+^2)}\right], \quad 2(1 + \Sigma_+), \quad 2(2 - \gamma).$$

The zero eigenvalue corresponds to the fact that $\mathcal{K}$ is a one–dimensional set of equilibrium points. The eigenvalue $2(1 + \Sigma_+)$ corresponds to the unphysical dimension. The subset $\mathcal{K}_s$ of $\mathcal{K}$ defined by $\frac{1}{2} < \Sigma_+ \leq 1$ has four positive eigenvalues, and hence is a local source in the extended state space (see Section 4.3.4). It follows that $\mathcal{K}_s$ *is a local source in the physical state space* $B(\tilde{h})$, *for each value of* $\tilde{h}$.

Certain points on the Kasner parabola $\mathcal{K}$ play a special role. Firstly, the invariant set $S(\tilde{h})$, given by (7.21), contains points of $\mathcal{K}$, which play an important role in the analysis of the exact Bianchi solutions in Chapter 9. These points are given by $3\Sigma_+^2 + \tilde{h}\tilde{\Sigma} = 0$, which, with (7.24), yields

$$\Sigma_+ = \pm\sqrt{\frac{-\tilde{h}}{3 - \tilde{h}}} = \pm\frac{1}{\sqrt{1 - 3\tilde{h}}}. \tag{7.25}$$

We denote the points with $\tilde{h} < 0$ by $K_\pm(\tilde{h})$, and the point with $\tilde{h} = 0$ by K_0. Secondly, the points which correspond to the Taub form of flat space–time are denoted by $T_1(\Sigma_+ = -1)$ and $T(\Sigma_+ = \frac{1}{2})$, and the points which correspond to LRS Kasner solutions are denoted by $Q_1(\Sigma_+ = 0)$ and $Q(\Sigma_+ = -\frac{1}{2})$. The special points are shown in Figure 7.1. Note that for $\tilde{h} = -1$, $K_+(\tilde{h})$ coincides with T and $K_-(\tilde{h})$ coincides with Q.

7.2.4 Plane wave equilibrium points

The plane wave equilibrium points form a curve $\mathcal{L}(\tilde{h})$ in the state space $B(\tilde{h})$, given by

$$\tilde{\Sigma} = -\Sigma_+(1 + \Sigma_+), \quad \Delta = 0, \quad \tilde{A} = (1 + \Sigma_+)^2,$$

$$N_+^2 = (1 + \Sigma_+)[\tilde{h}(1 + \Sigma_+) - 3\Sigma_+], \tag{7.26}$$

where Σ_+ is a parameter, which satisfies

$$-1 < \Sigma_+ \leq \frac{\tilde{h}}{3 - \tilde{h}}, \quad \text{if} \quad \tilde{h} < 0,$$

$$-1 < \Sigma_+ < 0, \qquad \text{if} \quad \tilde{h} \geq 0. \tag{7.27}$$

In the cases $\tilde{h} < 0$, $\tilde{h} = 0$ and $\tilde{h} > 0$, we have denoted this curve by $\mathcal{L}(VI_h)$, $\mathcal{L}^\pm(IV)$ and $\mathcal{L}^\pm(VII_h)$, respectively, so as to indicate the Bianchi type, in

Table 7.3. It follows from (7.13)–(7.15) that $q = -2\Sigma_+$.

The eigenvalues of the plane wave equilibrium points may be found explicitly, using the constraint equation (7.16) to eliminate $\tilde{\Sigma}$. The eigenvalues are:

$$0, \quad -4\Sigma_+ - (3\gamma - 2), \quad -2[(1 + \Sigma_+) \pm 2iN_+],$$

where N_+ is given by (7.26). It follows that *the part of the plane wave curve, denoted by $\mathcal{A}(\tilde{h})$, which satisfies*

$$\Sigma_+ > -\tfrac{1}{4}(3\gamma - 2), \tag{7.28}$$

is a local sink, since all eigenvalues, apart from the single zero one, have negative real part (see Section 4.3.4). If $\gamma \leq \tfrac{2}{3}$, the set $\mathcal{A}(\tilde{h})$ is empty, for all $\tilde{h} \in \mathbb{R}$. If $\tfrac{2}{3} < \gamma < 2$, $\mathcal{A}(\tilde{h})$ is non–empty if and only if

$$\tilde{h} > \tilde{h}_c(\gamma), \tag{7.29}$$

where $\tilde{h}_c(\gamma)$ is defined by (7.23). This inequality guarantees that the interval (7.27) intersects the interval (7.28). As regards the vacuum subset $\Omega = 0$, further analysis (Hewitt & Wainwright 1993, page 115) shows that *the plane wave curve $\mathcal{L}(\tilde{h})$ is a local sink in the vacuum state space $B(\tilde{h})|_{\Omega=0}$.*

The projections of the plane wave curve $\mathcal{L}(\tilde{h})$ in the $\Sigma_+ N_+$–plane, for representative values of $\tilde{h}$, are shown in Figure 7.2. All curves intersect and have a common tangent at $\Sigma_+ = -1$, $N_+ = 0$, which is the Kasner point T_1, corresponding to the Taub form of flat space–time. If $\tilde{h} \geq 0$, the curves intersect the line $\Sigma_+ = 0$, the intercepts being the Milne points M (if $\tilde{h} = 0$) and $M^{\pm}$ (if $\tilde{h} > 0$). If $\tilde{h} < 0$, the curves intersect the line $N_+ = 0$, the point of intersection being the diagonal plane wave solution (see Table 9.1), denoted $D(\tilde{h})$, which belongs to the invariant set $S(\tilde{h})$ (see Figure 7.3). In the limit $\tilde{h} \to -\infty$ (i.e. $h \to 0^-$), $\mathcal{L}(\tilde{h})$ approaches T_1, and in the limit $\tilde{h} \to +\infty$ (i.e. $h \to 0^+$), $\mathcal{L}(\tilde{h})$ approaches the infinite lines of equilibrium points $\mathcal{L}_1^{\pm}(\text{VII}_0)$ in Chapter 6, which correspond to Taub flat space–time (see Table 6.3).

We note that the stability of the plane wave equilibrium points has also been studied by Bogoyavlensky (1985, pages 95–7), by Barrow & Sonoda (1986, pages 31–5, for the Bianchi VII_h case), and by Uggla & Rosquist (1988, pages 778–9).

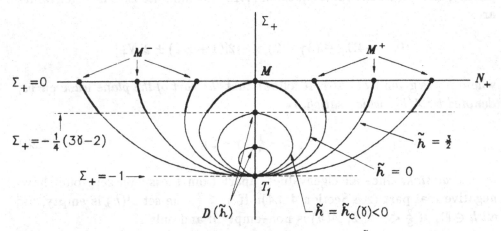

Fig. 7.2. The projection of the plane wave equilibrium set $\mathcal{L}(\tilde{h})$ in the $\Sigma_+ N_+$–plane, for different values of $\tilde{h}$. The local sink $\mathcal{A}(\tilde{h})$ lies between the lines $\Sigma_+ = -\frac{1}{4}(3\gamma-2)$ and $\Sigma_+ = 0$.

7.3 Asymptotic and intermediate evolution

In Section 7.2 we identified a local source and a local sink in the state space $B(\tilde{h})$, for both vacuum ($\Omega = 0$) and non–vacuum ($\Omega > 0$) models. We can now use the LaSalle Invariance Principle (Section 4.4.3) and the monotone functions in the Appendix in conjunction with the local stability results, to draw some conclusions about the global behaviour of the orbits and hence about the overall evolution of the models. We are able to give complete results for vacuum models ($\Omega = 0$), and non–vacuum models with $0 < \gamma < \frac{2}{3}$, and partial results for non–vacuum models with $\frac{2}{3} < \gamma < 2$.

7.3.1 Vacuum models

Proposition 7.1. In the vacuum state space $\underline{B(\tilde{h})}|_{\Omega=0}$, the Kasner arc $\overline{\mathcal{K}_s}$ is the past attractor and the plane wave arc $\mathcal{L}(\tilde{h})$ is the future attractor.

Proof See Hewitt & Wainwright (1993, page 115).   □

Interpretation: All vacuum models of types VI$_h$, IV and VII$_h$, except for a set of measure zero, are asymptotic at early times to a Kasner model and at late times to a plane wave model.

Comments:

(1) The qualifier 'except for a set of measure zero' is needed because models whose orbits lie in the stable manifold of one of the Milne points will not be asymptotic to a plane wave model. The Lukash solution (see Section 9.2.5) is an example of such a model.

(2) All models except those whose orbits lie in the unstable manifold of the Kasner point T (which corresponds to flat space–time), will have an initial curvature singularity, since the Weyl curvature invariants (1.35) will diverge.

(3) The result regarding the future asymptotic behaviour was obtained using different methods by Barrow & Sonoda (1986, page 32) and by Siklos (1991, page 1598).

7.3.2 Perfect fluid models.

The Kasner arc $\mathcal{K}_s$ is a local source for all values of the equation of state parameter $\gamma < 2$, and for all $\tilde{h}$. We have also seen that the local sink depends on γ and $\tilde{h}$, as follows:

if $\quad 0 < \gamma < \frac{2}{3}$, and $\quad \tilde{h} \in \mathbb{R}$, $\quad\quad F$ is a local sink,

if $\quad \frac{2}{3} < \gamma < 2$, and $\quad \tilde{h} < \tilde{h}_c(\gamma)$, $\quad P(\text{VI}_h)$ is a local sink,

if $\quad \frac{2}{3} < \gamma < 2$, and $\quad \tilde{h} > \tilde{h}_c(\gamma)$, $\quad \mathcal{A}(\tilde{h})$ is a local sink.

The proposition to follow deals with the case $0 < \gamma < \frac{2}{3}$.

Proposition 7.2. In the state space $B(\tilde{h})$, if $0 < \gamma < \frac{2}{3}$ then the Kasner arc $\overline{\mathcal{K}}_s$ is the past attractor and the flat FL equilibrium point F is the future attractor.

Proof See Hewitt & Wainwright (1993, page 115). $\quad\quad\quad\quad\quad\quad\quad\quad\quad\quad$ $\square$

Interpretation: If $0 < \gamma < \frac{2}{3}$, all models of types VI_h, IV and VII_h are asymptotic at late times to the flat FL model, and all models, except for a set of measure zero, are asymptotic at the big–bang to a Kasner model. The late–time behaviour in this case may be viewed as inflationary (see Section 8.3).

For values of γ which satisfy $\frac{2}{3} < \gamma < 2$ we do not have a complete analysis.

Conjecture: In the state space $B(\tilde{h})$, if $\frac{2}{3} < \gamma < 2$ then the local source is the past attractor and the local sink is the future attractor.

We have been able to provide some analytic evidence for this conjecture by proving that in each case there is an open set of orbits each of which is past asymptotic to the local source *and* future asymptotic to the local sink (see Hewitt & Wainwright 1993, page 116). The obstacle to providing a proof of this conjecture is the lack of a monotone function.

Interpretation: Assuming the validity of the conjecture, if $\frac{2}{3} < \gamma < 2$, all models of types VI$_h$, IV and VII$_h$, except for a set of measure zero, are asymptotic at the big–bang to a Kasner model, and are asymptotic at late times to a plane wave model, or, if $\tilde{h} < \tilde{h}_c(\gamma)$, to the Collins VI$_h$ model. In particular this implies that *all models are asymptotically self–similar*.

7.3.3 Intermediate evolution

As in Chapter 6, one can obtain information about the intermediate evolution by finding the heteroclinic sequences that join the past attractor to the future attractor. One obtains diagrams for the state space $B(\tilde{h})$, similar to Figure 6.12 but of somewhat greater complexity, depending on the value of $\tilde{h}$ (unpublished). The flat FL equilibrium point plays a central role, leading to flat quasi–isotropic epochs.

7.4 Summary

In this chapter we have given a qualitative description of the evolution of the Bianchi models of class B with non–tilted perfect fluid. We conclude this chapter by summarizing the main differences between the analysis and results for the class A and class B models (see Section 6.5). The first difference (see Section 7.1.1) is how we deal with the freedom in the choice of the spatial frame vectors. The second difference is that the (dimensionless) state space is defined by a constraint $g(\mathbf{y}) = 0$ in $\mathbf{R}^5$. The third difference is that the state space is a compact set, which leads to the existence of a future as well as a past attractor. The fourth difference is that our analysis suggests (but does not prove) that all models are asymptotically self–similar into the past and into the future.

Appendix to Chapter 7

Spatial curvature and Weyl curvature

The dimensionless spatial curvature variables are defined by

$$\mathcal{S}_+ = \frac{{}^3S_+}{H^2}, \quad \tilde{\mathcal{S}}_{AB} = \frac{{}^3\tilde{S}_{AB}}{H^2},$$

where ${}^3S_+$ and ${}^3\tilde{S}_{AB}$ are given by (1.146) and (1.147). It follows that

$$\mathcal{S}_+ = 2\tilde{N}, \quad \tilde{\mathcal{S}}_{AB}\tilde{\mathcal{S}}^{AB} = 24(\tilde{A} + N_+^2)\tilde{N}. \tag{7.30}$$

The dimensionless Weyl tensor components are defined by

$$\tilde{\mathcal{E}}_{AB} = \frac{\tilde{E}_{AB}}{H^2}, \quad \mathcal{E}_+ = \frac{E_+}{H^2}, \quad \tilde{\mathcal{H}}_{AB} = \frac{\tilde{H}_{AB}}{H^2}, \quad \mathcal{H}_+ = \frac{H_+}{H^2}.$$

It follows that

$$\mathcal{E}_+ \; = \; \Sigma_+(1 + \Sigma_+) - \tilde{\Sigma} + 2\tilde{N}, \quad \mathcal{H}_+ = 3\Delta,$$

$$\tilde{\mathcal{E}}_{AB}\tilde{\mathcal{E}}^{AB} \; = \; 6\Big[\tilde{\Sigma}(2\Sigma_+ - 1)^2 + 4\{\tilde{A}\Sigma_+ + \tilde{N}(\tilde{A} + 3\tilde{N} - 2\tilde{\Sigma})\} + 4\Delta^2$$

$$-4N_+\Delta(2\Sigma_+ - 1) + 4\tilde{h}\tilde{A}\tilde{N}\Big],$$

$$\tilde{\mathcal{H}}_{AB}\tilde{\mathcal{H}}^{AB} \; = \; 6\Big[9\Sigma_+^2\tilde{N} + 6\tilde{\Sigma}\tilde{N} + \tilde{\Sigma}\tilde{A} + 6\Delta^2$$

$$+12\Sigma_+ N_+\Delta + 4\tilde{h}\tilde{A}\tilde{\Sigma}\Big], \tag{7.31}$$

$$\tilde{\mathcal{E}}_{AB}\tilde{\mathcal{H}}^{AB} \; = \; 6\Big[2N_+\{(1 - 2\Sigma_+)\tilde{\Sigma} + 3\Sigma_+\tilde{N} + \Sigma_+\tilde{A}\}$$

$$+\Delta\{3(1 - 2\Sigma_+)\Sigma_+ + 4N_+^2 + \tilde{A}\}\Big].$$

Phase portraits

In Figure 7.3 we give the phase portraits for the invariant set $S(\tilde{h})$, as defined in (7.21) and Table 7.2, for various values of $\tilde{h} = 1/h$, and $\frac{2}{3} < \gamma < 2$. In case (a) there are two subcases depending on whether the eigenvalues of $P(\text{VI}_h)$ are complex, in which case $P(\text{VI}_h)$ is an attracting spiral, as in Figure 7.3(a), or real, in which case $P(\text{VI}_h)$ is an attracting node, as in Figure 9.3. We refer to Sections 7.2.1–7.2.4 for information about the equilibrium points in these figures.

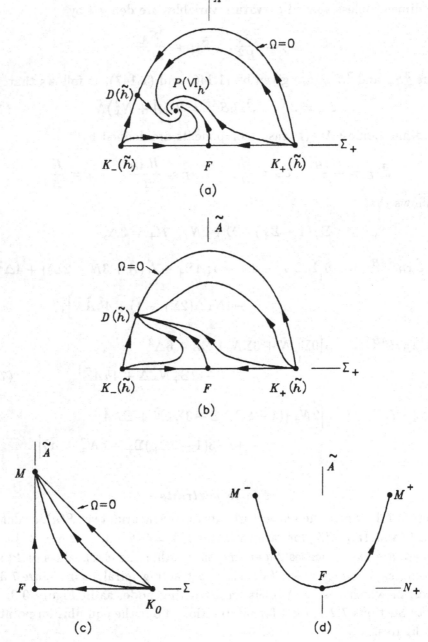

Fig. 7.3. Phase portraits for the invariant set $S(\tilde{h})$. In case (a), $\tilde{h} < \tilde{h}_c(\gamma) < 0$ (see (7.23)), in case (b), $\tilde{h}_c(\gamma) < \tilde{h} < 0$, in case (c) $\tilde{h} = 0$, and in case (d) $\tilde{h} > 0$.

Monotone functions

The DE (7.11) admits a number of monotone functions. We list the functions, their derivatives, and the invariant sets on which they are monotone.

(1) $Z_1 = (1 + \Sigma_+)^2 - \tilde{A}$, $Z_1' = -2(2 - q)Z_1 + 3(2 - \gamma)(1 + \Sigma_+)\Omega$,

$$S = B(\tilde{h})|_{\Omega=0}, \quad \text{or} \quad S = B(\tilde{h}) \quad \text{with} \quad \gamma = 2.$$

(2) $$Z_2 = \tilde{A}^2/N_+, \quad Z_2' = 3qZ_2,$$

$$S = S^{\pm}(\text{III}) \quad \text{with} \quad \gamma > \tfrac{2}{3}.$$

(3) Let

$$v = -\tfrac{1}{4}(3\gamma - 2), \quad m = \frac{-3vb^2(2 - \gamma)}{4(1 - b^2v^2)}, \quad b^2 = 1 - \frac{3}{\tilde{h}}.$$

Then

$$Z_3 = \frac{|\tilde{A}|^m \Omega^{1-m}}{(1 - b^2 v \Sigma_+)^2}, \quad Z_3' = \frac{4m(\Sigma_+ - v)^2}{v(1 - b^2 v \Sigma_+)} Z_3,$$

with $b^2 v^2 < 1$,

$$S = S(\text{VI}_h), \quad h = 1/\tilde{h}.$$

(4) *The density parameter:*
It follows from (5.20) and (5.21) that if $0 < \gamma \leq \tfrac{2}{3}$, Ω is monotone increasing on $B(\tilde{h})$.

8

Bianchi cosmologies: extending the scope

C. G. HEWITT

University of Waterloo

C. UGGLA

University of Stockholm

J. WAINWRIGHT

University of Waterloo

In Chapters 6 and 7 dynamical systems methods were used to give a detailed analysis of the ever–expanding Bianchi models with non–tilted perfect fluid as the source, apart from the so–called exceptional class B models (see Section 1.5.4). The methods used in Chapters 6 and 7 are more widely applicable, however, and in this chapter it is our intention to give a glimpse of their scope. We also mention other related qualitative investigations.

8.1 The exceptional Bianchi $VI^*_{-1/9}$ models

The exceptional class B Bianchi $VI^*_{-1/9}$ models with non–tilted perfect fluid source form a five–dimensional class of models and hence are of the same generality as the Bianchi VIII, IX models in class A, and are of greater generality than the Bianchi VII_h and non–exceptional VI_h models in class B. Despite their generality this class of models has apparently been regarded as an anomaly, and of little interest, perhaps because the group type is not compatible with the FL models (see Table 1.1). In any event, until recently no attempts were made to analyze the exceptional class.

The subclass for which the Abelian subgroup G_2 admits a hypersurface–orthogonal KVF (see Table 13.3) has a three–dimensional state space and has been analyzed in detail by Hewitt (1991a). In this paper it was proved that the models are asymptotically self–similar into the past and future. Specifically, the past attractor is the Kasner equilibrium point with indices $(p_\alpha) = \left(\frac{1}{3}(1 + \sqrt{3}), \frac{1}{3}(1 - \sqrt{3}), \frac{1}{3}\right)$ for all γ, while the future attractor depends on γ as follows† (the solutions are given in Tables 9.1 and 9.2):

---

† In the case $\gamma = \frac{10}{9}$, the proof is incomplete.

$0 < \gamma \le \frac{2}{3}$, the flat FL solution,

$\frac{2}{3} < \gamma < \frac{10}{9}$, the Collins VI$_{-\frac{1}{9}}$ perfect fluid solution,

$\gamma = \frac{10}{9}$, the 1–parameter set $\mathcal{L}(\text{VI}^*_{-\frac{1}{9}})$ of perfect fluid solutions,

$\frac{10}{9} < \gamma < 2$, the Collinson–French vacuum solution.

Thus dust models ($\gamma = 1$) are matter–dominated while radiation models are vacuum–dominated at late times. If $\gamma > \frac{2}{3}$, the flat FL universe is a saddle point with a two–dimensional stable manifold. It follows that an open subset of the models will have a flat quasi–isotropic epoch with $\Omega \approx 1$.

The general class is at present under investigation by one of the authors (Hewitt). The cosmological state space is a five–dimensional compact set (a subset of $\mathbb{R}^7$ with two constraints). There are three shear variables and two spatial curvature variables. In other words, compared to the non–exceptional class B models the group parameter h (a constant) is replaced by an additional shear variable. It appears that all models are asymptotically self–similar into the future, with the same future attractor as in the three–dimensional subset. The generic behaviour into the past is not asymptotically self–similar, since there is no equilibrium point, or set of equilibrium points, which acts as a local source (the Kasner equilibrium points are all saddle points). Since the state space is compact, however, a past attractor does exist, and it appears that the generic behaviour involves oscillations between Kasner states, governed by infinite heteroclinic sequences of Kasner equilibrium points in analogy with the Bianchi VIII and IX models.

8.2 Bianchi models with a two–fluid source

The formulation of the EFE with a perfect fluid source and linear equation of state as given in Chapters 6 and 7 can be generalized to the case of two such fluids, assumed to be non–interacting, and with both 4–velocities orthogonal to the group orbits. The stress–energy tensor is given in Chapter 2 (see (2.6), (2.17) and (2.20)). As mentioned there, a cosmological constant can be included as a special case, i.e. $\gamma_1 = 1$, $\gamma_2 = 0$ gives a model containing dust with a cosmological constant.

Following Coley & Wainwright (1992), the evolution equations in Chapters 6 and 7 can be generalized to the two–fluid case by introducing the variable

$$\chi = \frac{\mu_2 - \mu_1}{\mu_2 + \mu_1}, \tag{8.1}$$

which we shall refer to as the *transition variable*. It follows from the contracted Bianchi identities applied to each fluid that

$$\frac{d\chi}{d\tau} = \tfrac{3}{2}(\gamma_1 - \gamma_2)(1 - \chi^2), \tag{8.2}$$

where τ is the dimensionless time variable (see (5.10)). As described in Coley & Wainwright (1992) for the class A models, the remaining evolution equations have the same form as the single–fluid case, apart from the fact that the equation of state parameter γ (a constant in the single–fluid case) is replaced by the variable

$$\gamma = \tfrac{1}{2}(\gamma_1 + \gamma_2) - \tfrac{1}{2}(\gamma_1 - \gamma_2)\chi.$$

The same situation holds for the class B evolution equations, in both the exceptional and non–exceptional cases.

We assume that $\mu_1 \geq 0$, $\mu_2 \geq 0$, $\mu_1 + \mu_2 > 0$ and that $\gamma_1 > \gamma_2$. It follows from (8.1) and (8.2) that

$$-1 \leq \chi \leq 1,$$

and that χ is monotone increasing with

$$\lim_{\tau \to -\infty} \chi = -1, \qquad \lim_{\tau \to +\infty} = +1.$$

In other words, the models evolve from an initial state in which fluid one is dominant ($\chi = -1 \Leftrightarrow \mu_2 = 0$) to a final state in which fluid two is dominant ($\chi = +1 \Leftrightarrow \mu_1 = 0$).

In summary, the asymptotic behaviour of the two–fluid ever–expanding non–tilted Bianchi models is described by the asymptotic behaviour of the associated single–fluid models. The special case of the two–fluid LRS Bianchi II models (with a three–dimensional state space) is discussed in detail by Coley & Wainwright (1992). We also note that Weber (1987) has given a qualitative analysis of Bianchi models of type II (LRS), type V, type VI_0 ($n_\alpha{}^\alpha = 0$) and type VI_h ($n_\alpha{}^\alpha = 0$), with a non–tilted perfect fluid and cosmological constant $\Lambda > 0$ (or < 0), using the variables of Collins (1971).

8.3 Cosmic 'no–hair' theorems for Bianchi models

Bianchi models with a cosmological constant, or containing a perfect fluid with $0 < \gamma < \tfrac{2}{3}$ exhibit very simple future asymptotic behaviour, which may be described as inflationary in the sense that the models isotropize and the deceleration parameter q becomes negative. Models with a cosmological constant ($\gamma = 0$) correspond to exponential inflation i.e. $\ell(t) \sim e^{\lambda t}$, with

$\lambda > 0$, while models with $0 < \gamma < \frac{2}{3}$ correspond to power–law inflation i.e. $\ell(t) \sim t^r$, with $r > 1$ (see Barrow 1987b). Results of this type are referred to as cosmic 'no–hair' theorems.

The case of a cosmological constant is covered by a theorem proved by Wald (1983). The theorem does not require that the form of the stress–energy tensor be specified in detail, only that it satisfies the dominant and strong energy conditions given respectively by

(i) $T_{ab}t^b$ is (past) timelike or null for all (future) timelike t^a,
(ii) $(T_{ab} - \frac{1}{2}g_{ab}T_c{}^c)t^a t^b \geq 0$ for all timelike t_a.

Theorem 8.1 (Wald).   Suppose that the Einstein field equations with cosmological constant Λ hold, and that space–time admits a group G_3 of isometries, not of Bianchi type IX, acting on spacelike hypersurfaces. If

(i) $\Lambda > 0$,
(ii) T_{ab} satisfies the strong and dominant energy conditions, and
(iii) the normal congruence of the group orbits is expanding at some time,

then

(i) $\lim\limits_{t \to +\infty} H = \sqrt{\Lambda/3}, \quad \lim\limits_{t \to +\infty} \sigma = 0$,

where H is the Hubble scalar and σ the shear scalar of the normal congruence,
(ii) the spatial curvature of the group orbits tends to zero, and
(iii) the stress–energy tensor tends to zero.

Proof   See Wald (1983).   □

Since the limits in the conclusion of the theorem uniquely characterize the de Sitter solution, the theorem is interpreted as stating that the class of models is future asymptotic to the de Sitter universe. Note that the deceleration parameter q satisfies

$$\lim_{t \to +\infty} q = -1.$$

Comment: The proof does not require the full set of evolution equations, only the Raychauduri equation and the first integral applied to the group orbits and their normal congruence. In this situation, it is not necessary to use expansion–normalized variables, since the Hubble scalar does not tend to zero.

A similar, but less general theorem holds for Bianchi models with a perfect fluid which satisfies $0 < \gamma < \frac{2}{3}$.

Theorem 8.2. All non–tilted perfect fluid orthogonal Bianchi models, except for type IX, which satisfy $p = (\gamma - 1)\mu$ with $0 < \gamma < \frac{2}{3}$, are future asymptotic to the flat FL model with equation of state parameter γ.

Proof The proof is based on the evolution equation (5.21) for the density parameter, which can be written in the form

$$\Omega' = \left[-(3\gamma - 2)K + 3(2 - \gamma)\Sigma^2 \right] \Omega,$$

by using (5.19) and (5.20). Since we are excluding Bianchi type IX, we have $K \geq 0$, which implies $\Omega \leq 1$ by (5.19). The assumption $\gamma < \frac{2}{3}$ thus implies that $\Omega' \geq 0$, with equality if and only if $K = \Sigma = 0$, i.e. $\Omega = 1$. It follows that $\lim_{\tau \to +\infty} \Omega = 1$, and hence that $\lim_{\tau \to +\infty} \Sigma = 0$ and $\lim_{\tau \to +\infty} K = 0$. The latter condition implies that the spatial curvature variables (the expansion–normalized $n_{\alpha\beta}$ and a_α) are zero in the limit. For class B, this follows from (7.14) and (7.30)†, and for class A, we refer to the appendix in Coley & Wainwright (1992).

Comment. It follows from the proof and (5.20) that

$$\lim_{\tau \to +\infty} q = \tfrac{1}{2}(3\gamma - 2) < 0.$$

The discussion of two–fluid models in Section 8.2 leads to an immediate generalization of Theorem 8.2.

Corollary.‡ All two–fluid non–tilted Bianchi models, except for type IX, whose equation of state parameters satisfy

$$0 < \gamma_2 < \tfrac{2}{3} < \gamma_1 \leq 2,$$

are future asymptotic to the flat FL model with equation of state parameter γ_2.

† The same equations hold for the exceptional class B models.
‡ The limiting value $\gamma_2 = 0$, for which the second fluid is equivalent to a cosmological constant, is a special case of Theorem 8.1.

8.4 Tilted Bianchi models

The behaviour of tilted Bianchi perfect fluid models has been investigated by Belinskii, Khalatnikov and Lifshitz using piecewise approximations (see Belinskii *et al.* 1982) and by Jantzen using Hamiltonian methods (see Jantzen 1987), building on work by Misner and Ryan, (e.g. Misner *et al.* 1973, Ryan 1972). These investigations suggest that models of all Bianchi types, with the exception of types I and V, will in general exhibit an initial oscillatory regime (Jantzen 1987, page 132). However, in addition to the 'standard' type II bounces (see Section 10.3.1 for a detailed discussion) a tilted oscillating regime also involves 'centrifugal' bounces arising from the off–diagonal shear degrees of freedom. (The centrifugal bounce corresponds to a rotation of the Kasner axes in the terminology of Belinskii *et al.* 1982.) Thus, for example, the past attractor for the non–tilted models of types VIII and IX is expected to be replaced by a past attractor involving centrifugal bounces when including off–diagonal degrees of freedom. We note that in general a tilted perfect fluid will have three spatial velocity components v_α, which will give rise to three off–diagonal shear components. There exist special tilted perfect fluid cases for which some of the off–diagonal shear components are zero. In such cases one loses the possibility of a corresponding centrifugal bounce. Because of this one expects an initial Kasner state if $v_2 = v_3 = 0$ in types III–VII.§ In type II one also expects an initial Kasner state if one does not have the most general tilted perfect fluid with two non–zero velocity components.

The use of Hamiltonian methods makes it possible to obtain additional information about the initial behaviour. Thus, for example, the Hamiltonian analysis of Jantzen (1987) suggests that in all Bianchi models the centrifugal bounces can be described by the type II centrifugal bounce when one is sufficiently close to the initial singularity. However, the centrifugal bounces are expected to depend on the Bianchi type before reaching this stage (see Jantzen 1987, page 134).

Peresetsky (1985) has used dynamical systems methods in conjunction with the metric approach to analyze tilted models, and claims that the equilibrium point analysis together with the existence of a monotone function supports the existence of an initial oscillating regime in all general tilted models, with the exception of types I and V. Furthermore, Peresetsky also supports the claim that the initial behaviour of the type III–VII models with $v_2 = v_3 = 0$, and that of the type II model with only a single non–zero

---

§ We are excluding the exceptional class $VI^*_{-1/9}$, for which we expect an oscillatory singularity even in the non-tilted case (see Section 8.1).

component of the velocity, is given by Kasner states (*ibid*, page 191). He
also claims that the centrifugal bounces depend on the Bianchi type. This
does not contradict the Hamiltonian analysis which suggests that at first
there is such a dependence. This matter requires further investigation.

Dynamical systems methods can also be used, in conjunction with the orthonormal frame approach, to give a qualitative analysis of the evolution of
tilted Bianchi models, but relatively little progress has been made to date.
King & Ellis (1973) were the first to formulate the tetrad evolution equations for this class of models, generalizing the approach of Ellis & MacCallum
(1969). They used a group–invariant orthonormal frame, with e_0 equal to
the normal to the group orbits, and proved some general results about specific subclasses of models. As regards detailed qualitative analyses, Collins
& Ellis (1979) have given a complete analysis of the simplest class of tilted
models, the LRS Bianchi V models (see this paper for earlier references).
The cosmological state space is two–dimensional and the fluid has zero vorticity. Subsequently Hewitt & Wainwright (1992), using the approach of
Chapter 7, investigated a class of tilted models with a three–dimensional
state space (the Bianchi V models with zero vorticity, but not restricted to
be LRS), and gave a complete analysis of the evolution of the models with
non–extreme tilt. The models are all asymptotically self–similar.

The general class of tilted Bianchi V models is at present under investigation by two of the authors(Hewitt & Wainwright). The state space is a
five–dimensional compact set (a subset of $\mathbb{R}^7$, with two constraints). All
equilibrium points have been found, and have either zero vorticity or zero
matter density. Knowing the equilibrium points raises the possibility that
a detailed qualitative analysis of this class of tilted, rotating models can be
given.

To date, it has not proved possible to find all self–similar tilted perfect
fluid Bianchi models, that is, all solutions which correspond to equilibrium
points in the dimensionless state space. Some particular solutions have been
found, however:

(1) Hewitt (1991b) has found the general tilted Bianchi II self–similar
solution. The equation of state parameter satisfies $\frac{10}{7} < \gamma < 2$, and the fluid
is necessarily non–rotating.

(2) Rosquist (1983) and Rosquist & Jantzen (1985b) have found tilted
Bianchi VI_0 self–similar solutions with non–zero vorticity. One of the solutions has radiation ($\gamma = \frac{4}{3}$) as the source.

In addition Bradley (1988) has claimed that there are no self–similar Bianchi
solutions with tilted dust ($\gamma = 1$) as source. Lacking a complete catalogue

of equilibrium points, we do not expect to be able to give a detailed qualitative analysis of the general class of tilted Bianchi universes with vorticity, although some lower–dimensional subcases, for example rotating models with one hypersurface–orthogonal KVF (see Table 12.4) may be tractable. It should be possible, however, to analyze the local stability of the flat FL equilibrium point within the general class of tilted models, thereby gaining insight into the likelihood of flat quasi–isotropic epochs in these models. Likewise, an investigation of the local stability of the Kasner equilibrium points within the tilted state space would shed light on the structure of the past attractor for tilted models.

As regards tilted models with a positive cosmological constant and a perfect fluid with linear equation of state, Wald's theorem (Section 8.3) applies directly: if $\gamma > \frac{2}{3}$, equations (1.22) imply that the energy conditions in the theorem are satisfied. Thus all such tilted models, other than Bianchi type IX, are future asymptotic to the de Sitter universe in the sense of Wald's theorem. The theorem does not guarantee, however, that the tilt variable tends to zero.

It may also be possible to use the approach of Coley & Wainwright (1992), to investigate two–fluid models in which the 4–velocity of one fluid is tilted, for example models with radiation and dust, whose 4–velocities are not aligned.

8.5 Recollapsing models

The *recollapse conjecture*, namely the question of whether cosmological models whose space–sections have S^3 or $S^2 \times S^1$ topology will recollapse, has been studied since the 1930s (see Barrow *et al.* 1986 for some historical comments). The answer depends crucially on the conditions satisfied by the matter distribution, and examples exist for which recollapse does not occur (Barrow *et al.* 1986). Collins (1977) has proved a recollapse theorem for perfect fluid Kantowski–Sachs models, and Lin & Wald (1989) have proved such a theorem for non–tilted Bianchi IX models with a general fluid as source.

The evolution of recollapsing spatially homogeneous models with a non–tilted perfect fluid and linear equation of state can be analyzed using dynamical systems methods, as we now describe.

8.5.1 LRS Bianchi IX models with non–tilted perfect fluid

For recollapsing models, normalization with the Hubble variable H does not lead to a complete description of the evolution since the H–normalized variables diverge when $H \to 0$ at the point of maximal expansion. For the LRS type IX non–tilted perfect fluid models Uggla & von Zur–Mühlen (1990) found a different normalizing function, which led to a compact three–dimensional state space. Their analysis uses a dimensionless time variable $\hat{\tau}$ which takes on all real values, with $\hat{\tau} \to -\infty$ at the big bang and $\hat{\tau} \to +\infty$ at the big crunch. One of the state space variables is a normalized 'expansion' variable $\hat{\theta}$ (denoted by $\bar{q}_0$ in their paper), which satisfies $-1 < \hat{\theta} < 1$. This variable is monotonically decreasing along the orbits and satisfies $\lim_{\hat{\tau} \to -\infty} \hat{\theta} = +1$, $\lim_{\hat{\tau} \to +\infty} \hat{\theta} = -1$. The invariant sets $\hat{\theta} = +1$ and $\hat{\theta} = -1$ are part of the boundary of the state space and correspond to the LRS Bianchi type II models. These invariant sets can be identified with copies of the invariant set $S_1^+(\mathrm{II})$ in Chapter 6 (see Figure 6.5). We denote these invariant sets by $S_1^+(\mathrm{II})|_{\hat{\theta}>0}$ and $S_1^+(\mathrm{II})|_{\hat{\theta}<0}$, and note that the time direction is reversed in the invariant set with $\hat{\theta} < 0$. Since $\hat{\theta}$ is a monotone function, all of the equilibrium points are contained in the invariant sets $\hat{\theta} = \pm 1$, and can be read off from Figure 6.5. Thus the models are asymptotically self–similar into the past (big–bang) and future (big–crunch). It follows from Figure 6.5 that the past attractor is the Taub equilibrium point $T_1|_{\hat{\theta}>0}$, while the future attractor is the Taub equilibrium point $T_1|_{\hat{\theta}<0}$. Thus all models, except for a set of measure zero, are past asymptotic to the expanding Taub flat space–time at the big–bang and are future asymptotic to the contracting Taub flat space–time at the big crunch. The exceptional models correspond to a one–dimensional family of orbits starting at the expanding flat FL equilibrium point and a one–dimensional family of orbits ending at the contracting flat FL point, a single orbit starting at the expanding Collins–Stewart type II equilibrium point and a single orbit ending at the contracting Collins–Stewart type II equilibrium point (see Table 6.3).

An interesting feature of the dynamics is that there are orbits that start at the expanding flat FL equilibrium point and end at the contracting flat FL equilibrium point. The numerical investigations of Uggla & von Zur–Mühlen (1990) suggest strongly (see their Figure 5) that if the equation of state parameter γ equals $\frac{4}{3}$ there is a unique such orbit, namely the orbit which describes the closed FL models. For other values of γ in the interval $1 \le \gamma < 2$ there is at least one additional such orbit, with the number increasing as $\gamma \to 2$. This behaviour is of global nature and is not suggested by the analysis of the eigenvalues of the equilibrium points. These orbits are interesting since they describe models which have an isotropic big–bang and an isotropic big–crunch, but which are not FL models. In addition some of these orbits stay close to the closed FL orbit throughout their evolution. The closed FL orbit can also be part of a heteroclinic sequence, which means that there will be an open set of orbits which stay arbitrarily close to the closed FL orbit for a dimensionless time interval of arbitrary length.

It is also worth noting that one of the invariant sets in the boundary of the state space corresponds to the Kantowski–Sachs models (something which was not pointed out by Uggla and von Zur–Mühlen). This happens because the Kantowski–Sachs models can be obtained from the LRS type IX models by means of a Lie contraction.

8.5.2 Bianchi IX models with non–tilted perfect fluid

So far a satisfactory treatment of the full class of Bianchi IX models with non–tilted perfect fluid has not been given. Here we present a new approach which is a step towards a solution of this problem.

As in Section 6.1, we use the variables $(H, \mathbf{x})$, with $\mathbf{x}$ given by (6.3),

$$\mathbf{x} = (\sigma_+, \sigma_-, n_1, n_2, n_3), \tag{8.3}$$

with $n_1 > 0$, $n_2 > 0$, $n_3 > 0$ (see Table 1.1) to describe the physical state of the Bianchi IX models. We now permit H to assume all real values, so as to include the contracting epoch ($H < 0$). The generalized Friedmann equation (5.15) expresses the matter density μ in terms of the variables $(H, \mathbf{x})$ according to

$$\mu = 3H^2 - \sigma^2 + \tfrac{1}{2}\,{}^3R, \tag{8.4}$$

where ${}^3R$ is given by

$$ {}^3R = -\tfrac{1}{2}\left[n_1^2 + n_2^2 + n_3^2 - 2(n_1 n_2 + n_2 n_3 + n_3 n_1) \right], \tag{8.5}$$

(see (5.18) and (6.12)). The allowed region of the physical state space is given by the inequality $\mu \geq 0$.

We now introduce the new normalization factor D, defined by

$$D = \sqrt{H^2 + \tfrac{1}{4}(n_1 n_2 n_3)^{\frac{2}{3}}}. \tag{8.6}$$

It follows from (8.4) and (8.5) that $D > 0$ on the non–vacuum ($\mu > 0$) region of the physical state space. We note that $D = 0$ only at the points $H = 0$, $\sigma_\pm = 0$, $n_1 = 0$, $n_2 = n_3$ (cycle on $1, 2, 3$) on the vacuum boundary $\mu = 0$.

In analogy with (6.8), we define the dimensionless state by normalizing the physical state $(H, \mathbf{x})$ with D, giving

$$\tilde{\mathbf{y}} = (\tilde{H}, \tilde{\Sigma}_+, \tilde{\Sigma}_-, \tilde{N}_1, \tilde{N}_2, \tilde{N}_3), \tag{8.7}$$

where

$$\tilde{H} = \frac{H}{D}, \quad \tilde{\Sigma}_\pm = \frac{\sigma_\pm}{D}, \quad \tilde{N}_\alpha = \frac{n_\alpha}{D}. \tag{8.8}$$

On account of (8.6), these variables satisfy the constraint

$$\tilde{H}^2 + \tfrac{1}{4}(\tilde{N}_1\tilde{N}_2\tilde{N}_3)^{2/3} = 1. \tag{8.9}$$

In analogy with (5.10) we define a dimensionless time variable $\tilde{\tau}$ in terms of the normalization factor D by

$$\frac{dt}{d\tilde{\tau}} = \frac{1}{D}. \tag{8.10}$$

The evolution equation for D follows from (5.10), (5.11), (6.5) and (8.6), namely

$$\frac{dD}{d\tilde{\tau}} = -(1 + \tilde{q})\tilde{H}D, \tag{8.11}$$

where

$$\tilde{q} = \tilde{H}^2 q. \tag{8.12}$$

The evolution equation

$$\frac{d\tilde{\mathbf{y}}}{d\tilde{\tau}} = \tilde{\mathbf{f}}(\tilde{\mathbf{y}})$$

for the dimensionless state $\tilde{\mathbf{y}}$ can now be derived using (5.8), (6.5), (8.8), (8.10) and (8.11). We obtain

$$\begin{aligned}
\tilde{H}' &= -(1 - \tilde{H}^2)\tilde{q}, \\
\tilde{\Sigma}'_\pm &= -(2 - \tilde{q})\tilde{H}\tilde{\Sigma}_\pm - \tilde{S}_\pm, \\
\tilde{N}'_1 &= (\tilde{H}\tilde{q} - 4\tilde{\Sigma}_+)\tilde{N}_1, \\
\tilde{N}'_2 &= (\tilde{H}\tilde{q} + 2\tilde{\Sigma}_+ + 2\sqrt{3}\tilde{\Sigma}_-)\tilde{N}_2, \\
\tilde{N}'_3 &= (\tilde{H}\tilde{q} + 2\tilde{\Sigma}_+ - 2\sqrt{3}\tilde{\Sigma}_-)\tilde{N}_3,
\end{aligned} \tag{8.13}$$

where $'$ denotes differentiation with respect to $\tilde{\tau}$, and by (6.6),

$$\begin{aligned}
\tilde{S}_+ &= \tfrac{1}{6}\left[(\tilde{N}_2 - \tilde{N}_3)^2 - \tilde{N}_1(2\tilde{N}_1 - \tilde{N}_2 - \tilde{N}_3)\right], \\
\tilde{S}_- &= \tfrac{1}{2\sqrt{3}}\left[(\tilde{N}_3 - \tilde{N}_2)(\tilde{N}_1 - \tilde{N}_2 - \tilde{N}_3)\right].
\end{aligned} \tag{8.14}$$

It follows from (5.20), (8.9) and (8.12) that

$$\tilde{q} = \tfrac{1}{2}(3\gamma - 2)(1 - \tilde{V}) + \tfrac{3}{2}(2 - \gamma)\hat{\Sigma}^2, \tag{8.15}$$

where

$$\tilde{V} = \tfrac{1}{12}\left[\tilde{N}_1{}^2 + \tilde{N}_2{}^2 + \tilde{N}_3{}^2 - 2\tilde{N}_1\tilde{N}_2 - 2\tilde{N}_2\tilde{N}_3 - 2\tilde{N}_3\tilde{N}_1 + 3(\tilde{N}_1\tilde{N}_2\tilde{N}_3)^{2/3}\right],$$
(8.16)

and†

$$\hat{\Sigma}^2 = \tilde{\Sigma}_+^2 + \tilde{\Sigma}_-^2,$$
(8.17)

It is convenient to introduce the modified density parameter $\tilde{\Omega}$, defined in analogy with (5.16) by

$$\tilde{\Omega} = \frac{\mu}{3D^2}.$$

It follows from (6.8) and (8.8) that all quantities with a tilde (except for $\tilde{H}$) are related to the corresponding variables without a tilde by multiplication by an appropriate power of $\tilde{H}$:

$$\tilde{\Sigma}_\pm = \tilde{H}\Sigma_\pm, \quad \tilde{N}_\alpha = \tilde{H}N_\alpha, \quad \hat{\Sigma}^2 = \tilde{H}^2\Sigma^2, \quad \tilde{\Omega} = \tilde{H}^2\Omega.$$
(8.18)

The new variable $\tilde{V}$ has no direct counterpart, but is related to K (see (5.18) and (6.12)) by

$$\tilde{V} = \tilde{H}^2\left[K + \tfrac{1}{4}(N_1N_2N_3)^{2/3}\right],$$
(8.19)

as follows from (8.16). Equation (5.19) now can be written in the form

$$\hat{\Sigma}^2 + \tilde{V} + \tilde{\Omega} = 1.$$
(8.20)

There are two key points as regards the above choice of variables. Firstly the constraint (8.9) implies that

$$-1 \le \tilde{H} \le 1,$$
(8.21)

so that $\tilde{H}$ is bounded. Secondly, $\tilde{V}$ satisfies

$$\tilde{V} \ge 0,$$

as follows from (8.16). This inequality, in conjunction with (8.20), implies that the variables $\tilde{\Sigma}_\pm$ are also bounded. Unfortunately, the condition $\tilde{V} \le 1$ does not imply the $\tilde{N}_\alpha$ are bounded.

The evolution equations (8.13) have two attractive features. Firstly, the evolution equation for $\tilde{H}$ simplifies the dynamics significantly, as follows. Subject to the assumption $\tfrac{2}{3} < \gamma < 2$, (8.15) implies that $0 \le \tilde{q} \le 2$. In addition, the evolution equations imply that the orbits cut the surface

---

† To avoid a conflict of notation with $\tilde{\Sigma}$ in (7.10) we replace the tilde on Σ^2 by a hat.

defined by $\tilde{q} = 0$ transversely. It thus follows that $\tilde{H}$ is monotone decreasing along orbits in the invariant set $-1 < \tilde{H} < 1$. We conjecture that

$$\lim_{\tilde{\tau} \to -\infty} \tilde{H} = +1, \qquad \lim_{\tilde{\tau} \to +\infty} \tilde{H} = -1,$$

but have been unable to exclude the possibility of exceptional orbits for which $\tilde{H}$ approaches a value other than ± 1 and $\tilde{q}$ approaches 0.

The second feature arises from the fact that by (8.9), the conditions $\tilde{H} = \pm 1$ implies that at least one of the $\tilde{N}_\alpha$ are zero. Thus the invariant sets $\tilde{H} = \pm 1$ correspond to models of Bianchi types VII_0, II or I, and can be identified with the closure of the Bianchi invariant set

$$B^+(\mathrm{VII}_0) = B_1^+(\mathrm{VII}_0) \cup B_2^+(\mathrm{VII}_0) \cup B_3^+(\mathrm{VII}_0)$$

in Chapter 6 (see Table 6.1). It follows from (8.19) that on the invariant set $\tilde{H} = +1$, which we denote by $\overline{B^+(\mathrm{VII}_0)}\big|_{H>0}$, the evolution equations (8.13) reduce to the type VII_0 specializations of (6.9). Similarly, on $\tilde{H} = -1$, which we denote by $\overline{B^+(\mathrm{VII}_0)}\big|_{H<0}$, (8.19) reduces to the type VII_0 specialization of (6.9) *but with the direction of time reversed*, describing the fact that the model is contracting prior to the big crunch. The Mixmaster attractor (6.31) is contained in the invariant sets $\tilde{H} = \pm 1$, and, in accordance with the discussion in Section 6.4.3, we anticipate that it will be the past attractor in $\tilde{H} = +1$ and the future attractor in $\tilde{H} = -1$.

In conclusion, the variables (8.6) and (8.8) enable one to describe the complete evolution (expansion and recollapse) of non–tilted Bianchi IX universes, and provides a suitable framework for numerical studies of these models.

8.5.3 Kantowski–Sachs universes

The Kantowksi–Sachs universes are defined as admitting an isometry group G_4 which acts on spacelike hypersurfaces, but no subgroup G_3 which acts transitively on the hypersurfaces. Collins (1977) has proved that if the source is a perfect fluid with a physically reasonable equation of state, then all models start at a big–bang and recollapse to a final singularity. Collins (1977) has also given a qualitative analysis of these models with a perfect fluid source and $p = (\gamma - 1)\mu$, using expansion–normalized variables (his variables x, β' are related to Ω and Σ in Section 5.2 by $x = \Omega$, $4\beta'^2 = \Sigma^2$). His analysis is incomplete in that the variables are undefined at the instant of maximum expansion. As indicated at the end of Section 8.5.1, this problem is avoided by the variables of Uggla & von Zur–Mühlen (1990).

Weber generalized the analysis of Collins by including a cosmological constant (Weber 1984), and by considering a non–interacting mixture of dust and radiation (Weber 1986). Collins (1977) gives references to earlier work on the Kantowski–Sachs universes.

8.6 Other source terms: limitations

8.6.1 Non–linear equation of state

In Chapters 5–7 we assumed that the perfect fluid satisfied a linear equation of state $p = (\gamma - 1)\mu$, where γ is a constant. In a spatially homogeneous universe, a perfect fluid necessarily satisfies a barotropic equation of state $p = f(\mu)$. In this situation, one can *define* a dimensionless scalar γ by $\gamma - 1 = p/\mu$. The derivation of the dimensionless evolution equation (5.12) can still be carried out, but now $\mathbf{f}(\mathbf{y})$, which contains γ through (5.20), will depend on τ explicitly, so that the differential equation (5.12) is non–autonomous. Apart from the no–hair theorem of Wald (Theorem 8.1), the only asymptotic result of which we are aware concerning a general barotropic equation of state was proved recently by Rendall (1996), who showed that if the equation of state function $f(\mu)$ satisfies certain physically reasonable restrictions, then any Bianchi I perfect fluid universe has the same asymptotic behaviour as in the case of a linear equation of state (see the appendix in Rendall 1996). This paper also contains a qualitative analysis of Bianchi I models with a collisionless gas as source, described by the Vlasov equation.

8.6.2 Imperfect fluids

Bianchi and FL models with an imperfect fluid as source have also been studied qualitatively, in connection with the early universe. Early work (e.g. Belinskii & Khalatnikov 1976) used the first–order Eckhart (1940) theory, which, however, suffers from serious drawbacks concerning causality and stability (e.g. Maartens 1995, page 1456). In the 1970s a relativistic second–order theory was developed by Israel (1976) and Israel & Stewart (1979), which overcame these difficulties.† A simplified version of this theory, the so-called truncated Israel–Stewart theory, was first applied to Bianchi I models by Belinskii *et al.* (1979). Recently, this theory has also been applied, in conjunction with expansion–normalized variables, to FL (Coley & van den Hoogen 1995) and Bianchi I and V models (van den Hoogen & Coley 1995). The latter authors report the occurrence of a Hopf bifurcation in the class

---

† We refer to Maartens (1995, page 1456) for details and further references.

of Bianchi I models with non–zero anisotropic stress and zero bulk viscous pressure, leading to the existence of a periodic orbit, which is in fact a past attractor in the two–dimensional state space. Maartens (1995), however, has argued that the full (i.e. non–truncated) Israel–Stewart theory should be used in a cosmological context, and, in collaboration with Coley and van den Hoogen, has applied this theory to the FL universes, using expansion–normalized variables (Coley *et al.* 1996).

8.6.3 Magnetic fields

Bianchi cosmologies with a primordial magnetic field have been investigated sporadically since the 1960s. The three methods of qualitative analysis discussed in Section 5.1 have all been applied to these models. We refer to LeBlanc *et al.* (1995) for a review and other references. This paper also contains a detailed treatment of the Bianchi VI_0 models with magnetic field and non–tilted perfect fluid, generalizing the analysis of Section 6.3.3.

8.6.4 Scalar fields

Bianchi and FL models with a scalar field (1.24) as source have been studied extensively, starting in the 1980s, as examples of universes which undergo inflation. Before a detailed qualitative analysis can be undertaken, a specific form for the potential $V(\phi)$ has to be chosen. A common choice is

$$V(\phi) = V_0 e^{-\lambda\phi}, \qquad (8.22)$$

where V_0 and λ are constants, which corresponds to 'power–law inflation' (e.g. Halliwell 1987 and Kitada & Maeda 1993). One can verify that (8.22) is the only potential for which one can introduce expansion–normalized variables in analogy with the procedure in Section 5.2. Another common choice is the harmonic potential

$$V(\phi) = \tfrac{1}{2}m^2\phi^2, \qquad (8.23)$$

where m is a constant, which corresponds to 'chaotic inflation' (e.g. Chmielowski & Page 1988). Most of the qualitative analyses have dealt with the FL models (e.g. Belinskii *et al.* 1985, Halliwell 1987 and Belinskii *et al.* 1988). As regards Bianchi models, a general result has been proved by Kitada & Maeda (1993), who showed that all ever–expanding Bianchi universes, whose potential (8.22) satisfies $0 < \lambda < \sqrt{2}$ and whose matter content satisfies the dominant and strong energy conditions, isotropize as $t \to +\infty$ (*ibid.* theorem 1, page 710). This result can be regarded as a generalization of

Wald's theorem (Section 8.3). We also note that Belinskii & Khalatnikov (1989) have analyzed Bianchi I models for which the potential is given by (8.23).

9

Exact Bianchi cosmologies and state space

C. G. HEWITT
University of Waterloo

S. T. C. SIKLOS
Cambridge University

C. UGGLA
University of Stockholm

J. WAINWRIGHT
University of Waterloo

The goal in this chapter is to present a unified survey of known exact†
Bianchi cosmological solutions by describing them as orbits in the dimensionless state spaces of Chapters 6 and 7. We will restrict our considerations
to vacuum solutions and non–tilted‡ perfect fluid solutions with an equation of state $p = (\gamma - 1)\mu$, $\gamma =$ constant. We find that all known solutions
correspond to either

equilibrium points (and hence are self–similar; see Section 5.3.3)

or

heteroclinic orbits joining two equilibrium points
(and hence are asymptotically self–similar; see Section 5.3.3).

In this way we are able to give a simple description of the asymptotic behaviour of all known solutions, and to assess their relevance as potential
models of the real universe during some epoch in its evolution.

In the final section we describe the space of initial data for vacuum and
perfect fluid solutions, in order to determine the number of parameters in
the general solution for each Bianchi type.

9.1 Self–similar perfect fluid and vacuum solutions

Tables 9.1 and 9.2 provide a list of all Bianchi vacuum and perfect fluid
solutions which admit a similarity group H_4 acting simply transitively on

---

† By an exact solution we mean one in which the metric is either given explicitly or in terms of
  one quadrature. The time coordinate is not necessarily clock time.
‡ Self–similar tilted perfect fluid solutions are mentioned briefly in Section 8.4.

Table 9.1. *Self–similar vacuum solutions (see Theorem 9.1).*

Self–similar solution	Equilibrium points	Section
Kasner	$\mathcal{K}$	9.1.1
Taub flat space–time	T_α	9.1.6
Type VI_h plane wave	$\mathcal{L}$ (VI_h)	9.1.4
Diagonal plane wave	D (VI_h)	9.1.3
Type III flat space–time	D	9.1.6
Type IV plane wave	$\mathcal{L}^\pm$ (IV)	9.1.4
Milne flat space–time	M	9.1.6
Type VII_h plane wave	$\mathcal{L}^\pm$ (VII_h)	9.1.4
Milne flat space–time	$M^\pm$	9.1.6
Collinson–French	CF	9.1.5

space–time (see Section 1.2.3), and give the equilibrium points to which they correspond (see Tables 6.3 and 7.3). The solutions are characterized in the theorems stated at the end of this section (see Hsu & Wainwright 1986).

As regards the equation of state for the perfect fluid solutions, we note that first, spatial homogeneity implies that there must be an equation of state $p = p(\mu)$, and second that the existence of a homothetic vector field implies that the equation of state $p = p(\mu)$ must be of the form $p = (\gamma - 1)\mu$, γ = constant (Wainwright 1985). In addition, for all solutions except for the flat FL solution, the field equations imply that the permitted values of γ lie in the range $\frac{2}{3} \leq \gamma \leq 2$. The flat FL solution is valid for all $\gamma \neq 0$.

For each solution, we give the original reference, and mention subsequent papers which clarified the nature of the solutions. We use local coordinates (t, x, y, z), where the homogeneous spacelike hypersurfaces are given by $t =$ constant, and have unit normal $n = \partial/\partial t$. Thus in the perfect fluid solutions we have $u = \partial/\partial t$. The homothetic vector field X is also given for each line–element. In order to emphasize the common features of the solutions we have used the uniform notation of Hsu & Wainwright (1986).

9.1.1 *Bianchi I solutions*

The line–element is

$$ds^2 = -dt^2 + t^{2p_1} dx^2 + t^{2p_2} dy^2 + t^{2p_3} dz^2,$$

Table 9.2. *Self–similar, non–tilted perfect fluid solutions (see Theorem 9.2).*

Self–similar solution	Equilibrium points	Section
Flat FL	$P(\mathrm{I})$	9.1.1
Collins–Stewart II	$P_1^\pm$ (II)	9.1.2
Collins VI_0	$P_1^\pm$ (VI_0)	9.1.3
Collins VI_h	$P_1^\pm$ (VI_h)	9.1.3
Perfect fluid $\mathrm{VI}^*_{-1/9}$	$\mathcal{L}(\mathrm{VI}^*_{-1/9})$	9.1.5
Jacobs stiff fluid	$\mathcal{J}$	9.1.1
Closed FL $(\gamma = \frac{2}{3})$	$\mathcal{F}^\pm$ (IX)	9.1.7
Open FL $(\gamma = \frac{2}{3})$	$\mathcal{F}$ (V), $\mathcal{F}^\pm$ (VII_h)	9.1.7

and the homothetic vector field is

$$X = t\tfrac{\partial}{\partial t} + (1 - p_1)x\tfrac{\partial}{\partial x} + (1 - p_2)y\tfrac{\partial}{\partial y} + (1 - p_3)z\tfrac{\partial}{\partial z}.$$

(1) *Flat FL solution*

$$p_1 = p_2 = p_3 = \frac{2}{3\gamma},$$

$$\mu = \frac{4}{3\gamma^2}t^{-2}, \qquad 0 < \gamma \leq 2.$$

The dust solution ($\gamma = 1$) is the Einstein–de Sitter (1932) solution. The solution with arbitrary γ was first given by Harrison (1967). It has been shown in Chapters 6 and 7 (see also Section 15.2.1) that this solution, which corresponds to the equilibrium point F in Sections 6.2.1 and 7.2.1, is the future asymptote for all Bianchi I γ–law perfect fluid solutions, and determines a flat quasi–isotropic epoch for all other Bianchi types.

(2) *Kasner vacuum solutions*

$$p_1 + p_2 + p_3 = 1, \qquad p_1^2 + p_2^2 + p_3^2 = 1. \tag{9.1}$$

The Kasner exponents p_α can be conveniently expressed in terms of an angle ψ as in (6.17):

$$p_1 = \tfrac{1}{3}(1 - 2\cos\psi), \qquad p_{2,3} = \tfrac{1}{3}(1 + \cos\psi \pm \sqrt{3}\sin\psi). \tag{9.2}$$

These solutions were discovered by Kasner (1925), and were first studied as cosmological models by Lemaître (1933) and by Schücking and Heckmann in the 1950s (Heckmann & Schücking 1962). It has been shown in Chapters 6 and 7 that this one–parameter family of solutions, which corresponds to the Kasner equilibrium set $\mathcal{K}$†, essentially determines the past asymptotic behaviour of all Bianchi vacuum and almost all non–tilted γ–law perfect fluid models, either as individual equilibrium points or as part of the Mixmaster attractor (see Section 6.4.3).

(3) *Jacobs stiff perfect fluid solutions*

$$p_2 + p_2 + p_3 = 1, \qquad p_1^2 + p_2^2 + p_3^2 < 1$$

$$\mu = \tfrac{1}{2}(1 - p_1^2 - p_2^2 - p_3^2)t^{-2}, \qquad \gamma = 2.$$

This solution was first given by Jacobs (1968) and contains two essential parameters.

9.1.2 Bianchi II solutions

Collins–Stewart perfect fluid solution

The line–element is

$$ds^2 = -dt^2 + t^{2p_1}\left(dx + \frac{s}{2\gamma}zdy\right)^2 + t^{2p_2}(dy^2 + dz^2),$$

and the homothetic vector field is

$$X = t\tfrac{\partial}{\partial t} + (1 - p_1)x\tfrac{\partial}{\partial x} + (1 - p_2)\left(y\tfrac{\partial}{\partial y} + z\tfrac{\partial}{\partial z}\right),$$

with

$$p_1 = \frac{2 - \gamma}{2\gamma}, \qquad p_2 = \frac{2 + \gamma}{4\gamma}, \qquad s^2 = (2 - \gamma)(3\gamma - 2),$$

$$\mu = \frac{(6 - \gamma)}{4\gamma^2}t^{-2}, \qquad \tfrac{2}{3} < \gamma < 2.$$

This solution was first given by Collins & Stewart (1971) (see also Collins (1971), example 1(a), with $E = 0$). It has been shown that this solution is the future asymptote for all Bianchi II γ–law perfect fluid solutions (see Section 6.3.2), and is a past asymptote or determines a quasi–equilibrium epoch for more general Bianchi types. Earlier indications of its role as an intermediate equilibrium state were given by Doroshkevich *et al.* (1973,

---

† $\mathcal{K}$ is a circle in the class A state space (Figure 6.2) and a parabola in the class B state space (Figure 7.1).

page 741, equation (12)) and Bogoyavlensky & Novikov (1973, page 749, equation (2.3)).

9.1.3 Bianchi VI solutions with diagonal metric

The line–element is

$$ds^2 = -dt^2 + t^2 dx^2 + t^{2p_2} e^{2c_2 x} dy^2 + t^{2p_3} e^{2c_3 x} dz^2, \qquad (9.3)$$

and the homothetic vector field is

$$X = t\frac{\partial}{\partial t} + (1 - p_2)y\frac{\partial}{\partial y} + (1 - p_3)z\frac{\partial}{\partial z}.$$

The group parameter h is given by

$$h = -\left[\frac{c_2 + c_3}{c_2 - c_3}\right]^2.$$

(1) *Collins VI_0 perfect fluid solution*

This solution is given as the special case $r = 0$ in (2) below. It was first given by Collins (1971, example 2(a)) apart from the dust solution ($\gamma = 1$), which is due to Ellis & MacCallum (1969, page 125). We refer to Sections 6.2.1 and 6.3.3, and Table 6.3 for the role played by this solution.

(2) *Collins VI_h perfect fluid solution*

$$p_{2,3} = \frac{2 - \gamma \pm rs}{2\gamma}, \qquad c_{2,3} = \frac{r(2 - \gamma) \pm s}{2\gamma},$$

where

$$s^2 = (2 - \gamma)(3\gamma - 2), \qquad 0 < r < 1.$$

The parameter r satisfies

$$h = \frac{-(2 - \gamma)^2 r^2}{s^2}.$$

The energy density and equation of state parameter are

$$\mu = \frac{(2 - \gamma)(1 - r^2)}{\gamma^2}t^{-2}, \qquad \tfrac{2}{3} < \gamma < \frac{2(1 - h)}{1 - 3h}.$$

This solution was first given by Collins (1971, example 3(a)(i)). The limit $r = 0$ gives the Collins VI_0 perfect fluid solution and the limit $r = 1$ gives the diagonal vacuum plane wave solution below, with different parameters. We refer to Sections 7.2.2 and 7.3.2 for the role played by this solution.

(3) *Diagonal vacuum plane wave solution*

$$c_2 = p_2, \qquad c_3 = p_3, \qquad p_{2,3} = r \pm \sqrt{r(1-r)}, \qquad 0 < r < 1.$$

The parameter r is determined by the group parameter h according to

$$h = -r/(1-r).$$

This solution corresponds to the equilibrium point $D(\tilde{h})$ on the arc $\mathcal{L}(\tilde{h})$ in Figure 7.2. This solution was first given by Lifshitz & Khalatnikov (1963, page 232) and has been found by other authors. We refer to Wainwright (1983) for details.

9.1.4 Vacuum plane wave solutions

The line–element is

$$ds^2 = -dt^2 + t^2 dx^2 + t^{2r} e^{2rx} \left[e^{\beta}(Ady + Bdz)^2 + e^{-\beta}(Cdy + Adz)^2 \right], \quad (9.4)$$

where $A^2 - BC = 1$, and A, B, C are functions of

$$v = b(x + \ln t), \quad b = \text{constant},$$

and r, b, β are constants. The homothetic vector field is

$$X = t\frac{\partial}{\partial t} - \frac{\partial}{\partial x} + y\frac{\partial}{\partial y} + z\frac{\partial}{\partial z}.$$

(1) *Bianchi* VI_h *plane wave solutions*

$$A = \cosh v, \quad B = C = \sinh v,$$

where b and β are determined in terms of r and the group parameter $h < 0$ according to

$$b^2 \cosh^2 \beta = r(1-r), \quad b^2 = -\frac{r^2}{h}.$$

It follows that r takes on values in the range

$$0 < r \leq \frac{-h}{1-h}.$$

This one–parameter family of solutions corresponds to the arc $\mathcal{L}(\tilde{h})$ of equilibrium points, with $\tilde{h} < 0$, which is shown in Figure 7.2. This arc is an ellipse minus one point, the flat Kasner point T_1, which is given by $r = 0$. The value $r = -h/(1-h)$ implies $\beta = 0$, which gives the diagonal plane wave solution.† This solution corresponds to the equilibrium point $D(\tilde{h})$ on the arc $\mathcal{L}(h)$ (see Figure 7.2).

(2) *Bianchi IV plane wave solutions*

$$A = 1, \quad B = v, \quad C = 0, \quad b^2 = 4r(1-r),$$

where r is an arbitrary parameter satisfying $0 < r < 1$. This one–parameter family of solutions corresponds to the arc $\mathcal{L}(\tilde{h})$, with $\tilde{h} = 0$, which is shown in Figure 7.2. The value $r = 0$ gives the flat Kasner point T_1 and the value $r = 1$ gives the Milne point M. This family of solutions was first given by Harvey & Tsoubelis (1977).

(3) *Bianchi plane wave VII$_h$ solutions*

$$A = \cos v, \quad B = -C = \sin v,$$

where b and β are determined in terms of a parameter r, $0 < r < 1$, and the group parameter $h > 0$ according to

$$b^2 \sinh^2 \beta = r(1-r), \quad b^2 = \frac{r^2}{h}.$$

This one–parameter family of solutions corresponds to the arc $\mathcal{L}(\tilde{h})$, with $\tilde{h} > 0$, shown in Figure 7.2. The value $r = 0$ gives the flat Kasner point T_1 and $r = 1$ gives the Bianchi VII$_h$ version of the Milne model, i.e., the points $M^\pm$. This family of solutions was first given by Doroshkevich *et al.* (1973, page 741, equation (12)).

Comment: These vacuum plane wave solutions were derived systematically by Siklos (1981) as algebraically special spatially homogeneous solutions, and his derivation led him to recognize the solutions as particular plane gravitational waves. We refer to Siklos (1981, 1984) for more information.

9.1.5 *Bianchi VI$^*_{-1/9}$ solutions (the exceptional case)*

The line–element is

$$ds^2 = -dt^2 + t^2 dx^2 + t^{\frac{2}{5}} \left[\exp\left(-\tfrac{\sqrt{6}}{5} rx \right) dy + bt^{\frac{4}{5}} dx \right]^2 + t^{\frac{6}{5}} \exp\left(\tfrac{4\sqrt{6}}{5} rx \right) dz^2, \tag{9.5}$$

† See Section 9.1.3 (3). The line–elements are related by the coordinate transformation

$$y' = \frac{1}{\sqrt{2}}(y+z), \quad z' = \frac{1}{\sqrt{2}}(y-z).$$

where
$$b^2 = \tfrac{9}{4}r^2 - 1,$$

and the homothetic vector field is
$$X = t\tfrac{\partial}{\partial t} + \tfrac{4}{5}y\tfrac{\partial}{\partial y} + \tfrac{2}{5}z\tfrac{\partial}{\partial z}.$$

(1) *Perfect fluid* $VI^*_{-1/9}$ *solutions*
$$\mu = \tfrac{27}{25}(1 - r^2)t^{-2}, \qquad \gamma = \tfrac{10}{9},$$

with
$$\tfrac{2}{3} < r < 1.$$

This solution was first given by Wainwright (1984, page 666, equation (20)). In the limit $r = \tfrac{2}{3}$ it reduces to the Collins VI_h perfect fluid solution with $\gamma = \tfrac{10}{9}$ (Section 9.1.3 (2)), and in the limit $r = 1$, it gives the vacuum $VI^*_{-1/9}$ solution below.

(2) *Collinson–French vacuum* $VI^*_{-1/9}$ *solution*
$$r = 1, \qquad b^2 = \tfrac{5}{4}.$$

This solution was first given by Robinson & Trautman (1962), but was not presented as a spatially homogeneous solution. It was found in a different form by Collinson & French (1967). The spatial homogeneity of the solution was recognized by Ellis & MacCallum (1969). The form given by Siklos (1984, equation (2.15)) is the closest to ours. See also Siklos (1981, page 401, equation (4.10)).

9.1.6 Spatially homogeneous forms of flat space–time

(1) *Taub form of flat space–time*
$$ds^2 = -dt^2 + t^2 dx^2 + dy^2 + dz^2, \qquad (9.6)$$

$$X = t\tfrac{\partial}{\partial t} + y\tfrac{\partial}{\partial y} + z\tfrac{\partial}{\partial z}.$$

The Kasner solution reduces to this line–element when $(p_\alpha) = (1, 0, 0)$.

(2) *Bianchi III form of flat space–time*
$$ds^2 = -dt^2 + t^2(dx^2 + e^{2x}dy^2) + dz^2, \qquad (9.7)$$

$$X = t\tfrac{\partial}{\partial t} + z\tfrac{\partial}{\partial z}.$$

The diagonal plane wave solution reduces to this line–element when $b = \frac{1}{2}$, so that $h = -1$.

(3) *Milne form of flat space–time*

$$ds^2 = -dt^2 + t^2\left[dx^2 + e^{2x}(dy^2 + dz^2)\right], \qquad (9.8)$$

$$X = t\frac{\partial}{\partial t}.$$

The Bianchi VII_h vacuum plane wave solution reduces to this line–element when $\beta = 0$ and $r = 1$.

Comment: These three line–elements describe the only possible ways of choosing a family of homogeneous spacelike hypersurfaces with expanding normals in flat space–time. These line–elements describe the asymptotic states of various Bianchi models, the Taub form primarily as $t \to 0^+$ and the Milne and type III forms as $t \to +\infty$ (see Chapters 6 and 7).

9.1.7 FL solutions with $\gamma = \frac{2}{3}$

The line–element is

$$ds^2 = -dt^2 + \left(\tfrac{t}{b}\right)^2\left[dr^2 + f(r)^2(d\theta^2 + \sin^2\theta d\phi^2)\right], \qquad (9.9)$$

where

$$f(r) = \sin r, \quad \sinh r \quad \text{for} \quad k = +1, -1,$$

respectively, and b is a positive constant. The homothetic vector field is

$$X = t\frac{\partial}{\partial t}.$$

The matter density and pressure are

$$\mu = 3(1 + kb^2)t^{-2}, \quad p = -\tfrac{1}{3}\mu.$$

In the case $k = -1$, we require $b < 1$. The limiting case $b = 1$ gives the Milne form of flat space–time with spherical spatial coordinates.

We conclude this section by summarizing the results in the following theorems.

Theorem 9.1. Table 9.1 contains all spatially homogeneous solutions of the vacuum Einstein field equations which admit a similarity group H_4 acting simply transitively on space–time.

Theorem 9.2. Table 9.2 contains all spatially homogeneous solutions of the Einstein field equations with a perfect fluid as source, which admit a similarity group H_4 acting simply transitively on space–time, subject to the following conditions:

(i) the fluid 4–velocity is orthogonal to the homogeneous hypersurfaces,

(ii) the matter density is positive.

In connection with these theorems we should note that we cannot claim to have found all Bianchi vacuum solutions and non–tilted Bianchi perfect fluid solutions which admit a homothetic vector field, since the maximal similarity group H_r may not have a simply transitive subgroup H_4. This is analogous to the fact that not all homogeneous space–times admit a simply transitive G_4 of isometries (see Kramer *et al.* 1980, chapter 10).

9.2 Evolving vacuum solutions

In this section we give the known Bianchi vacuum solutions which are not self–similar. As mentioned at the beginning of the chapter, all known solutions correspond to heteroclinic orbits, i.e., orbits which join two equilibrium points. In each case the past asymptote is a Kasner equilibrium point or the Taub form of flat space–time T_α, while the future asymptote is either a Kasner point, the diagonal plane wave point $D(\bar{h})$, or one of the representations of flat space–time, T_α, D or M (see Table 9.1).

The various classes are summarized in Table 9.3. All solutions, except for the Bianchi VII$_h$ case, are the general solution for the indicated invariant set, and all are known in explicit form. As regards the number of essential parameters† listed in the table, we make the following observation. Because of the scale invariance of the EFE, any non–self–similar solution with metric **g** determines a one–parameter family of conformally related (but inequivalent) solutions $\lambda^2\mathbf{g}$, where $\lambda > 0$ is a constant. Such a one–parameter family of solutions corresponds to one ordinary orbit in the dimensionless state space. Thus, for each class of solutions in Table 9.3, one of the parameters is a conformal scaling parameter, denoted λ. One drawback of writing one of the parameters as a scaling parameter λ is that one cannot set $\lambda = 0$ to obtain limiting cases. Thus sometimes it is desirable to reparameterize a solution by rescaling the coordinates, as in the Taub solution in Section 9.2.1.

---

† Essential parameters means parameters in addition to the group parameter h.

Table 9.3. *Evolving vacuum solutions.*

Class	Invariant Set	Parameters
Bianchi II	$B_1^+(\text{II})$	2
Bianchi V	$B(\text{V})$	1
Bianchi VI$_0$ ($n_\alpha{}^\alpha = 0$)	$S_1^+(\text{VI}_0)$	1
Bianchi VI$_h$ ($n_\alpha{}^\alpha = 0$)	$S^+(\text{VI}_h)$	1
Bianchi VII$_h$	$B^+(\text{VII}_{4/11})$	1
Bianchi VIII (LRS)	$S_1^+(\text{VIII})$	2
Bianchi IX (LRS)	$S_1^+(\text{IX})$	2

9.2.1 Taub solutions (Bianchi II)

The line–element is

$$ds^2 = -A^2 dt^2 + t^{2p_1} A^{-2}(dx + 4p_1 bz dy)^2 + t^{2p_2} A^2 dy^2 + t^{2p_3} A^2 dz^2, \quad (9.10)$$

with

$$A^2 = 1 + b^2 t^{4p_1}, \quad p_1 + p_2 + p_3 = 1, \quad p_1^2 + p_2^2 + p_3^2 = 1.$$

This is a two–parameter family of solutions, labelled by b and the one degree of freedom in the choice of the constants p_α. The solutions were discovered by Taub (1951). We refer to the above form of the Taub solutions as the *Kasner form*, since it contains the Kasner solutions as a limiting case, namely $b = 0$, and makes clear the Kasnerian asymptotic behaviour of the solutions.

In state space, these solutions correspond to the Taub orbits in the invariant set $B_1^+(\text{II})$, which are past asymptotic to a Kasner equilibrium point in the arc $\mathcal{K}_2 \cup \mathcal{T}_1 \cup \mathcal{K}_2$ and future asymptotic to a point in $\mathcal{K}_1$ (see Figure 6.6). If $p_1 > 0$, t approximates clock time as $t \to 0^+$ since then $A \approx 1$. It follows from Figure 6.2 that the solutions with $p_2 < 0$ are past asymptotic to $\mathcal{K}_2$, those with $p_3 < 0$ are past asymptotic to $\mathcal{K}_3$ and those with $p_2 = 0 = p_3$ are past asymptotic to the equilibrium point $\mathcal{T}_1$. If $p_1 < 0$, t approximates clock time as $t \to +\infty$ since then $A \approx 1$. This means that the solutions with $p_1 < 0$ are isometric to those with $p_1 > 0$; one is simply using a t coordinate for which their future asymptotic behaviour is in a manifestly Kasner form. The parameter b labels the conformally related solutions which correspond to a single orbit in state space.

9.2.2 Joseph solution (Bianchi V)

This one–parameter family of solutions is given as the special case $b = 0$ of (9.12) below, and was discovered by Joseph (1966). It corresponds to the orbit in the vacuum boundary of the Bianchi V state space $B(\text{V})$, as shown in Figure 7.3(c). The solution is thus past asymptotic to the Kasner solution with $\cos\psi = 0$ in (9.2), and is future asymptotic to the Milne form of flat space–time.

9.2.3 Ellis–MacCallum solution (Bianchi $VI_0, n_\alpha{}^\alpha = 0$)

The line–element is

$$\lambda^2 ds^2 = t^{-\frac{1}{2}}e^{t^2}(-dt^2 + dx^2) + t(e^{2x}dy^2 + e^{-2x}dz^2). \qquad (9.11)$$

This one–parameter family of solutions corresponds to the heteroclinic vacuum orbit in the invariant set $S_1^+(\text{VI}_0)$ (see Figure 6.10). This orbit is past asymptotic to the LRS Kasner equilibrium point Q_1, with exponents $\left(-\frac{1}{3}, \frac{2}{3}, \frac{2}{3}\right)$, and is future asymptotic to the Taub equilibrium point T_1. This solution was discovered by Ellis & MacCallum (1969).

9.2.4 Ellis–MacCallum solution (Bianchi $VI_h, n_\alpha{}^\alpha = 0$)

The line–element is

$$\lambda^2 ds^2 = \sinh 2t \left[A^b(-dt^2 + dx^2) + Ae^{2(1+b)x}dy^2 + A^{-1}e^{2(1-b)x}dz^2\right], \qquad (9.12)$$

with

$$A = (\sinh 2t)^b(\tanh t)^{\sqrt{3+b^2}}.$$

The constant b is related to the group parameter h by $b^2 = -\frac{1}{h}$. There are two inequivalent classes of solutions corresponding to $b > 0$ and $b < 0$. These solutions were discovered by Ellis & MacCallum (1969). The Joseph solution arises as the special case $b = 0$.

This family of solutions corresponds to the orbits in the vacuum boundary of the two–dimensional invariant set $S(\text{VI}_h)$ (see Figure 9.2). The solutions with $b > 0$ correspond to the orbit $K_-(\tilde{h}) \to D(\tilde{h})$ and the solutions with $b < 0$ correspond to the orbit $K_+(\tilde{h}) \to D(\tilde{h})$. The Kasner exponents of the equilibrium points $K_\pm(\tilde{h})$ are given by (9.24). In Section 9.3.5(1) we obtain an alternative representation of this family of solutions which makes the asymptotic behaviour more apparent.

9.2.5 Lukash solution (Bianchi VII$_{4/11}$)

The line–element is

$$\lambda^2 ds^2 = (\sinh 2t)^{-\frac{3}{8}} e^{\frac{11}{4}t}(-dt^2 + dx^2) + e^{2x}\sinh 2t\Big[f(\cos vdy + \sin vdz)^2$$
$$+ f^{-1}(-\sin vdy + \cos vdz)^2\Big], \tag{9.13}$$

with

$$f = (\tanh t)^{\frac{1}{2}}, \qquad v = \tfrac{\sqrt{11}}{2}(t - x).$$

The solution is past asymptotic to the Kasner solution with exponents $(p_1, p_2, p_3) = \left(-\frac{3}{13}, \frac{12}{13}, \frac{4}{13}\right)$, and is future asymptotic to the Milne solution, represented by the equilibrium point M^+. Thus, in the class B state space, this family of solutions is represented by a heteroclinic orbit in the vacuum boundary which joins a point on the Kasner equilibrium set to the Milne point M^+. This solution was discovered by Lukash (1975, page 796).

9.2.6 NUT solutions (LRS Bianchi VIII)

The line–element is

$$\lambda^2 ds^2 = -\frac{Y^4}{X^2}dt^2 + X^2(dx + 2hdz)^2 + \frac{Y^2}{X^2}(dy^2 + f^2dz^2), \tag{9.14}$$

with

$$X^2 = (\cosh 2t)^{-1}, \quad Y = [\sinh(t + t_0)]^{-1}$$
$$h = \cosh y, \qquad f = \sinh y,$$

and $\lambda > 0$ and t_0 are arbitrary constants. We refer to Ellis & MacCallum (1969, pages 133–4) for the history of these solutions and of the Taub–NUT† solutions in Section 9.2.7.

9.2.7 Taub–NUT solutions (LRS Bianchi IX)

The line–element is given by (9.14), with

$$X^2 = (\cosh 2t)^{-1}, \quad Y = [\cosh(t + t_0)]^{-1}$$
$$h = \cos y, \qquad f = \sin y,$$

and $\lambda > 0$ and t_0 are arbitrary constants.

† The abbreviation NUT stands for Newman–Unti–Tamburino.

9.3 Evolving non–tilted perfect fluid solutions

In this section we give the known Bianchi perfect fluid solutions with linear equation of state, which are not self–similar. In all cases the coordinates are comoving, i.e. the fluid 4–velocity **u** is parallel to $\partial/\partial t$. We restrict our considerations to solutions for which the equation of state parameter γ satisfies $0 < \gamma < 2$, partly because solutions with a stiff equation of state ($\gamma = 2$) are of less interest physically than those with $\gamma = 1$ or $\frac{4}{3}$ and partly for reasons of brevity. We remark, however, that all known solutions with $\gamma = 2$, apart from the Jacobs self–similar solutions, correspond to hetero-clinic orbits which are past and future asymptotic to equilibrium points on the Jacobs disc or on the Kasner circle. Most of these solutions can be generated from the corresponding vacuum solutions by a standard technique (Wainwright *et al.* 1979, Wainwright & Marshman 1979).

All of the known solutions with $0 \leq \gamma < 2$, except for the Bianchi II solutions, form two–parameter families, and correspond to the heteroclinic orbits in one of the two–dimensional invariant sets encountered in Chapters 6 and 7. The phase portraits, together with the identification of the equilibrium points in Tables 9.1 and 9.2, enable one to infer immediately the asymptotic behaviour of the solutions. The various classes are summa-rized in Table 9.4. All solutions listed in Table 9.4, except for the Bianchi II family, are the general solution for the indicated class, subject to the restric-tion on γ. The parameter k is related to the group parameter h according to $k = 1/\sqrt{1 - 3h}$. The format column indicates whether the solution is known explicitly, or up to a quadrature. The remark concerning conformal scaling parameters at the beginning of Section 9.2 applies here also.

9.3.1 Bianchi I

The general solution of Bianchi type I for dust ($\gamma = 1$) was discovered by Robinson (1961) and independently by Heckmann & Schücking (1962), and generalized to arbitrary γ by Jacobs (1968). These solutions correspond to the orbits in the two–dimensional invariant set $B(\text{I})$, shown in Figure 6.4. We give a new form of the solutions in which the asymptotic behaviour is manifestly clear.

The line–element is

$$ds^2 = -A^{2(\gamma-1)}dt^2 + t^{2p_1}A^{2q_1}dx^2 + t^{2p_2}A^{2q_2}dy^2 + t^{2p_3}A^{2q_3}dz^2, \qquad (9.15)$$

where

$$A^{2-\gamma} = \alpha + m^2 t^{2-\gamma},$$

Table 9.4. *Evolving non–tilted perfect fluid solutions with equation of state*
$$p = (\gamma - 1)\mu.$$

Class	Restriction	Format	Parameters
Bianchi I	$0 \leq \gamma < 2$	explicit	2
Bianchi II	$\frac{2}{3} < \gamma < 2$	explicit	1
Bianchi III $(n_\alpha{}^\alpha = 0)$	$\gamma = 1$	explicit	2
	$\gamma = \frac{4}{3}$	explicit	2
Bianchi V	$\gamma = \frac{4}{3}$	explicit	2
	$0 \leq \gamma \leq 2$	quadrature	2
Bianchi VI$_h$ $(n_\alpha{}^\alpha = 0)$	$k = \frac{1}{4}(3\gamma - 2)$	explicit	2
	$k = \frac{1}{8}(3\gamma + 2)$	quadrature	2
	$k = \frac{1}{2}(4 - 3\gamma)$	quadrature	2
	$k = \frac{1}{2}(3\gamma - 4)$	quadrature	2
	$k = \frac{4}{5}, \gamma = \frac{6}{5}$	explicit	2

the constants p_1, p_2 and p_3 satisfy the Kasner constraints (9.1) and

$$q_\alpha = \tfrac{2}{3} - p_\alpha.$$

The matter density and pressure are

$$\mu = \frac{4m^2}{3t^\gamma A^\gamma}, \qquad p = (\gamma - 1)\mu \qquad 0 \leq \gamma < 2.$$

If $m = 0$, we obtain the Kasner vacuum solutions (note that we can rescale the coordinates so that $\alpha = 1$). If $m \neq 0$, we can rescale the coordinates so that $m = 1$. The solutions thus depend on two arbitrary parameters, namely α and the angle ψ which labels the Kasner indices p_α in (9.2). We can, without loss of generality,† assume $\alpha \geq 0$. Note that $\alpha = 0$ gives the flat FL model, but with t not representing clock time.

As $t \to 0^+$, the line–element assumes the Kasner form with exponents p_α, and as $t \to +\infty$ the line–element approaches the flat FL form, in agreement with the evolution shown in Figure 6.4.

---

† If $\alpha < 0$, one can rewrite the line–element with A as the time coordinate, using the fact that $A^{\gamma-1} dt = t^{\gamma-1} dA$ when $m = 1$.

9.3.2 Bianchi II

This solution is a simplified form of a solution first given by Collins (1971, example 1a). The line–element is

$$ds^2 = -A^{2(\gamma-1)}dt^2 + t^{2p_1}A^{2p_1}(dx + bzdy)^2 + t^{2p_2}A^{2p_3}dy^2 + t^{2p_3}A^{2p_2}dz^2, \quad (9.16)$$

where

$$A^{2-\gamma} = \alpha + m^2 t^{2-\gamma},$$

$$b^2 = \tfrac{1}{4}(3\gamma - 2)(2 - \gamma)m^2,$$

and the p_α are given by (9.2) with

$$\cos\psi = \tfrac{1}{8}(3\gamma - 2). \quad (9.17)$$

The p_α thus satisfy the Kasner conditions (9.1). The density and pressure are given by

$$\mu = \frac{(6 - \gamma)m^2}{4t^\gamma A^\gamma}, \qquad p = (\gamma - 1)\mu, \qquad \tfrac{2}{3} < \gamma < 2.$$

If $m = 0$, we obtain the Kasner vacuum solutions. If $m \neq 0$, we can set $m = 1$. The solution thus depends on one arbitrary parameter α. We can, without loss of generality, assume $\alpha \geq 0$ (see Section 9.3.1). If $\alpha = 0$, we obtain the Collins–Stewart type II self–similar solution (see Table 9.2), but with t not representing clock time.

As $t \to 0^+$, the line–element assumes a Kasner–like form, with exponents (9.17). As $t \to +\infty$, it follows that $A \approx t$ and the line–element is approximated by the self–similar line–element with $\alpha = 0$. Thus, in the Bianchi II state space $B_1^+(\mathrm{II})$, this solution is represented by an orbit which joins the Kasner equilibrium point with exponents (9.17) to the Collins–Stewart type equilibrium point $P_1^+(\mathrm{II})$. This orbit is shown in Figure 6.7, and lies in the plane $\Sigma_+ = \tfrac{1}{8}(3\gamma - 2)$.

9.3.3 Bianchi III

The general solution with $n_\alpha{}^\alpha = 0$ (and hence LRS) was found in explicit form for the cases $\gamma = 1$ and $\gamma = \tfrac{4}{3}$ by Kantowski (1966) [See Kantowski & Sachs 1966 for the case $\gamma = 1$.] In the case $\gamma = 1$, one exceptional solution, namely (1)(iii) below, was not given by the above authors, but was found by Evans (1978). These classes of solutions correspond to the orbits in the two–dimensional invariant set $S(\mathrm{VI}_h)$ with $h = -1$ and $\gamma = 1$ or $\tfrac{4}{3}$. They are thus contained as special cases of the Bianchi VI_h solutions in Section 9.3.5, but are given here separately in a simpler form.

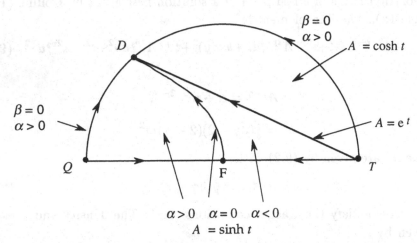

$\beta = 0$
$\alpha > 0$
$A = \cosh t$

D

$\beta = 0$
$\alpha > 0$

$A = e^t$

Q

F

T

$\alpha > 0 \quad \alpha = 0 \quad \alpha < 0$
$A = \sinh t$

Fig. 9.1. LRS Bianchi III ($n_\alpha^{\ \alpha} = 0$) solutions with $\gamma = 1$.

(1) $\gamma = 1$

The line–element is

$$\lambda^2 ds^2 = A^4(-dt^2 + dx^2 + e^{4x}dy^2) + A^{-2}B^2 dz^2, \qquad (9.18)$$

where A and B are given in three subcases by

(i) $A = \cosh t, \quad B = \alpha \sinh t + \beta^2(t \sinh t - \cosh t),$
(ii) $A = \sinh t, \quad B = \alpha \cosh t + \beta^2(t \cosh t - \sinh t),$
(iii) $A = e^t, \quad B = e^t(\alpha + \beta^2 t),$

and λ, α, β are constants. The matter density and pressure are

$$\mu = \frac{4\lambda^2\beta^2}{A^3 B}, \quad p = 0.$$

If $\beta = 0$, we obtain the vacuum limits. Otherwise, β^2 can be made equal to 1 by rescaling z.

　　The qualitative behaviour of the solutions can be inferred from the skeleton of the orbits in the invariant set $S(VI_h)$, with $h = -1$, shown in Figure 9.1. All solutions are future asymptotic to the Bianchi III form of flat space–time, represented by the equilibrium point D. If $A \to$ finite $\neq 0$ and $B \to 0$ at the initial singularity, then the solutions are past asymptotic to T (*pancake* singularity). This occurs for all solutions in cases (i) and (iii), and, if $\alpha < 0$, in case (ii). If $A \to 0$ and $B \to$ finite $\neq 0$, the solutions are past asymptotic to Q (*cigar* singularity). This occurs in case (ii) if $\alpha > 0$. A special one–parameter family of solutions are past asymptotic to F (*isotropic singularity*). This occurs in case (ii) if $\alpha = 0$, since then $A \sim t$ and $B \sim t^3$ as $t \to 0^+$.

(2) $\gamma = \frac{4}{3}$

The line–element is

$$ds^2 = -B^2 dt^2 + B^4(dx^2 + e^{4\beta x} dy^2) + t^2 B^{-2} dz^2, \tag{9.19}$$

where

$$B^2 = \alpha + 4\beta^2 t^2 + m^2 t^{2/3},$$

and α, β and m are constants. The matter density and pressure are

$$\mu = \frac{4m^2}{3t^{\frac{4}{3}} B^4}, \qquad p = \frac{1}{3}\mu.$$

If $m = 0$, we obtain the vacuum limit. Otherwise m^2 can be made equal to 1 by rescaling the coordinates.

The interpretation of the asymptotic behaviour is similar to that in Figure 9.1, and is shown in Figure 9.2, subject to the specialization $\tilde{h} = -1$, so that $K_-(\tilde{h})$ is Q, $K_+(\tilde{h})$ is T, and $D(\tilde{h})$ is D.

9.3.4 Bianchi V

The general Bianchi V solution for dust was given up to a quadrature by Heckmann & Schücking (1962), and generalized to arbitrary γ by Ellis & MacCallum (1969). The general solution for radiation $\left(\gamma = \frac{4}{3}\right)$ was given in explicit form by Ruban (1977). These solutions correspond to orbits in the two–dimensional invariant set $B(V)$ (see Figure 7.3(c)).

(1) $\gamma = \frac{4}{3}$

The line–element is

$$\lambda^2 ds^2 = AB\left[-dt^2 + dx^2 + e^{2x}\left\{\left(\tfrac{A}{B}\right)^{\sqrt{3}} dy^2 + \left(\tfrac{A}{B}\right)^{-\sqrt{3}} dz^2\right\}\right], \tag{9.20}$$

where

$$A = \sinh t, \qquad B = \alpha \cosh t + m^2 \sinh t,$$

and $\lambda > 0$, m are constants. The density and pressure are

$$\mu = \frac{3\lambda^2 m^2}{A^2 B^2}, \qquad p = \frac{1}{3}\mu.$$

If $m = 0$, we obtain the Joseph vacuum solution (see Section 9.2.4). Otherwise, we can rescale the coordinates to set $m = 1$. Without loss of generality, we can assume $\alpha \geq 0$. If $\alpha = 0$ we obtain the open FL radiation model.

As $t \to 0^+$, the line–element approaches a Kasner–like form with $\cos \psi = 0$ in (9.2). As $t \to +\infty$, $A/B \sim$ constant, and the line–element approaches an isotropic form, corresponding to the Milne model. This asymptotic behaviour can also be inferred from the orbits in the two–dimensional invariant set $B(V)$ with $\gamma = \frac{4}{3}$, which are past asymptotic to the Kasner equilibrium point with $\cos \psi = 0$, and future asymptotic to the Milne equilibrium point M (see Figure 7.3(c)).

(2) $0 \leq \gamma \leq 2$

The line–element is

$$ds^2 = -N^2 dt^2 + t^2 \left[dx^2 + e^{2rx} \left(e^{2g(t)} dy^2 + e^{-2g(t)} dz^2 \right) \right], \qquad (9.21)$$

where

$$N^2 = \left[m^2 t^{-3\gamma+2} + 3s^2 t^{-4} + r^2 \right]^{-1}, \quad g'(t) = 3st^{-3} N,$$

and m, s and r are constants. The density and pressure are

$$\mu = \frac{3m}{t^{3\gamma}}, \qquad p = (\gamma - 1)\mu, \qquad 0 \leq \gamma \leq 2.$$

This family is the general Bianchi V solution. Of the three constants, two are essential since by rescaling the coordinates one can set any one of the constants, if non–zero, equal to 1. By setting the constants to zero one also obtains as special cases the solutions which correspond to the orbits in the boundary of the Bianchi V state space as shown in Figure 7.3(c):

$r = 0$ gives the Bianchi I solutions with $\psi = \frac{\pi}{2}$.
$m = 0$ gives an alternate form of the Joseph vacuum solution
$s = 0$ gives the open FL solutions.

It follows from Figure 7.3(c) that the solutions with $r > 0$ and $s > 0$ are past asymptotic to the Kasner solution with $\cos \psi = 0$ in (9.2) and are future asymptotic to the Milne form of flat space–time.

9.3.5 Bianchi VI$_h$

Five general classes of solutions with $n_\alpha{}^\alpha = 0$ have been found, as listed in Table 9.4. These solutions correspond to orbits in the two–dimensional invariant set $S(\mathrm{VI}_h)$, for various values of h and γ (see Table 7.2 and Section 7.1.3).

The line–element for each class has the form

$$\lambda^2 ds^2 = -A^{2a_0} B^{2b_0} dt^2 + A^{2a_1} B^{2b_1} dx^2 + A^{2a_2} B^{2b_2} e^{2c_2 x} dy^2 + A^{2a_3} B^{2b_3} e^{2c_3 x} dz^2,$$
(9.22)

involving two functions of time, A and B, a constant conformal factor λ, two sets of constant exponents a_i and b_i, $i = 0, 1, 2, 3$, which depend only on a constant k which is related to the group parameter according to

$$k = \frac{1}{\sqrt{1 - 3h}},$$
(9.23)

and two constants c_2, c_3 which determine the group type. The constants a_α and b_α, $\alpha = 1, 2, 3$, are related to the Kasner exponents for the Kasner equilibrium points which lie in the invariant set $S(\mathrm{VI}_h)$ (see (6.17) and (7.25)),

$$
\begin{aligned}
K_+(\tilde{h}) &: \quad p_1 = \tfrac{1}{3}(1 - 2k), \quad p_{2,3} = \tfrac{1}{3}\left(1 + k \mp \sqrt{3}\sqrt{1 - k^2}\right), \\
K_-(\tilde{h}) &: \quad q_1 = \tfrac{1}{3}(1 + 2k), \quad q_{2,3} = \tfrac{1}{3}\left(1 - k \pm \sqrt{3}\sqrt{1 - k^2}\right).
\end{aligned}
$$
(9.24)

These constants satisfy

$$p_\alpha + q_\alpha = \tfrac{2}{3}.$$

(1) $k = \tfrac{1}{4}(3\gamma - 2)$, with $\tfrac{2}{3} < \gamma < 2$.

The functions A and B are

$$A = t, \quad B = \alpha + m^2 t + \beta^2 \left(\frac{1 - k}{1 + k}\right) t^{(1+k)/(1-k)},$$

where α, β and m are arbitrary constants. The other constants are

$$a_0 = b_0 = \frac{4k - 1}{4(1 - k)}, \quad a_\alpha = \frac{3p_\alpha}{4(1 - k)}, \quad b_\alpha = \frac{3q_\alpha}{4(1 - k)},$$
(9.25)

$$c_{2,3} = \frac{\beta}{2(1 - k)}\left(\sqrt{1 - k^2} \pm \sqrt{3}k\right).$$

The density and pressure are

$$\mu = \frac{3\lambda^2 m^2}{4(1 - k)^2}(AB)^{-2b_1}, \quad p = (\gamma - 1)\mu.$$

For a given γ, the solution is determined by four arbitrary constants $\lambda > 0$, α, β and m, only two of which are essential. If $\beta = 0$, we obtain the Bianchi I perfect fluid solutions of Section 9.3.1, in terms of a different time coordinate.

In particular, the flat FL model is given by $\alpha = 0 = \beta$. If $m = 0$, we obtain the Bianchi VI_h vacuum solutions of Section 9.2.4 in terms of a different time coordinate. If $\beta \neq 0$ and $m \neq 0$, we can set $\beta = 1$, $m = 1$ by rescaling the coordinates. We note that the choice $k = \frac{1}{2}$ (i.e. $\gamma = \frac{4}{3}$) gives the Bianchi III radiation solution in Section 9.3.3.

Three subclasses of solutions $\alpha > 0$, $\alpha = 0$ and $\alpha < 0$ occur. The relation of these subclasses to the orbits in the invariant set $S(VI_h)$ is shown in Figure 9.2. If $\alpha > 0$ the singularity occurs when $A = 0$ (i.e. $t = 0$) and is Kasner–like with exponents p_α, and if $\alpha < 0$ the singularity occurs when $B = 0$ (at some time $t = t_s > 0$) and is Kasner–like with exponents q_α. The special case $\alpha = 0$ gives a one–parameter family of solutions with an isotropic singularity (the orbit is past asymptotic to the flat FL equilibrium point).

(2) $k = \frac{1}{8}(3\gamma + 2)$, $0 < \gamma < 2$

There are three different forms for the functions A and B:

$$(i) \quad A = \cosh t, \quad B = [\alpha + \beta^2 \int (\cosh^r \bar{t} / \sinh^2 \bar{t}) d\bar{t}\,] \sinh t,$$

$$(ii) \quad A = \sinh t, \quad B = [\alpha + \beta^2 \int (\sinh^r \bar{t} / \cosh^2 \bar{t}) d\bar{t}\,] \cosh t,$$

$$(iii) \quad A = e^t, \quad B = \begin{cases} e^t \left[\alpha + \frac{\beta^2}{r-2} e^{(r-2)t}\right], & r \neq 2, \\ \\ e^t(\alpha + \beta^2 t), & r = 2, \end{cases} \tag{9.26}$$

where α and β are arbitrary constants and

$$r = \frac{2k}{1 - k}. \tag{9.27}$$

The other constants are given by (9.25). The density and pressure are

$$\mu = \frac{3\lambda^2}{4(1 - k)^2}(AB)^{-2b_0}, \quad p = (\gamma - 1)\mu.$$

For a given γ, the solution is determined by three arbitrary constants $\lambda > 0$, α and β, only two of which are essential. If $\beta = 0$, we obtain the Bianchi I perfect fluid solutions of Section 9.3.1 with a different time coordinate, with the flat FL model corresponding to choice (iii) for A and B. Otherwise we can set $\beta = 1$ by rescaling the constants and x, y and z. Various subclasses of solutions arise depending on the choice (i), (ii) or (iii) and the sign of α. The relation of these subclasses to the orbits in the invariant set $S(VI_h)$ for $\frac{2}{3} < \gamma < 2$ (and hence $\frac{1}{2} < k < 1$ and $r > 2$), is shown in Figure 9.3. We

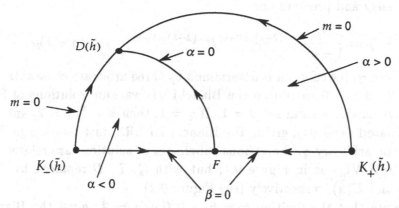

Fig. 9.2. Bianchi VI_h ($n_\alpha{}^\alpha = 0$) solutions with $k = \frac{1}{4}(3\gamma - 2)$.

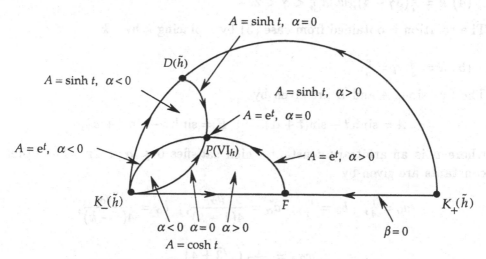

Fig. 9.3. Bianchi VI_h ($n_\alpha{}^\alpha = 0$) solutions with $k = \frac{1}{8}(3\gamma + 2)$.

note that the solutions with $A = e^t$ and $\alpha > 0$ have an isotropic singularity.

(3) $k = \frac{1}{2}(4 - 3\gamma)$, with $\frac{2}{3} < \gamma < \frac{4}{3}$.
The functions A and B are given by (9.26) and (9.27). The constants in the line–element (9.22) are given by

$$a_0 = a_1, \quad b_0 = b_1, \quad a_\alpha = \frac{3q_\alpha}{2(1-k)}, \quad b_\alpha = \frac{3p_\alpha}{2(1+k)},$$

$$c_{2,3} = \frac{1}{\sqrt{1-k^2}} \left(\sqrt{1-k^2} \pm \sqrt{3}k \right).$$

The density and pressure are

$$\mu = \frac{3\lambda^2\beta^2}{1-k^2} A^{-(2-k)/(1-k)} B^{-(2-k)/(1+k)}, \quad p = (\gamma-1)\mu.$$

For a given γ, the solution is determined by three arbitrary constants, $\lambda > 0$, α and β. If $\beta = 0$ we obtain the Bianchi VI_h vacuum solutions of Section 9.2.4. Otherwise we can set $\beta = 1$. If $\gamma = 1$, then $k = \frac{1}{2}$, $r = 2$, and B can be evaluated explicitly, giving the Bianchi III LRS dust solution in Section 9.3.3. For arbitrary γ, the various subclasses of solutions are related to the orbits in $S(VI_h)$ as in Figure 9.1, but with Q, T, D replaced by $K_-(\tilde{h})$, $K_+(\tilde{h})$ and $D(\tilde{h})$, respectively (see Figure 9.2).

We note that the limiting case $k = 0$ (i.e. $\gamma = \frac{4}{3}$) gives the Bianchi V radiation solution in Section 9.3.4.

(4) $k = \frac{1}{2}(3\gamma - 4)$, *with* $\frac{4}{3} < \gamma < 2$.

The solution is obtained from case (3) by replacing k by $-k$.

(5) $k = \frac{4}{5}$, $\gamma = \frac{6}{5}$.

The functions A and B are given by

$$A = \sinh t + \sin(t + \alpha), \qquad B = \sinh t - \sin(t + \alpha),$$

where α is an arbitrary constant which satisfies $0 \leq \alpha < 2\pi$. The other constants are given by

$$a_0 = \frac{1}{4}, \quad b_0 = \frac{11}{4}, \quad a_\alpha = \frac{p_\alpha}{4(1-k)}, \quad b_\alpha = \frac{3q_\alpha}{4(1-k)},$$

$$c_{2,3} = \frac{1}{2\sqrt{2}} \left(\sqrt{3} \pm 4 \right).$$

The density and pressure are

$$\mu = \frac{25\lambda^2}{8} A^{-\frac{3}{2}} B^{-\frac{9}{2}}, \qquad p = \frac{1}{5}\mu.$$

This class of solutions depends on two essential parameters $\lambda > 0$ and α and corresponds to the orbits in Figure 7.3(a), with $h = -\frac{3}{16}$ and $\gamma = \frac{6}{5}$. The value $\alpha = \pi$ gives the orbit which joins F and $P(VI_h)$. The trigonometric time dependence corresponds to the fact that $P(VI_h)$ is a spiral local sink for these values of h and γ.

We conclude this section by giving the origins of these solutions. The solution in case (1) was discovered by Collins (1971, example 3(a)(ii)), and given in a simplified form in Wainwright (1984, page 669). In the form given above the parameters and coordinates have been redefined to give a unified formulation of all the Bianchi VI_h solutions. Of the remaining solutions, case (5) was discovered by Uggla (1990) and cases (2), (3) and (4) by Uggla & Rosquist (1990). We note that the simple structure of the solutions in this section, as exemplified by the form of the line–element (9.22), can be explained using the so–called Killing tensor symmetries (see Uggla *et al.* 1995). The existence of invariant algebraic curves in the dimensionless state space also plays an important role in determining whether exact solutions can be found (Hewitt 1991c).

9.4 The space of initial data

In general, counting functions to give a measure of the space of initial data is unreliable. Even when all the functions are analytic, care must be taken to compare sets of functions at the same level of differentiation (Siklos 1996). For spatially homogeneous metrics, there is no problem of this sort, since the initial data set consists only of (constant) parameters. The counting is straightforward. The initial data are the invariant triad components of $g_{\alpha\beta}$ and $\dot{g}_{\alpha\beta}$, plus the structure constants† $\hat{n}^{\alpha\beta}$ and $\hat{a}_\beta$, and, for perfect fluid with given equation of state, the energy density and components of the (unit) velocity vector. These 25 parameters determine the evolution. The triad freedom on the initial surface is a general linear transformation (nine parameters); the Jacobi identities ($\hat{n}^{\alpha\beta}\hat{a}_\beta = 0$) supply three constraints and the Einstein $T_{0\alpha}$ and T_{00} equations supply a further four constraints. There is also the freedom to choose the initial hypersurface (one parameter, corresponding to $t_0 \to t_0 + T$), all of which leaves, in general, just eight free parameters. The state space for spatially homogeneous perfect fluid solutions of the EFE is therefore eight dimensional.

In order to calculate the dimension of the space of initial data for a given Bianchi type we have to investigate the structure constants and the Einstein constraint equations individually. We need to know how many parameters remain in the structure constants (for example, none in the case of Bianchi I) and whether the constraint equations become degenerate.

The space occupied by the parameters $\hat{n}^{\alpha\beta}$ and $\hat{a}_\beta = 0$ is nine–dimensional, though we are only interested in the six–dimensional subspace defined by

---

† See Section 1.5.1 for definitions and notation.

the Jacobi identities. For most Bianchi types, the structure constants comprise six free parameters including, for class B, the group parameter h. The simple way to find the number of free parameters for each group type is to count them explicitly. Thus types IX and VIII have six, which is the number of independent elements of the rank 3 symmetric matrix $\hat{n}^{\alpha\beta}$. Types VII$_h$ and VI$_h$ also have six. In these cases, the matrix $\hat{n}^{\alpha\beta}$ is rank 2 and hence is defined by two orthogonal vectors (five components), and the vector $\hat{a}_\beta$, which must be orthogonal to these (by the Jacobi identities), supplies one more parameter. When h is given a fixed value, there remain just five free parameters; this includes the cases $h = 0$, and $h = \infty$ which is Bianchi IV. For Bianchi V, only $\hat{a}_\beta$ is non–zero so there are only three free parameters. For Bianchi II, $\hat{a}_\beta = 0$ and $\hat{n}^{\alpha\beta}$ has rank 1 so there are again three free parameters.

The constraint equations are written out in Chapter 1 (equations (1.92) and (1.93)). The $(0,0)$ constraint (1.92) does not depend on the structure constants and therefore cannot be trivially satisfied for any group type; it always provides one constraint.

In the case of a tilted perfect fluid, none of the constraint equations becomes degenerate when the Bianchi type is specialized, because the equations can just be regarded as defining the energy density μ and heat flux q_α (see (1.22)) in terms of other parameters. For a tilted perfect fluid, therefore, the only variation in the dimension of the state space for different Bianchi types is due to the reduction of parameters in the structure constants described above.

In vacuum or for a non–tilted fluid, the $(0,\alpha)$ constraints (1.93) are linear in the structure constants and are satisfied identically for some Bianchi types. In Bianchi I all the structure constants are zero, so that all three constraints are trivially satisfied. For types II and VI$_h$ with $h = -1/9$ two of the three constraints are linearly independent, reducing the number of constraints by one. The easiest way of showing this is to write out the constraints in components, as in the appendix of Ellis & MacCallum (1969). Type VI$_{-1/9}$ is peculiar, because for a non–tilted perfect fluid, the class of solutions is a whole parameter larger than the class of solutions for any other value of the invariant parameter h. The additional parameter is due to the presence of an extra shear mode and when the shear mode is present, the solution is said to be of type VI$^*_{-1/9}$ (see Section 1.5.4). Note that the shear modes are only 'extra' when the left sides of the constraint equations orthogonal to $\hat{a}_\beta$ vanish.

Table 9.5 gives the dimension of the space of initial data for each Bianchi type. These dimensions equal the number of arbitrary parameters in the

Table 9.5. *The number of arbitrary constants in the general solution for each Bianchi type. The second column is for vacuum (add one parameter for non–tilted perfect fluid with given equation of state). The third column is for tilted perfect fluid with given equation of state. The group parameter h is regarded as fixed.*

Group types	Vacuum	Tilted fluid
I	1	–
II	2	5
VI_0, VII_0	3	7
VIII, IX	4	8
IV	3	7
V	1	5
VI_h ($h \neq -1/9$)	3	7
$VI_{-\frac{1}{9}}$	4	7
VII_h	3	7

general solution for each type, and also equal the dimensions of the Bianchi state spaces for the expansion–normalized variables used in Chapters 6 and 7.

10

Hamiltonian cosmology

C. UGGLA

University of Stockholm

Not all spatially homogeneous models admit a Hamiltonian formulation. However, many do and for these models the Hamiltonian formulation provides a useful tool that complements and sheds light on the use of dynamical systems methods. The advantage of the Hamiltonian approach is that all the dynamical content is encoded in a single function. As discussed in Section 5.1, this often enables one to make conjectures about the evolution of the models without any detailed calculations, by using potential diagrams. The Hamiltonian also enables one to find monotone functions, which are not easily discovered using the detailed evolution equations.

10.1 The Hamiltonian formulation of Einstein's equations

10.1.1 The variational principle

It is well known that one can derive the vacuum Einstein field equations from the variational principle

$$I = \int L_G \, d^4 x = \int \sqrt{-g} R \, d^4 x, \qquad (10.1)$$

where $g = \det g_{ij}$ and R is the scalar curvature (e.g. Wald 1984, pages 453–4). The variation, δg_{ij}, is taken over a four–dimensional region with a three–dimensional boundary which contributes a term that is removed by setting $\delta g_{ij} = 0$ on the boundary. One obtains

$$\delta I = \int \frac{\delta L_G}{\delta g_{ij}} \, \delta g_{ij} \, d^4 x = \int \sqrt{-g} \, G^{ij} \delta g_{ij} \, d^4 x = 0, \qquad (10.2)$$

yielding the vacuum EFE equations.

If one tries to apply the above variational principle to the Bianchi models, by expressing the metric in a group–invariant frame while integrating over the group orbits, one encounters a problem. If one restricts oneself to spatially homogeneous variations that satisfy $\delta g_{ij} = 0$ on the boundary, then $\delta g_{ij} = 0$ on every group orbit and thus it follows that $\delta g_{ij} = 0$ everywhere (MacCallum 1979, pages 552–3). It turns out, however, that for spatially homogeneous variations the boundary term is identically zero for Bianchi models of class A and some models of class B, even if $\delta g_{ij} \neq 0$ on the boundary. Hence the above variational principle yields the correct vacuum EFE only for these models.

For many source terms it is possible to generalize the action I in (10.1), to one of the form

$$I = \int [\sqrt{-g}R + L_{\text{source}}(g_{ij}, \phi^A, \partial_i \phi^A)] d^4x , \qquad (10.3)$$

where ϕ^A are the degrees of freedom associated with the source. Variation with respect to g_{ij} gives

$$\delta I = \int \sqrt{-g}(G^{ij} - T^{ij}) \delta g_{ij} d^4x = 0 ,$$

where T^{ij} is the stress–energy tensor. Variation with respect to ϕ^A leads to the source equations. Note that there are no derivatives of g_{ij} in the source Lagrangian. This means that the source does not contribute to the boundary term when variations are taken with respect to g_{ij}. Hence potential complications only arise from the vacuum part, when going over to the spatially homogenous case.

10.1.2 The gravitational Hamiltonian

We consider Bianchi models of class A, and as in Section 1.5.2 we introduce a frame $\{\mathbf{n}, \mathbf{E}_\alpha\}$, where the $\mathbf{E}_\alpha$ are time–independent, group–invariant vector fields, and $\mathbf{n}$ is the unit normal to the group orbits. The line–element has the form (1.89), i.e.

$$ds^2 = -N(\tilde{t})^2 d\tilde{t}^2 + g_{\alpha\beta}(\tilde{t}) \mathbf{W}^\alpha \mathbf{W}^\beta, \qquad (10.4)$$

where the 1–forms $\mathbf{W}^\alpha$ are dual to the vector fields $\mathbf{E}_\alpha$. For class A models, the non–zero structure constants are given by

$$C^1{}_{23} = \hat{n}_1, \quad C^2{}_{31} = \hat{n}_2, \quad C^3{}_{12} = \hat{n}_3, \qquad (10.5)$$

where the $\hat{n}_\alpha$ equal 0, +1 and −1, as in Table 1.1. For vacuum, and for various sources, the spatial metric in (10.4) is diagonal. We restrict our considerations to this case, and write

$$g_{\alpha\beta} = \text{diag}\,(e^{2\beta_1}, e^{2\beta_2}, e^{2\beta_3}). \qquad (10.6)$$

It is useful to make the following transformation of the metric variables† (Misner 1969b, page 1323):

$$\beta_1 = \beta^0 - 2\beta^+, \quad \beta_2 = \beta^0 + \beta^+ + \sqrt{3}\beta^-, \quad \beta_3 = \beta^0 + \beta^+ - \sqrt{3}\beta^-. \quad (10.7)$$

Expressed in the β^A variables ($A = 0, +, -$), the variational principle (10.1) leads to the following gravitational Lagrangian (e.g. Jantzen 1987):

$$L_G = T_G - U_G,$$

where

$$\begin{aligned} T_G &= 6N^{-1}e^{3\beta^0}[-(\dot{\beta}^0)^2 + (\dot{\beta}^+)^2 + (\dot{\beta}^-)^2], \\ U_G &= Ne^{\beta^0}V^*(\beta^\pm), \end{aligned} \quad (10.8)$$

and

$$V^*(\beta^\pm) = \tfrac{1}{2}e^{4\beta^+}\,h_-{}^2 + \hat{n}_1 e^{-2\beta^+}h_+ + \tfrac{1}{2}(\hat{n}_1)^2 e^{-8\beta^+}, \quad (10.9)$$

with

$$h_\pm = \hat{n}_2 e^{2\sqrt{3}\beta^-} \pm \hat{n}_3 e^{-2\sqrt{3}\beta^-}.$$

In (10.8) an overdot denotes $d/d\tilde{t}$. We note that the potential V^* is related to the Ricci scalar ^3R of the spatial metric according to

$$V^*(\beta^\pm) = -e^{-2\beta^0}\,{}^3R. \quad (10.10)$$

The advantage of using the β^A variables is that the kinetic part of the Lagrangian becomes 'conformally flat'. In addition, β^0 is related to the length scale function for the metric (10.4) according to

$$\ell = \ell_0 e^{\beta^0}, \quad (10.11)$$

where ℓ_0 is the value of ℓ at some arbitrary reference time, and the Hubble scalar H is

$$H = N^{-1}\dot{\beta}^0, \quad (10.12)$$

(see (5.6)).

It follows from the variational principle that the Lagrangian equations with respect to β^A are equivalent to the $\alpha\alpha$–field equations (the $\alpha\beta$–equations, with $\alpha \neq \beta$, are identically satisfied). Since $g_{00} = -N^2$ and L_G does not depend on the derivatives of N, it follows from (10.2) that

$$\frac{\delta L_G}{\delta N} = -2N\frac{\delta L_G}{\delta g_{00}} = -2N\sqrt{-g}\,G^{00} = 2\sqrt{^3g}\,n^i n^j G_{ij} = 0, \quad (10.13)$$

---

† In Misner's articles, $-\beta^0$ is denoted by Ω, which is used there as a time variable. See the Appendix to Chapter 11 for the relations between different time variables.

where $^3g = \det(g_{\alpha\beta})$, and n^i is the unit normal to the group orbits. The explicit form (10.8) implies that

$$\frac{\delta L_G}{\delta N} = \frac{\partial L_G}{\partial N} = -N^{-1}(T_G + U_G). \qquad (10.14)$$

The transition to the Hamiltonian formulation is accomplished by a Legendre transformation (e.g. Gelfand & Fomin 1963, Chapter 4). Equation (10.8) leads to

$$\mathcal{H}_G = T_G + U_G , \qquad T_G = \tfrac{1}{24}N e^{-3\beta^0}[-(p_0)^2 + (p_+)^2 + (p_-)^2] , \qquad (10.15)$$

where

$$p_0 = -12e^{3\beta^0} N^{-1}\dot{\beta}^0 , \qquad p_\pm = 12e^{3\beta^0} N^{-1}\dot{\beta}^\pm . \qquad (10.16)$$

It follows from (10.13)–(10.15), (10.6) and (10.7) that

$$\mathcal{H}_G = -2N e^{3\beta^0} n^i n^j G_{ij} = 0 , \qquad (10.17)$$

for vacuum models.

There are a number of sources with diagonal stress–energy tensor that give rise to Lagrangians of the type

$$L = L_G + L_{\text{source}}, \qquad (10.18)$$

in the diagonal class A Bianchi case. This Lagrangian leads to a Hamiltonian that, in analogy with the vacuum case, has the form

$$\mathcal{H} = \mathcal{H}_G + \mathcal{H}_{\text{source}} = 0, \qquad (10.19)$$

where $\mathcal{H}_G$ is given by (10.15), and

$$\mathcal{H}_{\text{source}} = 2N e^{3\beta^0} n^i n^j T_{ij}. \qquad (10.20)$$

In the Bianchi case one can often solve the equations of motion for the source variables, and hence express them in terms of metric variables and constants of the motion. As an example, consider a non–tilted perfect fluid with stress–energy tensor (1.20) and equation of state $p = (\gamma - 1)\mu$, where γ is a constant. The contracted Bianchi identity (1.50), with (5.6), can be integrated to give $\mu = \mu_0 \ell^{-3\gamma}$, or equivalently by (10.11),

$$\mu = \mu_0 e^{-3\gamma\beta^0} . \qquad (10.21)$$

Since the fluid is non–tilted, we have $n^i = u^i$, implying that $\mathcal{H}_{\text{source}}$, as given by (10.20), depends only on the gravitational field variables β^A, N. One can thus think of $\mathcal{H}_{\text{source}}$ as an additional potential in the Hamiltonian, given by

$$\mathcal{H}_{\text{source}} = U_{\text{fluid}} = 2N\mu_0 e^{-3(\gamma-1)\beta^0} , \qquad (10.22)$$

as follows from (1.20) and (10.21).

A cosmological constant and a magnetic field can be treated in a similar manner to the perfect fluid leaving only a potential contribution to the total Hamiltonian, which takes the form

$$\mathcal{H} = T_G + U_G + U_{\text{source}} = 0 \, , \tag{10.23}$$

where U_{source}, given by (10.20), can be expressed in terms of the gravitational variables.

A cosmological constant in the EFE will be considered as an additional term $-\Lambda g_{ab}$ in the total stress–energy tensor T_{ab}. Inserting this term in equation (10.20) gives rise to the potential

$$U_\Lambda = 2Ne^{3\beta^0}\Lambda \, . \tag{10.24}$$

The simplest magnetic field is aligned with one of the axes, leading to a diagonal stress–energy tensor. If we choose the first axis to be the preferred axis, we obtain

$$U_{\text{mag}} = 2Nb^2 e^{-\beta^0 - 4\beta^+} \, , \tag{10.25}$$

where b is a constant (e.g. Uggla *et al.* 1995).

For a scalar field ϕ, with stress–energy tensor (1.24), one cannot, in general, integrate the scalar field equation and reduce the problem to one only involving gravitational variables. However, scalar fields do have a variational formulation (10.3), which leads to a Hamiltonian (10.19) with the source contribution given by (10.20). Thus, in contrast with the above sources, a scalar field adds another dependent variable $\beta^\dagger = \sqrt{6}\phi$ and contributes both to the kinetic and the potential part of the Hamiltonian (Uggla *et al.* 1995):

$$T_{\text{sc}} = 6N^{-1}e^{3\beta^0}(\dot{\beta}^\dagger)^2 = \tfrac{1}{24}Ne^{-3\beta^0}(p_\dagger)^2 \, , \qquad U_{\text{sc}} = 2Ne^{3\beta^0}V_{\text{sc}}(\beta^\dagger) \, , \tag{10.26}$$

where $V_{\text{sc}}(\beta^\dagger)$ is the scalar field potential.

Finally we note that if a number of non–interacting sources are present, the corresponding potentials are simply added together.

In summary, we have shown that for diagonal Bianchi class A cosmologies the EFE are equivalent to Hamilton's equations

$$\dot{\beta}^A = \frac{\partial H}{\partial p_A}, \qquad \dot{p}_A = -\frac{\partial \mathcal{H}}{\partial \beta^A}, \tag{10.27}$$

and the constraint $\mathcal{H} = 0$, where $\mathcal{H}$ is of the form (10.19) with (10.15) and (10.20).

10.2 Relationship with the orthonormal frame approach

In this section we relate the Hamiltonian variables (p_A, β^A) to the variables $(\Sigma_\pm, N_\alpha)$ that were used in Chapter 6 to study the class A models with non-tilted perfect fluid. We begin by using Hamilton's equations to derive reduced evolution equations in terms of normalized variables, which we then relate to the DE (6.9).

10.2.1 Reduced differential equations

The Hamiltonian for class A non-tilted perfect fluid models is given by (10.15), (10.19) and (10.22):

$$\mathcal{H} = \tfrac{1}{12}e^{-3\beta^0}N[\tfrac{1}{2}\{-(p_0)^2 + (p_+)^2 + (p_-)^2\} \\ + 12e^{4\beta^0}V^*(\beta^\pm) + 24\mu_0 e^{3(2-\gamma)\beta^0}] = 0 , \qquad (10.28)$$

where $V^*(\beta^\pm)$ is given by (10.9). We restrict our consideration to expanding models, which implies $p_0 < 0$ on account of (10.12) and (10.16). We shall use β^0 as the time variable, so that $\dot{\beta}^0 = 1$. By (10.16) this implies that

$$N = -12e^{3\beta^0}p_0^{-1}, \qquad (10.29)$$

or equivalently

$$N = H^{-1}, \qquad (10.30)$$

by (10.12) and (10.16). On account of (5.9) and (10.11), β^0 can be identified, up to an additive constant, with the time variable τ in Section 5.2. With this choice of gauge in (10.28), Hamilton's equations (10.27) assume the form

$$\frac{d\beta^0}{d\tau} = 1, \qquad \frac{d\beta^\pm}{d\tau} = -p_0^{-1}p_\pm,$$

$$\frac{dp_0}{d\tau} = (2-q)p_0, \qquad \frac{dp_\pm}{d\tau} = p_0 \mathcal{S}_\pm, \qquad (10.31)$$

where

$$\mathcal{S}_\pm = \frac{12e^{4\beta^0}}{p_0^2}\frac{\partial V^*(\beta^\pm)}{\partial \beta^\pm}, \qquad (10.32)$$

$$q = 2(1-K) - \tfrac{3}{2}(2-\gamma)\Omega, \qquad (10.33)$$

and

$$K = \frac{24e^{4\beta^0}V^*(\beta^\pm)}{p_0^2}, \qquad \Omega = \frac{48\mu_0 e^{3(2-\gamma)\beta^0}}{p_0^2}. \qquad (10.34)$$

It follows from (10.10), (10.21), (10.29), (10.30), (5.16) and (5.18)–(5.20) that q, K and Ω are precisely the deceleration parameter, the curvature parameter and the density parameter as introduced in Section 5.2.

We now introduce non–canonical variables $(\Sigma_\pm, \tilde{N}_\alpha)$ according to

$$\Sigma_\pm = \frac{p_\pm}{(-p_0)}, \qquad \tilde{N}_\alpha = \frac{12e^{2\beta_\alpha}}{(-p_0)}, \tag{10.35}$$

where the β_α are given in terms of β^0 and $\beta^\pm$ by (10.7). Equations (10.31) assume the form

$$\begin{aligned}
\Sigma_\pm' &= -(2-q)\Sigma_\pm - \mathcal{S}_\pm, \\
\tilde{N}_1' &= (q - 4\Sigma_+)\tilde{N}_1, \\
\tilde{N}_2' &= (q + 2\Sigma_+ + 2\sqrt{3}\Sigma_-)\tilde{N}_2, \\
\tilde{N}_3' &= (q + 2\Sigma_+ - 2\sqrt{3}\Sigma_-)\tilde{N}_3,
\end{aligned} \tag{10.36}$$

where $'$ denotes $d/d\tau$. It follows from (10.7), (10.32) and (10.34) that

$$\mathcal{S}_+ = \tfrac{1}{6}[(\hat{n}_2\tilde{N}_2 - \hat{n}_3\tilde{N}_3)^2 - \hat{n}_1\tilde{N}_1(2\hat{n}_1\tilde{N}_1 - \hat{n}_2\tilde{N}_2 - \hat{n}_3\tilde{N}_3)], \tag{10.37}$$

$$\mathcal{S}_- = \tfrac{1}{2\sqrt{3}}(\hat{n}_3\tilde{N}_3 - \hat{n}_2\tilde{N}_2)(\hat{n}_1\tilde{N}_1 - \hat{n}_2\tilde{N}_2 - \hat{n}_3\tilde{N}_3),$$

and

$$\begin{aligned}
K = \tfrac{1}{12}[&(\hat{n}_1\tilde{N}_1)^2 + (\hat{n}_2\tilde{N}_2)^2 + (\hat{n}_3\tilde{N}_3)^2 \\
&- 2\hat{n}_1\hat{n}_2\tilde{N}_1\tilde{N}_2 - 2\hat{n}_2\hat{n}_3\tilde{N}_2\tilde{N}_3 - 2\hat{n}_3\hat{n}_1\tilde{N}_3\tilde{N}_1].
\end{aligned} \tag{10.38}$$

The Hamiltonian constraint (10.28) can be written in the form

$$\Sigma^2 + K + \Omega = 1, \tag{10.39}$$

where

$$\Sigma^2 = \Sigma_+^2 + \Sigma_-^2. \tag{10.40}$$

It follows from (10.33) that

$$q = \tfrac{1}{2}(3\gamma - 2)(1 - K) + \tfrac{3}{2}(2 - \gamma)\Sigma^2. \tag{10.41}$$

10.2.2 Orthonormal frame approach

With the line–element (10.4) we associate the orthonormal frame $\{e_a\}$ defined by

$$e_0 = \frac{1}{N}\frac{\partial}{\partial \hat{t}}, \qquad e_\alpha = e^{-\beta_\alpha}E_\alpha, \tag{10.42}$$

where the E_α are the time–independent frame vectors dual to the 1–forms W^α.

First, evaluation of the commutator $[\mathbf{e}_0, \mathbf{e}_\alpha]$ using (1.11), (1.61), (1.107), (6.2) and (10.16) yields

$$H = -\tfrac{1}{12}e^{-3\beta^0}p_0, \quad \sigma_\pm = \tfrac{1}{12}e^{-3\beta^0}p_\pm. \tag{10.43}$$

Second, evaluation of $[\mathbf{e}_\alpha, \mathbf{e}_\beta]$ using (1.11), (1.62), (1.88), (1.103) and (10.5) yields

$$n_\alpha = \hat{n}_\alpha\, e^{-3\beta^0 + 2\beta_\alpha} \quad \text{(no sum)}. \tag{10.44}$$

It follows from (10.35) and (10.43) that $\Sigma_\pm = \sigma_\pm/H$, showing that $\Sigma_\pm$ as defined by (10.35) coincide with $\Sigma_\pm$ as defined by (6.8). Further, (10.35) and (10.44) imply that the variables $N_\alpha = n_\alpha/H$ (see (6.8)) are related to the $\tilde{N}_\alpha$ by

$$N_\alpha = \hat{n}_\alpha \tilde{N}_\alpha \quad \text{(no sum)}. \tag{10.45}$$

It follows that *(10.36)–(10.41) coincide with the evolution equations (6.9)–(6.14)* in Chapter 6. It is worth noting that the variables $\Sigma_\pm$ are in fact the derivatives of the metric variables $\beta^\pm$:

$$\Sigma_\pm = \frac{d\beta^\pm}{d\tau}, \tag{10.46}$$

as follows from (10.31) and (10.35).

There is one difference between (10.36) and (6.9). When a structure constant $\hat{n}_\alpha$ is zero in (10.34) the corresponding $\tilde{N}'_\alpha$ equation decouples, while in (6.9) the corresponding N'_α equation reduces to '0 = 0', since $N_\alpha = 0$. The decoupled equation is in fact redundant, since, if desired, the metric functions $\beta^\pm$ can be determined by quadrature from $\Sigma_\pm$, using (10.46).

10.2.3 Self–similar solutions in β^A–space

We now discuss the equilibrium points of the DE (10.36), given by

$$\Sigma_\pm = \text{constant}, \quad \tilde{N}_\alpha = \text{constant}, \tag{10.47}$$

from the Hamiltonian point of view. As shown in Section 5.3.3, the equilibrium points correspond to self–similar solutions. It follows from (10.46) that each equilibrium point (self–similar solution) determines a straight line in β^A–space, which we can write in parametric form as

$$\beta^\pm = \Sigma_\pm \tau + \beta_0^\pm, \quad \beta^0 = \tau, \tag{10.48}$$

where $\Sigma_\pm$ and $\beta_0^\pm$ are constants. The form of the kinetic energy term in (10.28) suggests that we introduce the Lorentzian metric

$$(\eta_{AB}) = \text{diag}\,(-1, 1, 1)$$

in β^A–space. It follows from (10.40) and (10.46) that

$$\frac{d\beta^A}{d\tau}\frac{d\beta^B}{d\tau}\,\eta_{AB} = -1 + \Sigma^2. \tag{10.49}$$

It is known that all equilibrium points of the class A perfect fluid DE (10.36) with $\gamma \neq \frac{2}{3}$ are of Bianchi types I–VII$_0$ (see Table 6.3, or Jantzen & Rosquist 1986). Since $K \geq 0$ for these Bianchi types, it follows from (10.39) that $\Sigma \leq 1$, with equality only for the Kasner solutions. Thus, by (10.49), *the Kasner solutions are described by null lines and all other self–similar solutions are described by timelike lines in β^A–space.* For the non–Kasner equilibrium points, the dependence of the p_A on $\beta^0 = \tau$ follows from (10.34) and (10.35). Since the $\tilde{N}_\alpha$ and Ω are constant and at least one is non–zero, we obtain

$$p_A \sim e^{C\beta^0}, \tag{10.50}$$

where C is a constant.

For each non–Kasner equilibrium point, we can now perform a Lorentz transformation

$$\beta^A = L^A{}_B \bar\beta^B, \quad p_A = \bar p_B (L^{-1})^B{}_A, \tag{10.51}$$

so that the $\bar\beta^0$–axis is the timelike line in question. The self–similar solution is then described by

$$\bar\beta^\pm = \bar\beta_0^\pm, \quad \bar p_\pm = 0, \quad \bar p_0 \propto e^{\bar C \bar\beta^0}, \tag{10.52}$$

where $\bar\beta_0^\pm$ and $\bar C$ are constants and $\bar\beta^0$ is the parameter.

It is now convenient to write the Hamiltonian (10.28) in the so–called Taub gauge,† determined by

$$N = 12e^{3\beta^0}. \tag{10.53}$$

We obtain

$$\mathcal{H}_{\text{taub}} = \tfrac{1}{2}(-p_0^2 + p_+^2 + p_-^2) + U_{\text{taub}} = 0, \tag{10.54}$$

where

$$U_{\text{taub}} = 12e^{4\beta^0} V^*(\beta^\pm) + 24\mu_0 e^{3(2-\gamma)\beta^0}. \tag{10.55}$$

After the Lorentz transformation, the Hamiltonian (10.54) has the form‡

$$\mathcal{H}_{\text{taub}} = \tfrac{1}{2}(-\bar p_0^2 + \bar p_+^2 + \bar p_-^2) + U_{\text{taub}} = 0, \tag{10.56}$$

where U_{taub} now depends on the variables $\bar\beta^A$.

---

† Also called the supertime gauge (Misner 1972, page 450; Jantzen 1987, page 91), but first used by Taub (1951), without the factor 12.

‡ This is an abuse of notation. More precisely, one should write $\bar U_{\text{taub}}(\bar\beta^A)$, with $\bar U_{\text{taub}}(\bar\beta^A) = U_{\text{taub}}(\bar\beta^A)$, since the potential has the same values but a different functional form.

For the solution (10.52), (10.56) implies that $\bar{p}_0^2 = 2U_{\text{taub}}$. For consistency with (10.52), the transformed potential must have the form

$$U_{\text{taub}} = e^{2\bar{C}\bar{\beta}^0} V(\bar{\beta}^\pm). \tag{10.57}$$

In addition, since $\dot{\bar{p}}_\pm = 0$ for the solution (10.52), Hamilton's equations imply $\partial U_{\text{taub}}/\partial \bar{\beta}^\pm = 0$ when $\bar{\beta}^\pm = \bar{\beta}_0^\pm$. It follows from (10.57) that

$$\frac{\partial V(\bar{\beta}_0^\pm)}{\partial \bar{\beta}^\pm} = 0. \tag{10.58}$$

Thus, *self-similar solutions arise if the Taub potential (10.55) can be transformed into the form (10.57)* (i.e. $\bar{\beta}^0$–dependence is 'exponentially factored'), *and the potential* $V(\bar{\beta}^\pm)$ *has a critical point* $\bar{\beta}_0^\pm$.

For the special Bianchi types of class A (i.e. I, II, VI$_0$ and VII$_0$), the Taub potential (10.55) can be transformed into the form (10.57) (see Rosquist *et al.* 1990). For type I (i.e. $V^*(\beta^\pm) = 0$) no change is needed. For the others, we perform a boost in the β^+–direction:

$$\beta^0 = \Gamma(\bar{\beta}^0 + v\bar{\beta}^+), \quad \beta^+ = \Gamma(v\bar{\beta}^0 + \bar{\beta}^+), \quad \beta^- = \bar{\beta}^-,$$

$$p_0 = \Gamma(\bar{p}_0 - v\bar{p}_+), \quad p_+ = \Gamma(-v\bar{p}_0 + \bar{p}_+), \quad p_- = \bar{p}_-, \tag{10.59}$$

where $\Gamma = (1 - v^2)^{-1/2}$. For type II, we choose

$$v = \tfrac{1}{8}(3\gamma - 2), \tag{10.60}$$

and obtain (10.57) with

$$\bar{C} = \tfrac{3}{2}\Gamma(2 - \gamma), \tag{10.61}$$

and

$$V(\bar{\beta}^\pm) = V_g + V_{\text{fluid}} = 6e^{-r\bar{\beta}^+} + 24\mu_0 e^{s\bar{\beta}^+},$$

$$r = \tfrac{3}{2}\Gamma(6 - \gamma), \quad s = \tfrac{3}{8}\Gamma(2 - \gamma)(3\gamma - 2). \tag{10.62}$$

For types VI$_0$ and VII$_0$ we choose

$$v = -\tfrac{1}{4}(3\gamma - 2), \tag{10.63}$$

and obtain (10.57) with $\bar{C}$ given by (10.61), and

$$V(\bar{\beta}^\pm) = V_g + V_{\text{fluid}} = 6h_-^2 e^{r\bar{\beta}^+} + 24\mu_0 e^{-s\bar{\beta}^+},$$

$$r = 3\Gamma(2 - \gamma), \quad s = \tfrac{3}{4}\Gamma(2 - \gamma)(3\gamma - 2), \tag{10.64}$$

where h_- is given by (10.9). One can now verify that for types II and VI$_0$, but not VII$_0$, V($\bar{\beta}^\pm$) has critical points, implying the existence of self–similar solutions of types II and VI$_0$, but not VII$_0$ (see Table 6.3). Since $\bar{p}_\pm = 0$ for a self–similar solution (see (10.52)), it follows from (10.59) and (10.35) that

$$\Sigma_+ = v, \quad \Sigma_- = 0$$

for the self–similar solutions†, with v given by (10.60) or (10.63). These expressions agree with the results for the equilibrium points P_1^+(II) and P_1^+(VI$_0$) in Section 6.2.1.

In Section 10.3.2, we shall use the potentials $V(\bar{\beta}^\pm)$ to describe the evolution of perfect fluid models of types II and VI$_0$ at late times.

10.3 Qualitative Hamiltonian cosmology

As mentioned in Section 5.1, one can use Hamiltonian methods to give a heuristic analysis of the generic asymptotic behaviour of Bianchi models near the singularity and at late times, without doing lengthy, detailed calculations. The essential information is contained in the potential function in the Hamiltonian. In this section we illustrate this process by considering the non–tilted Bianchi models of class A near the singularity (all types) and at late times (types I, II and VI$_0$). As a byproduct of the late–time analysis we derive some new monotone functions for Bianchi types II, VI$_0$ and VII$_0$.

10.3.1 Behaviour at early times

The usual analysis near the singularity deals with the vacuum case, since it is argued that in a typical model the matter does not affect the evolution in that regime (see Section 6.5 for details).

We use the time gauge defined by (10.29), so that the evolution equations are given by (10.31), with $\mu_0 = 0$ for vacuum. One can regard these equations as being derived from a *time–dependent reduced* Hamiltonian $\mathcal{H}_{red} = -p_0$, obtained by solving the Hamiltonian constraint $\mathcal{H} = 0$ for p_0, and writing $\beta^0 = \tau$. From (10.28) we obtain

$$\mathcal{H}_{red} = \sqrt{(p_+)^2 + (p_-)^2 + 24e^{4\tau}V^*(\beta^\pm)}, \tag{10.65}$$

where V^* is given by (10.9). Hamilton's equations read

$$\frac{d\beta^\pm}{d\tau} = \frac{\partial\mathcal{H}_{red}}{\partial p_\pm}, \quad \frac{dp_\pm}{d\tau} = -\frac{\partial\mathcal{H}_{red}}{\partial\beta^\pm}, \quad \frac{d\mathcal{H}_{red}}{d\tau} = \frac{\partial\mathcal{H}_{red}}{\partial\tau}, \tag{10.66}$$

† For type I, (10.58) is trivially satisfied, and so we obtain a self–similar solution with $v = 0$, giving $\Sigma_+ = 0$ (the flat FL solution).

giving

$$\frac{d\beta^\pm}{d\tau} = \frac{p_\pm}{\mathcal{H}_{\text{red}}}, \quad \frac{dp_\pm}{d\tau} = -\frac{12e^{4\tau}}{\mathcal{H}_{\text{red}}}\frac{\partial V^*}{\partial \beta^\pm}, \quad \frac{d\mathcal{H}_{\text{red}}}{d\tau} = \frac{48e^{4\tau}}{\mathcal{H}_{\text{red}}}V^*. \tag{10.67}$$

Following Misner, we now describe the evolution in terms of the motion of the 'universe point' (β^+, β^-) in the $\beta^\pm$-plane under the influence of a time–dependent potential

$$U = 24e^{4\tau}V^*(\beta^\pm).$$

By (10.65) and (10.67) the velocity $(d\beta^+/d\tau, d\beta^-/d\tau)$ of the point satisfies

$$\left(\frac{d\beta^+}{d\tau}\right)^2 + \left(\frac{d\beta^-}{d\tau}\right)^2 = 1 - \frac{24e^{4\tau}}{\mathcal{H}_{\text{red}}^2}V^*(\beta^\pm). \tag{10.68}$$

For Bianchi I models we have $V^*(\beta^\pm) = 0$. Equation (10.67) implies $p_\pm$ are constant and the universe point moves with constant unit velocity in the $\beta^\pm$-plane, representing a Kasner vacuum solution. In order to analyze the motion for more general Bianchi types it is convenient to use a 'moving wall' approximation for the potential, as introduced by Misner. We first illustrate the idea for the Bianchi II models. Without loss of generality we choose $\hat{n}_1 = 1$, $\hat{n}_2 = \hat{n}_3 = 0$. By (10.9),

$$V^*(\beta^\pm) = \tfrac{1}{2}e^{-8\beta^+}. \tag{10.69}$$

The level curves $V^* = C$ are straight lines $\beta^+ = $ constant. Because of the exponential dependence we can approximate the potential by an infinitely steep wall located at a specific level curve $V^* = C$, for a value of C sufficiently large, with the potential effectively zero to the right of the wall. The value C should be thought of as the value of the potential which is large enough to significantly affect the motion of the universe point (Jantzen 1987, page 115). The full potential in (10.65) is, however, time–dependent,

$$U = 12e^{4(\tau - 2\beta^+)}, \tag{10.70}$$

and hence the wall that represents the potential will move with speed $\tfrac{1}{2}$ in the *negative* β^+-direction as the model evolves *towards* the singularity $(\tau \to -\infty)$.

When the universe point is to the right of the wall and not close to the wall the potential is effectively zero (Bianchi I approximation) and the universe point moves in a straight line with speed 1. If this line makes an angle θ_{initial} with the β^+-axis, where $-\frac{\pi}{3} < \theta_{\text{initial}} < \frac{\pi}{3}$, then the universe point will reach the moving wall, bounce† off it, and then continue in a straight line with

† We refer to Jantzen (1987, page 116) for further details about the bounce process.

angle θ_{final}, corresponding to a new Kasner solution. The relation between θ_{final} and θ_{initial}, the 'bounce law', can be deduced from the Hamiltonian (Misner 1970, pages 63–4):

$$\sin \theta_{\text{final}} = \frac{3 \sin \theta_{\text{initial}}}{5 - 4 \cos \theta_{\text{initial}}}. \qquad (10.71)$$

In this way the Hamiltonian method describes a BKL Kasner–to–Kasner transition, the bounce representing a Taub Bianchi II vacuum solution. We refer to Jantzen (1987, page 118) for details. There are two other equivalent representations of the type II models, determined by $\hat{n}_1 = \hat{n}_3 = 0$, $\hat{n}_2 = 1$ and $\hat{n}_1 = \hat{n}_2 = 0$, $\hat{n}_3 = 1$ in the potential (10.9). The corresponding potential walls are obtained by a rotation through $\frac{2\pi}{3}$ or $\frac{4\pi}{3}$.

We now consider the most general models, those of types VIII and IX. The level curves of the potential $V^*(\beta^\pm)$, given by (10.9), are shown schematically in Figure 10.1, for all Bianchi types. We refer to Jantzen (1987, page 86) for an accurate representation.

Note that for large values of V^*, the level curves for models of types VIII and IX are approximately triangular in shape, with the sides of the triangles being approximated by level curves of the three type II potentials. The channels (respectively corners) that occur at the 'vertices' of the triangles are approximated by the level curves of the type VII_0 (respectively VI_0) potential. Since one can approximate each of the type II potential walls as being infinitely steep, one can approximate the type VIII and IX potentials by a triangular box with infinitely high walls, located at the level curve $V^* = C$, for C sufficiently large, with V^* effectively zero inside the triangle. The presence of the $e^{4\tau}$ time dependence in the potential (see (10.55) with $\beta^0 = \tau$) means that each wall of the triangular box will move outwards with a speed $\frac{1}{2}$ perpendicular to the wall as time evolves into the past. For evolution close to the singularity (τ large and negative), the potential box will be large and for an arbitrary initial state within the box, the potential will be effectively zero (type I approximation) and hence the universe point will initially move in a straight line with speed 1. Eventually the point will reach one of the potential walls, bounce off it and continue in a straight line. Since the three walls form a bounded, though expanding region in the $\beta^\pm$-plane, this process will continue indefinitely, leading to the infinite sequence of Kasner states referred to in Section 5.1. A single Kasner state is called a *Kasner epoch*, and a sequence of Kasner epochs in which the bounces occur off only two of the walls is called a *Kasner era*. Thus the claim is that the evolution of type VIII and IX models into the past near the initial singularity can be approximated by a sequence of Kasner epochs, which can

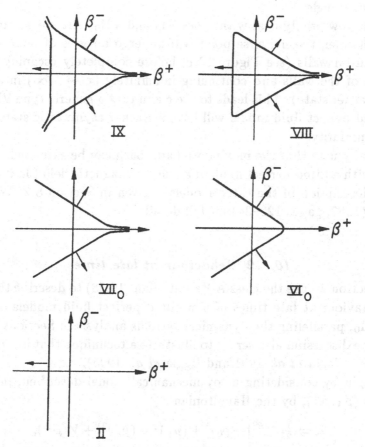

Fig. 10.1. Level curves for the potential $V^*(\beta^\pm)$ in the $\beta^\pm$-plane for class A models. The arrows describe the directions in which the potential walls move as one approaches the initial singularity.

be grouped into eras. This pattern of evolution can, however, be disrupted if the universe point interacts with one of the channels or corners. In this event, the evolution will be approximated by a vacuum VII_0 or VI_0 model, respectively. Since the vertices of the triangle move outward with speed 1 (Misner 1970, page 63, Figure 3), one expects that the universe point will eventually leave the influence of the channel/corner, and resume its Kasner motion. This expectation has been confirmed by numerical experiments and analytic approximations (Misner 1970, pages 64–5, Khalatnikov & Pokrovski 1972, and Jantzen 1987, page 115 and pages 123–5). One also expects that this anomalous behaviour occurs less frequently as $\tau \to -\infty$, since the size of the channels/corners becomes increasingly small compared to the size of the whole triangle.

We can now briefly deal with types VI_0 and VII_0. After emerging from a corner/channel, the universe point will undergo a series of bounces off the two potential walls (see Figure 10.1) before completely escaping from the influence of the walls and continuing indefinitely ($\tau \rightarrow -\infty$) in a straight line (a Kasner state). This leads to the claim that a generic type VI_0 or VII_0 non–tilted perfect fluid model will have a Kasner asymptotic state near the initial singularity.

We finally note that the moving wall approach can be extended to Bianchi models with a tilted perfect fluid, or an electromagnetic field (Jantzen 1986). A brief description of the tilted models is given in Section 8.2. We refer to Jantzen (1987, pages 129–34) for full details.

10.3.2 Behaviour at late times

In this section we use the class A Hamiltonian (10.28) to describe the asymptotic behaviour at late times of non–tilted perfect fluid models of types I, II and VI_0, paralleling the dynamical systems analysis in Sections 6.3.2 and 6.3.3. The discussion also serves to illustrate a technique that is more widely applicable (Uggla *et al.* 1991 and Uggla *et al.* 1995).

We begin by considering a toy mechanical model described, in terms of variables (p_A, β^A), by the Hamiltonian

$$\mathcal{H} = \tfrac{1}{2}e^{-c\beta^0}[-(p_0)^2 + (p_+)^2 + (p_-)^2] + V(\beta^\pm), \qquad (10.72)$$

and the constraint, $\qquad \mathcal{H} = 0,$

where c is a positive constant (Uggla *et al.* 1991).

Introducing an 'energy' E and a 'mass' m according to

$$E = \tfrac{1}{2}e^{-c\beta^0}(p_0)^2, \quad m = e^{c\beta^0}, \qquad (10.73)$$

the constraint assumes the form

$$\tfrac{1}{2m}[(p_+)^2 + (p_-)^2] + V(\beta^\pm) = E. \qquad (10.74)$$

In our applications, $V(\beta^\pm)$ will have a positive lower bound $V_{\min}$. The constraint thus implies that

$$E \geq V(\beta^\pm) \geq V_{\min} > 0. \qquad (10.75)$$

With cosmological applications in mind, we consider 'expanding' solutions, which satisfy the conditions

$$\dot{\beta}^0 > 0, \quad p_0 < 0.$$

It follows from the Hamiltonian equations for β^0 and p_0 that

$$\dot{E} = -\tfrac{1}{2}cm^{-2}(-p_0)[(p_+)^2 + (p_-)^2] = -cm^{-1}(-p_0)[E - V(\beta^{\pm})], \quad (10.76)$$

Thus, for expanding solutions, E is monotone decreasing and m is monotone increasing. Hamilton's equations for $\beta^{\pm}$ and $p_{\pm}$ lead to

$$m\ddot{\beta}^{\pm} = -c(-p_0)\dot{\beta}^{\pm} - \frac{\partial V}{\partial \beta^{\pm}}.$$

This equation suggests that one can view the Hamiltonian (10.72) as describing a particle moving in the $\beta^{\pm}$-plane under the influence of a potential $V(\beta^{\pm})$. The increase in the mass m introduces an effective coefficient of friction which dissipates the energy E. Our (heuristic) conclusion is that the particle will continue to move in the potential well defined by (10.75) and the energy will continue to decrease until E approaches the value $V_{\min}$.

The Hamiltonian $\mathcal{H}$, as given by (10.28), is not of the form (10.72) in general. For Bianchi type I models, however, with the choice of lapse $N = 12e^{3(\gamma-1)\beta^0}$, $\mathcal{H}$ is of the form (10.72) with $c = 3(2-\gamma)$ and $V = 24\mu_0$. Thus, if $\gamma < 2$, the dynamics is that of a free particle with increasing mass m, losing energy by friction, and eventually coming to rest with $\beta^{\pm} = $ constant. This asymptotic state at late times corresponds to the flat FL model, in agreement with Section 6.3.1 (see Figure 6.4).

For Bianchi II models the boost (10.59)–(10.60) and the choice of lapse $N = 12e^{3\beta^0 - 3\Gamma(2-\gamma)\bar{\beta}^0}$ transforms $\mathcal{H}$ to a Hamiltonian of the form (10.72), namely

$$\mathcal{H} = \tfrac{1}{2}e^{-3\Gamma(2-\gamma)\bar{\beta}^0}[-(\bar{p}_0)^2 + (\bar{p}_+)^2 + (\bar{p}_-)^2] + V(\bar{\beta}^{\pm}), \quad (10.77)$$

where $V(\bar{\beta}^{\pm})$ is given by (10.62). In this case V depends only on $\bar{\beta}^+$. The potential is thus a 'trough', whose base is the line $\bar{\beta} = \bar{\beta}^+_{\text{crit}}$, $\bar{\beta}^- \in \mathbb{R}$, where $V(\bar{\beta}^+_{\text{crit}}) = V_{\min}$. Motion of the universe point along the base of the trough (i.e. $\ddot{\bar{\beta}}^+ = 0$, which implies $\bar{p}_+ = 0$ and hence $\Sigma_+ = \tfrac{1}{8}(3\gamma - 2)$, by (10.35), (10.59) and (10.60)) corresponds to special solutions. If the point is actually at rest ($\dot{\bar{\beta}}^{\pm} = 0$ and hence $\bar{p}^{\pm} = 0$) the solution is self-similar (see Section 10.2.3, in particular (10.52)), and is in fact the Collins–Stewart solution (Section 9.1.2). Otherwise, one obtains the Collins type II evolving solution (Section 9.3.2), in which case $\dot{\bar{\beta}}^- > 0$ and $\ddot{\bar{\beta}}^- < 0$ (or $\dot{\bar{\beta}}^- < 0$ and $\ddot{\bar{\beta}}^- > 0$), and the universe point eventually comes to rest. An arbitrary solution is described by the universe point moving in the trough while losing energy and eventually coming to rest at the base of the trough. This heuristic analysis corresponds to the phase portrait of the invariant set $B_1^+(\text{II})$ in Figure 6.7.

For type VI_0 models, the boost (10.59) with (10.63), and the above choice of lapse, leads to the Hamiltonian (10.77), where $V(\bar{\beta}^{\pm})$ is given by (10.64). This potential has closed level curves and a unique minimum point, and thus defines a 'bowl'. An arbitrary solution is described by the universe point moving in the bowl while losing energy and eventually coming to rest at the bottom of the bowl. As before, this state of rest corresponds to a self–similar solution, the Collins VI_0 solution (Section 9.1.3).

In this way, by analyzing the potential $V(\beta^{\pm})$ in the transformed Hamiltonian (10.77), one can give a simple heuristic argument suggesting that non–tilted type I, type II and type VI_0 universes are asymptotically self–similar.

Appendix to Chapter 10

The monotone function E, defined by (10.74), gives rise to the monotone functions Z_7 (type VI_0) and Z_4 (type II) in the Appendix to Chapter 6, when transformed into the dimensionless variables $(\Sigma_{\pm}, N_{\alpha})$.

By (10.72) and (10.73),

$$E = \frac{\bar{p}_0^2 V}{\bar{p}_0^2 - \bar{p}_+^2 - \bar{p}_-^2}.$$

Using invariance under a Lorentz transformation and (10.35), (10.39) and (10.40),

$$\bar{p}_0^2 - \bar{p}_+^2 - \bar{p}_-^2 = p_0^2(\Omega + K).$$

Writing $V = V_g + V_{\text{fluid}}$, equations (10.34), (10.35), (10.55) and (10.57) imply

$$\Omega V = (\Omega + K) V_{\text{fluid}}.$$

Using these equations and (10.59) we obtain

$$E = \Gamma^2 (1 - v\Sigma_+)^2 \Omega^{-1} V_{\text{fluid}}.$$

Finally, the detailed form of V_g and V_{fluid}, as given by (10.62) and (10.64), leads to

$$V_{\text{fluid}} \propto \left(\frac{\Omega}{N_1^2}\right)^m, \quad V_{\text{fluid}} \propto \left(\frac{\Omega}{|N_2 N_3|}\right)^m,$$

for type II and VI_0, respectively, where $m = s/(r + s)$, and we ignore multiplicative constants. After simplifying the exponent, one obtains (6.47) and (6.42).

11
Deterministic chaos and Bianchi cosmologies

D. W. HOBILL
University of Calgary

It is well known that solutions of non–linear differential equations in three and higher dimension can display apparently random behaviour referred to as deterministic chaos, or simply, chaos. The associated dynamical system is then referred to as being chaotic. It was recognized some years ago that the oscillatory approach to the past or future singularity of Bianchi IX vacuum models displays random features (e.g. Belinskii *et al.* 1970, which we shall refer to as BKL, and Barrow 1982b), and hence is a potential source of chaos. This oscillatory behaviour is also believed to occur in other classes of models, provided that the Bianchi type and/or source terms are sufficiently general (see Sections 8.1 and 8.4). The goal in this chapter is to address the question of whether the dynamical systems which describe the evolution of Bianchi models are chaotic.

Historically, both Poincaré and Birkhoff in the late nineteenth and early twentieth centuries were aware that non–linear DEs could admit complicated aperiodic or quasi–periodic solutions. The modern development of a theory of chaotic dynamical systems was stimulated in a large part by two papers, namely Lorenz (1963), a numerical simulation of a three–dimensional DE, and Smale (1967), a theoretical analysis of discrete dynamical systems. The field developed rapidly once computer simulations of dynamical systems became widely available. Despite a lengthy history, complete agreement on a definition of chaotic dynamical system has not been reached. This state of affairs is due partly to the conflicting aims and needs of pure mathematicians and applied researchers. Another feature of the discipline is that it is extremely difficult to prove that a dynamical system governed by a particular set of DEs, whose solutions display apparently random behaviour in numerical simulations, is chaotic in terms of a rigorous mathematical definition. These features of the discipline are reflected in the controversy that has arisen over whether or not the vacuum Bianchi IX models are chaotic.

In Section 11.1, we thus introduce some of the concepts that arise in discussions of chaotic dynamical systems. This section and Chapter 4 provide the mathematical background for the present chapter.

As discussed in Section 5.1, three methods have been used to investigate the oscillatory singularity. In reaching our main goal in this chapter, we shall describe the connections between these approaches and compare their suitability for calculating numerical solutions of the field equations.

11.1 Chaotic dynamical systems

We consider both continuous and discrete dynamical systems, as discussed in Chapter 4. Recall from Section 4.1.3 that a continuous dynamical system is a flow ϕ_t, $t \in \mathbb{R}$, determined by a DE $\mathbf{x}' = \mathbf{f}(\mathbf{x})$ on $\mathbb{R}^n$, while a discrete dynamical system is the set of iterates g^m of a map $g : X \to X$, where m is an arbitrary integer if g is invertible, and $m \geq 0$ if g is not invertible. In our applications the set X will either be the unit interval $[0,1]$ or the unit circle S^1.

11.1.1 A definition of chaotic dynamical system

The most important feature of a chaotic dynamical system (a flow φ_t or a map g) is that long–term predictions of the evolution from given initial data are impossible despite the deterministic nature of the equations. This property means that orbits arbitrarily close initially will eventually diverge and become effectively uncorrelated, while remaining bounded. The definition to follow formalizes this idea (Wiggins 1990, page 608, and Glendinning 1994, page 299).

Definition 11.1. Let A be a compact invariant set of a flow φ_t on $\mathbb{R}^n$. The flow φ_t has *sensitive dependence on initial conditions (SDIC)* on A means that there exists a $\delta > 0$ such that for all $\mathbf{x} \in A$ and for all $\varepsilon > 0$, there exists a $\mathbf{y} \in A$ and $t > 0$ such that

$$\| \mathbf{x} - \mathbf{y} \| < \varepsilon \quad \text{and} \quad \| \varphi_t(\mathbf{x}) - \varphi_t(\mathbf{y}) \| > \delta. \qquad (11.1)$$

Comment: For a map g, replace '$t > 0$' by '$m > 0$' and '$\| \varphi_t(\mathbf{x}) - \varphi_t(\mathbf{y}) \|$' by '$\| g^m(\mathbf{x}) - g^m(\mathbf{y}) \|$'.

The next concept, topological transitivity, ensures that under the action of the flow all points in the set A (except possibly for a set of measure zero) come arbitrarily close to all other points in A. In writing this definition $\varphi_t(U)$ denotes the image of the set U under the map φ_t.

Definition 11.2. A flow φ_t on $\mathbb{R}^n$ is *topologically transitive* on a compact invariant set A means that for all open sets $U, V \subset A$ there exists a $t > 0$ such that $\varphi_t(U) \cap V \neq \phi$.

Comments:

(i) For a map g replace '$t > 0$' by '$m > 0$' and '$\varphi_t(U) \cap V \neq \phi$' by '$g^m(U) \cap V \neq \phi$'.

(ii) If A contains an orbit that is dense in A (i.e. the closure of the orbit equals A), then the flow or map is topologically transitive on A.

Following Wiggins (1990, page 608) we now define a chaotic dynamical system.

Definition 11.3: A flow φ_t (or map g) is *chaotic* on a compact invariant set A means that

(i) φ_t (respectively g) has sensitive dependence on initial conditions on A, and

(ii) φ_t (respectively g) is topologically transitive on A.

Comment: The above definition is the weakest criterion for a chaotic dynamical system that is given in the literature. It makes no statement about how rapidly orbits separate. Some authors require that the separation be exponentially rapid:

$$\|\varphi_t(\mathbf{x}_1) - \varphi_t(\mathbf{x}_0)\| \approx e^{\lambda t} \|\mathbf{x}_1 - \mathbf{x}_0\|, \tag{11.2}$$

for neighbouring points $\mathbf{x}_0, \mathbf{x}_1$, where λ is a positive constant. One might also wish to require that the set on which φ_t is chaotic is an attractor (see Section 4.4.1) in order to ensure that the chaotic behaviour will be observable. We refer to Wiggins (1990, pages 608–15) for a discussion of the merits of various criteria and some illustrative examples. He mentions that numerical experiments of chaotic dynamical systems arising in applications suggest that exponential separation does not always occur (*ibid*, remark 3 on page 615). One can also argue that a requirement such as exponential separation, which depends on the choice of time variable, is not appropriate in a theory such as general relativity in which there is no preferred time variable. The definition of Wiggins depends only on the metric and topological properties of the state space. We refer to Glendinning (1994, pages 291–302) and to Kirchgraber & Stoffer (1989), for further discussion of these matters within the context of discrete dynamical systems. Knudsen (1994) gives further arguments in favour of the definition used by Wiggins.

We now state a theorem which gives a simple sufficient condition for a map on an interval $[a, b]$ to be chaotic (Glendinning 1994, page 300). The definition of chaos used in this reference is stronger than that of Wiggins. We note that a point of period n of a map g is a point x that satisfies $g^n(x) = x$ and $g^k(x) \neq x$ for $k = 1, \ldots, n - 1$.

Theorem 11.1. Suppose that g maps the interval $[a, b]$ into itself. If g is continuous and has a point of period three, then g is chaotic on $[a, b]$.

11.1.2 Lyapunov exponents

The Lyapunov exponents of a dynamical system provide a measure of how rapidly neighbouring orbits converge or diverge. To make this precise let x_0 be a reference point and x_1 a point in the neighbourhood of x_0. We wish to describe the long–term evolution of the vector separating the orbits that pass through x_0 and x_1. By the linear approximation,

$$\varphi_t(x_1) - \varphi_t(x_0) \approx D\varphi_t(x_0)(x_1 - x_0).$$

provided that $x_1 - x_0$ is sufficiently close to $\mathbf{0}$. Thus the derivative matrix of the flow, $D\varphi_t(x_0)$, describes the evolution of the relative separation of the orbits through x_0 and x_1. The Lyapunov exponents of the flow at the point x_0 are defined in terms of the asymptotic behaviour of the eigenvalues of this matrix (e.g. Parker & Chua 1989, pages 67 and 305). We refer to Lichtenberg & Lieberman (1983, page 262) and Wiggins (1990, page 603) for other introductory discussions of Lyapunov exponents.

Definition 11.4. Given a DE $x' = f(x)$ in $\mathbb{R}^n$ with a flow φ_t, let $d_i(t)$, where $i = 1, 2, \ldots, n$, be the eigenvalues of the derivative matrix $D\varphi_t(x_0)$ at a given point x_0. The *Lyapunov exponents* of the DE/flow at x_0 are defined by

$$\lambda_i = \lim_{t \to +\infty} \frac{1}{t} \ln |d_i(t)|, \qquad (11.3)$$

whenever the limit exists.

Comment: A positive Lyapunov exponent is an indication that the dynamical system has SDIC with an *exponential rate of separation*. We refer to Parker & Chua (1989, pages 70–2) for examples of Lyapunov exponents in a variety of situations.

For an arbitrary point $\mathbf{x_0}$ one cannot calculate the derivative matrix, $D\varphi_t(\mathbf{x_0})$ explicitly. However, it can be calculated numerically in a simple manner, since it is the unique solution to the linear, non–autonomous matrix DE (the variational equation),

$$\mathcal{M}'(t) = D\mathbf{f}(\varphi_t(\mathbf{x_0}))\mathcal{M}(t), \tag{11.4}$$

with the initial condition

$$\mathcal{M}(0) = I, \tag{11.5}$$

where I is the identity matrix. Since $\varphi_t(\mathbf{x_0})$ is the solution of the initial value problem:

$$\mathbf{x}' = \mathbf{f}(\mathbf{x}), \quad \mathbf{x}(0) = \mathbf{x_0} \tag{11.6}$$

one can integrate the combined system (11.4)–(11.6) simultaneously to obtain $\mathcal{M}(t) = D\varphi_t(\mathbf{x_0})$ (see Parker & Chua 1989, pages 305–6).

The actual numerical calculation of the spectrum of Lyapunov exponents is not straightforward, however, since $\mathcal{M}(t)$ can increase without bound. An algorithm which circumvents this difficulty has been introduced by Benettin *et al.* (1980), and implemented numerically by Wolf *et al.* (1985).

11.2 The metric approach

11.2.1 The BKL equations

We consider the Bianchi models of class A, for which the source is a non–tilted perfect fluid with equation of state $p = (\gamma - 1)\mu$ and 4–velocity $\mathbf{u}$. As in Section 1.5.2 we introduce a frame $\{\mathbf{u}, \mathbf{E}_\alpha\}$, where the $\mathbf{E}_\alpha$ are time–independent, group–invariant vector fields. The line–element has the form (1.89) in terms of an arbitrary time variable $\tilde{t}$, i.e.

$$ds^2 = -N(\tilde{t})^2 d\tilde{t}^2 + g_{\alpha\beta}(\tilde{t})\mathbf{W}^\alpha\mathbf{W}^\beta, \tag{11.7}$$

where the 1–forms $\mathbf{W}^\alpha$ are dual to the vector fields $\mathbf{E}_\alpha$. The structure constants $C^\alpha{}_{\mu\nu}$ are given by (10.5). In this situation the metric $g_{\alpha\beta}$ can be chosen to be diagonal (MacCallum 1973, Theorem 2, page 126). As in Section 10.1.2 we write

$$g_{\alpha\beta} = \text{diag}\,(e^{2\beta_1}, e^{2\beta_2}, e^{2\beta_3}). \tag{11.8}$$

The length scale function ℓ is given by

$$\ell^3 = e^{\beta_1 + \beta_2 + \beta_3}. \tag{11.9}$$

We specify the time variable $\tilde{t}$ by choosing

$$N = \ell^3, \tag{11.10}$$

and denote $\tilde{t}$ by T, which is thus related to clock time t by†

$$\frac{dt}{dT} = \ell^3. \tag{11.11}$$

In order to write the field equations concisely, we define the quantities R_α, $\alpha = 1, 2, 3$, by

$$R_1 = \tfrac{1}{2}\left[\hat{n}_1^2\, e^{4\beta_1} - (\hat{n}_2\, e^{2\beta_2} - \hat{n}_3\, e^{2\beta_3})^2\right] \tag{11.12}$$

with R_2, R_3 defined by cycling on 1, 2, 3. The field equations are (see the Appendix):

$$\dot{\beta}_\alpha = k_\alpha, \quad \dot{k}_\alpha = -R_\alpha + \tfrac{1}{2}(2 - \gamma)\mu\ell^6, \tag{11.13}$$

where μ is given by

$$\mu\ell^6 = \tfrac{1}{2}(R_1 + R_2 + R_3) + (k_1 k_2 + k_2 k_3 + k_3 k_1), \tag{11.14}$$

and an overdot denotes d/dT. The first equation in (11.13) should be viewed as defining the 'momenta' k_α. We refer to the field equations in the form (11.13)–(11.14) as the *BKL equations*, although BKL give them for vacuum, $\mu = 0$, and write them as second–order DEs (see BKL page 534, equations (3.5) and (3.6)).

Comment: The state vector for the DE (11.13)–(11.14) is

$$\mathbf{x} = (\beta_1, \beta_2, \beta_3, k_1, k_2, k_3) \in \mathbb{R}^6,$$

and the state space for expanding models is the subset of $\mathbb{R}^6$ defined by the inequalities

$$k_1 + k_2 + k_3 > 0, \quad \mu \geq 0, \tag{11.15}$$

(see ((11.29)). As the initial singularity is approached, $T \to -\infty$, i.e. the solutions of the DE can be extended arbitrarily into the past. Thus the DE (11.13)–(11.14) on $\mathbb{R}^6$ defines a negative semi–flow (see Section 4.1.3) on the state space. The state space is, however, unbounded and, since orbits do not remain in a compact set into the past, there is no past attractor (see Section 4.4.1). Thus the BKL equations are not satisfactory from a dynamical systems point of view. Furthermore, because of the lack of compactness, one cannot use the concept of SDIC. However, one can numerically calculate Lyapunov exponents for the BKL equations.

† The relation between T and the dimensionless time variable τ used in Chapters 5–7 can be inferred from (11.39).

11.2.2 The piecewise approximation method

The BKL equations (11.13)–(11.14) form the basis for the piecewise approximation method described in Section 5.1, in particular, the BKL analysis of the oscillatory approach to the singularity. The following discussion, in which we assume $\mu = 0$, is intended to give an idea of the method.

We begin by noting that the Kasner solutions are obtained from the BKL equations when $\hat{n}_\alpha = 0$ and $\mu = 0$. It follows from (11.13) that $\beta_\alpha = k_\alpha T$, where the k_α are constants which, on account of (11.12) and (11.14), satisfy

$$(k_1 + k_2 + k_3)^2 = k_1^2 + k_2^2 + k_3^2. \tag{11.16}$$

It follows from (11.9) and (11.11), after absorbing constants, that

$$e^{\beta_\alpha} = t^{p_\alpha}, \tag{11.17}$$

where the Kasner exponents p_α are given by

$$p_\alpha = \frac{k_\alpha}{k_1 + k_2 + k_3}. \tag{11.18}$$

Suppose that over some time interval, $e^{\beta_\alpha} \ll 1$ for $\alpha = 1, 2, 3$. Equation (11.13) with $\mu = 0$ implies that the $\dot{k}_\alpha$ are close to zero, and hence that the universe will be close to a Kasner state. One is effectively setting $\hat{n}_\alpha = 0$, $\alpha = 1, 2, 3$, in (11.13). The restrictions (11.15) and (11.16) imply that in general, one k_α is negative and two are positive, say

$$k_1 < 0, \quad k_2, k_3 > 0. \tag{11.19}$$

These conditions, which yield $\dot{\beta}_1 < 0$, and $\dot{\beta}_2$, $\dot{\beta}_3 > 0$ (into the future), imply that a Kasner state is unstable, both into the past and into the future. Into the past, $\dot{\beta}_1 > 0$, and $\dot{\beta}_2$, $\dot{\beta}_3 < 0$, and so e^{β_1} will become significant while e^{β_2} and e^{β_3} will remain insignificant. Thus, over some time interval, the evolution will be approximated by a Bianchi II model ($\tilde{n}_1 = 1, \tilde{n}_2 = \tilde{n}_3 = 0$), a Taub vacuum model (see BKL, equation (3.12), page 535; also our Section 9.2.1, where a Kasner–like time variable with $N = \ell^3 t^{-1}$ is used). Analysis of the Taub model shows that as time evolves into the past, it approaches a Kasner state which differs from the initial one, with $k_2 < 0$ and $k_1, k_3 > 0$, say.[†] This state is likewise unstable, and so the above process is repeated. Arguments such as these lead to the claim that into the past, the evolution of a type IX or VIII model is approximated by a sequence of Kasner states, with the transitions being approximated by Taub vacuum solutions (BKL, pages 533–8; see Khalatnikov & Lifshitz 1970 and Landau & Lifshitz 1975, pages 392–3 for a brief description).

† This asymptotic behaviour can be inferred from Figure 6.6.

11.2.3 The discrete dynamical system

BKL use a variable u to parameterize the Kasner exponents according to

$$p_1 = -\frac{u}{1+u+u^2}, \quad p_2 = \frac{1+u}{1+u+u^2}, \quad p_3 = \frac{u(1+u)}{1+u+u^2}, \qquad (11.20)$$

with $1 \leq u \leq \infty$. BKL (see pages 535-7) show that the Kasner transitions are given by

$$u_{\text{final}} = f(u_{\text{initial}}),$$

where

$$f(u) = \begin{cases} u - 1, & \text{if} \quad u \geq 2 \\ \dfrac{1}{u-1}, & \text{if} \quad 1 < u < 2. \end{cases} \qquad (11.21)$$

One can form the iterates f^n, for $n > 0$, but in order to obtain a discrete dynamical system, it is necessary to describe the transitions by a map g of a finite closed interval into itself. Following Ma & Wainwright (1992), we let $\tilde{u} = 1/u$ and then

$$\tilde{u}_{\text{final}} = g(\tilde{u}_{\text{initial}}),$$

where

$$g(\tilde{u}) = \begin{cases} \dfrac{\tilde{u}}{1-\tilde{u}}, & \text{if} \quad 0 \leq \tilde{u} \leq \frac{1}{2} \\ \dfrac{1-\tilde{u}}{\tilde{u}}, & \text{if} \quad \frac{1}{2} \leq \tilde{u} \leq 1, \end{cases} \qquad (11.22)$$

(see also Rugh 1994, page 377, who refers to g as the Farey map). The iterates $\{g^m\}$, $m \geq 0$, then define a discrete dynamical system on the interval $[0, 1]$. Since g is continuous on $[0, 1]$, we can invoke Theorem 11.1 to prove that g is a chaotic map. All that has to be done is to observe (Ma 1988) that $\tilde{u} = \frac{1}{6}(\sqrt{13} - 1)$ is a periodic point of least period 3 of g. We shall refer to g as the *compactified BKL map*.

11.2.4 Numerical solutions

Numerical solutions of the DE (11.13) and constraint (11.14), with $\mu = 0$ have been calculated by a number of authors, confirming that the evolution is approximated by a sequence of Kasner states (Hobill *et al.* 1991, Figure 2 on page 1162, and Rugh 1994). Rugh (1994) extracts the values of the BKL parameter u from a numerical solution, thereby verifying that the transition law defined by (11.21), or equivalently by (11.22), is verified to a high degree of accuracy (*ibid*, Figure 5 and Table 1, on pages 374-5). These authors and others confirm that the Lyapunov exponents associated with this DE are actually zero (Francisco & Matsas 1988, Burd & Tavakol† 1993, and Rugh 1994, page 383).

11.3 Hamiltonian method

As discussed in Section 10.3.1, the Hamiltonian method describes the evolution of a Bianchi class A universe in terms of the motion of the universe point in the $\beta^{\pm}$-plane, subject to a time–dependent potential (see (10.65) and (10.67). When the universe point is not under the influence of the potential, it moves in a straight line in the $\beta^{\pm}$-plane, corresponding to a Kasner state. This straight line can be characterized by the angle θ that it makes with the β^{+}-axis. A transition between Kasner states is described by the bounce law (10.71), which reads

$$\sin \theta_{\text{final}} = \frac{3 \sin \theta_{\text{initial}}}{5 - 4 \cos \theta_{\text{initial}}}, \tag{11.23}$$

for the case $|\theta| \leq \frac{\pi}{3}$. The angle θ is related to the BKL parameter u according to

$$u = \frac{2 - \cos \theta - \sqrt{3} \sin \theta}{2 \cos \theta - 1}, \tag{11.24}$$

for $-\frac{\pi}{3} < \theta \leq 0$ (Berger 1991, page 1388). It follows that *the bounce law (11.23) corresponds to the BKL transition law (11.21).*

Numerical solutions of the Hamiltonian DE have been calculated by a number of authors. Misner (1970, page 65) reports some numerical experiments and gives pictures showing the trajectory of the universe point in the $\beta^{\pm}$-plane (*ibid*, Figures 7 and 8). Berger (1991) has performed extensive studies of numerical solutions, and has extracted the values of the BKL parameter u from a numerical solution, again verifying the transition law defined by (11.21) to a high degree of accuracy (*ibid*, Table 1, page 1393).

11.4 Expansion–normalized variables

In Chapter 6 the evolution of the class A Bianchi cosmologies was analyzed using dimensionless variables, defined by normalizing with the Hubble scalar H. The analysis in Section 6.4 suggests that the oscillatory approach to the singularity in universes of types VIII and IX can be described by a past attractor in the dimensionless state space. In addition, the analysis leads to

---

† These authors essentially define SDIC to mean that there is a positive Lyapunov exponent, whereas we make a distinction between these two requirements.

a discrete dynamical system, generated by the Kasner map (see Section 6.4.1 and Figure 6.13). In this section, we discuss the numerical experiments that support the above conclusion, and then use the properties of the Kasner map to draw some tentative conclusions about the continuous dynamical system.

11.4.1 Numerical experiments

The state vector for the non–tilted class A Bianchi models is

$$\mathbf{y} = (\Sigma_+, \Sigma_-, N_1, N_2, N_3) \in \mathbb{R}^5,$$

when $\Sigma_\pm$ describe the anisotropy in the Hubble flow, and the N_α describe the spatial curvature of the group orbits. The state space is the subset of $\mathbb{R}^5$ defined by the inequality $\Omega \geq 0$, where Ω is the density parameter, given by (6.14). The evolution equation

$$\frac{d\mathbf{y}}{d\tau} = \mathbf{f}(\mathbf{y}), \qquad (11.25)$$

is given by the DE (6.9), where τ is the dimensionless time variable defined by (5.9). The Kasner solutions arise as a circle $\mathcal{K}$ of equilibrium points of this DE:

$$\Sigma_+^2 + \Sigma_-^2 = 1, \quad N_1 = N_2 = N_3 = 0, \qquad (11.26)$$

where $\Sigma_\pm$ are constants (see Section 6.2.2).

Numerical solutions of the DE (11.25) have been studied by Ma (1988), Burd *et al.* (1990), and Creighton & Hobill (1994). For evolution into the past ($\tau \to -\infty$) it is found that $\Sigma_\pm$ and the N_α remain bounded. There is a difficulty, however, that when the universe is close to a Kasner state (11.26), the N_α are very close to zero, leading to a rapid loss of accuracy. This difficulty is circumvented by introducing variables Z_α defined by

$$Z_\alpha = \log |N_\alpha|. \qquad (11.27)$$

First, the numerical solutions corresponding to initial conditions with $N_1 N_2 N_3 \neq 0$ and $\Omega > 0$ provide evidence for the limits (6.32), namely that the density parameter Ω and the function

$$\Delta = (N_1 N_2)^2 + (N_2 N_3)^2 + (N_3 N_1)^2$$

tend to zero. These limits justify the existence of the Mixmaster attractor (6.30) or (6.31). The stage of the evolution during which Δ and/or Ω differ appreciably from zero is called the *transient* stage. The stage during which the orbit is close to the attractor is called the *asymptotic stage*.

Second, the numerical solutions show that in the asymptotic stage, the projection of the orbit in the $\Sigma_\pm$–plane is determined, to a high degree of accuracy, by the action of the Kasner map, as in Figure 6.13 (see also Creighton & Hobill 1994, Figure 4, for a comparison between a numerical solution and the corresponding exact representation by the Kasner map).

Third, as the evolution progresses (into the past), one finds that the Kasner epochs become successively longer (in term of τ–time). This effect is illustrated in Figure 11.1, which shows the shear parameter Σ as a function of $\log \tau$. In the asymptotic stage the graph is a horizontal line, $\Sigma = 1$ (the Kasner value), punctuated by downward spikes corresponding to the rapid transition between Kasner states. Similarly, the graphs of the N_α show a horizontal line $N_\alpha = 0$ (the Kasner value) punctuated by spikes (see Burd *et al.* 1990, Figures 3–5). One can monitor the gravitational field by plotting the expansion–normalized Weyl curvature scalars $\mathcal{E}$ and $\mathcal{H}$, defined by

$$\mathcal{E}^2 = \tfrac{1}{2}\mathcal{E}_{\alpha\beta}\mathcal{E}^{\alpha\beta}, \quad \mathcal{H}^2 = \tfrac{1}{2}\mathcal{H}_{\alpha\beta}\mathcal{H}^{\alpha\beta},$$

where $\mathcal{E}_{\alpha\beta}$ and $\mathcal{H}_{\alpha\beta}$ are defined by (6.33). These graphs are shown in Figures 11.2 and 11.3. One sees from Figure 11.3 that the magnetic Weyl curvature generates the transitions between Kasner states. This result is to be expected since $\mathcal{H}_{\alpha\beta} = 0$ for the Kasner solutions, while $\mathcal{H}_{\alpha\beta} \neq 0$ for the Taub solutions which describe the transitions (see Section 6.3.2 and (6.37)).

11.4.2 The Kasner map

As shown in Section 6.4.1, the Kasner map (represented geometrically in Figure 6.13) determines the sequence of Kasner states as the singularity is approached, and provides another way of using a discrete dynamical system to approximate the evolution. The three maps, the compactified BKL map (11.22), the bounce law (11.23) and the Kasner map of course determine the same sequence of Kasner states. The bounce law is in fact an analytic representation of the Kasner map, as follows from Figure 6.13 using trigonometry[†], and, as mentioned in Section 11.3, the bounce law determines the BKL transition law (11.21). The Kasner map, however, contains more information, since it also encodes the order of the Kasner exponents p_α for each Kasner state. Another difference is that the Kasner map determines the infinite heteroclinic sequences on the Mixmaster attractor (see Section 6.4.2), that approximate typical orbits when they are close to the attractor (the asymptotic stage of the evolution). One can thus use the

---

† On account of (10.46), the angle θ in (11.23) is the polar coordinate angle in the $\Sigma_\pm$–plane.

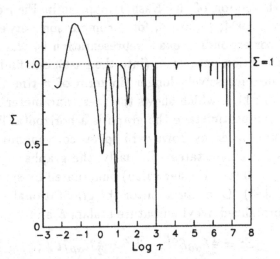

Fig. 11.1. The graph of the dimensionless shear parameter Σ as a function of $\log \tau$, for a model of Bianchi type IX, with τ increasing into the past.

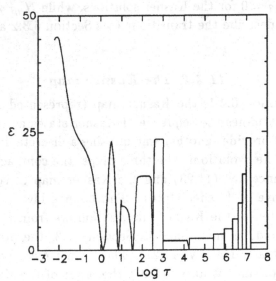

Fig. 11.2. The graph of the dimensionless electric Weyl scalar $\mathcal{E}$ as a function of $\log \tau$, using the same initial conditions as in Figure 11.1.

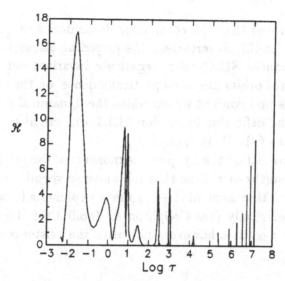

Fig. 11.3. The graph of the dimensionless magnetic Weyl scalar $\mathcal{H}$ as a function of $\log \tau$, using the same initial conditions as in Figure 11.1.

Kasner map to obtain information about the continuous dynamical system generated by the DE (11.25).

Since the compactified BKL map is chaotic, and the Kasner map encodes the same information, one expects that the Kasner map is chaotic. This expectation is confirmed by the fact that it is topologically equivalent to a simple map on the circle, defined by $f(z) = z^{-2}$, $|z| = 1$ (Smith & Wainwright, unpublished), that is known to be chaotic. We can now draw two conclusions:

(1) the Kasner map has SDIC and hence the infinite heteroclinic sequences have SDIC in the sense that sequences that start arbitrarily close will not remain close;

(2) a typical orbit of the Kasner map is dense on the Kasner circle $\mathcal{K}$ (i.e. the map is topologically transitive on $\mathcal{K}$), and hence a typical heteroclinic sequence will be dense on the Mixmaster attractor.

Since numerical solutions follow the heteroclinic sequences, it is plausible that

(1) the continuous dynamical system has SDIC into the past, on a negatively invariant set containing the attractor;

(2) a typical orbit of the DE (11.25) comes arbitrarily close to each point

on the attractor†; one could say that a typical orbit is 'asymptotically dense' on the attractor.

It should be noted that the continuous dynamical system is not chaotic on the attractor itself.† Nevertheless, the preceding discussion suggests that the attractor induces SDIC on a negatively invariant set containing the attractor, and that orbits are 'asymptotically dense' on the attractor. Thus, although there is no invariant set on which the dynamical system is chaotic in the sense of the definition in Section 11.1.1, one might argue that it does satisfy the intent of the definition.

We finally comment on the Lyapunov exponents of the DE (11.25). Due to the increasing lengths of τ–time that the universe spends close to a Kasner state, one expects that none of the Lyapunov exponents is positive and this is confirmed numerically (see Creighton & Hobill 1994, Figure 3). One of the exponents is negative, however, indicating the existence of an attractor.

11.5 Summary

The discussion in this chapter can be summarized in the following conclusions.

(1) Numerical solutions, using each of the three approaches (i.e. BKL, Hamiltonian, and expansion–normalized variables), confirm that the associated discrete dynamical systems approximate the sequence of Kasner states to a high degree of accuracy.

(2) It appears that Lyapunov exponents are not particularly useful for analyzing whether the continuous dynamical systems are chaotic, since the exponents depend on the choice of time variable.

(3) Expansion–normalized variables have the advantage that the resulting DE defines a semi–flow on a *compact* negatively invariant set. Thus one can use the concepts of SDIC and topological transitivity, on which the formal definition of a chaotic dynamical system is based. The numerical solutions support the existence of a past attractor and the associated infinite heteroclinic sequences that are generated by the Kasner map. Since the Kasner map is chaotic we are able to argue that the continuous dynamical system has SDIC.

A final mention should be made of the possible physical significance of the infinite number of Kasner epochs that are predicted to occur as the initial singularity is approached (see Zel'dovich & Novikov 1983, pages 547–

† i.e. the α–limit set of a typical initial state is the whole attractor
† Note that all orbits on the attractor are either Taub heteroclinic orbits or Kasner equilibrium points.

8). Classical general relativity, on which the analysis is based, is only valid after the Planck time (given by $t_p = (\hbar G/c^5)^{1/2} \approx t^{-43}$s, where $\hbar$ is Planck's constant; e.g. Coles & Lucchin 1995, page 111), corresponding to the Planck length ($\ell_p \equiv ct_p \approx 10^{-33}$cm). The characteristic scale today is of order $\ell_0 \approx 10^{28}$cm. Thus, by (5.30), the change in τ is

$$\Delta\tau = \log(\ell_0/\ell_p) \approx 140,$$

implying that one can typically expect fewer than 10 Kasner epochs to occur (e.g. Figure 11.1).

This result casts doubt on the physical significance of the oscillatory approach to the singularity. Nevertheless, it is useful to study the singularities implied by the classical theory, both in terms of understanding that theory in its own right, and in order to be able to make comparisons with the results of a future quantum theory.

Appendix to Chapter 11
Orthonormal frame variables and the BKL equations

We use the orthonormal frame (10.42). The rate of expansion tensor (1.30) is diagonal, and

$$\Theta_\alpha = k_\alpha \ell^{-3}, \tag{11.28}$$

as follows from (1.58) with $\ell_\alpha = e^{\beta_\alpha}$, and (11.11). This implies

$$H = \tfrac{1}{3}(k_1 + k_2 + k_3)\ell^{-3}. \tag{11.29}$$

Equations (10.44) and (10.11) lead to

$$n_\alpha = \hat{n}_\alpha \ell^{-3} e^{2\beta_\alpha}. \tag{11.30}$$

Using (1.94), (1.95) and (11.30), it follows that the frame components of the spatial Ricci tensor, $^3R_{\alpha\beta} = {}^3S_{\alpha\beta} + \tfrac{1}{3}{}^3R\delta_{\alpha\beta}$, are given by

$$^3R_{\alpha\beta} = \mathrm{diag}\,(R_1, R_2, R_3)\,\ell^{-6}, \tag{11.31}$$

where the quantities R_α are defined by (11.12). The identity

$$3H^2 - \sigma^2 = \Theta_1\Theta_2 + \Theta_2\Theta_3 + \Theta_3\Theta_1,$$

shows that (11.14) is the generalized Friedmann equation (1.57). For the models under consideration, (1.65)–(1.67) give an evolution equation for $\Theta_{\alpha\beta}$:

$$\partial_0\Theta_{\alpha\beta} = -3H\Theta_{\alpha\beta} - {}^3R_{\alpha\beta} - \tfrac{1}{2}(2 - \gamma)\mu\, h_{\alpha\beta}. \tag{11.32}$$

This equation, in conjunction with (11.28), (11.31) and (11.11) gives (11.13).

Table 11.1. *Basic variables in the three approaches to class A Bianchi universes.*

BKL	Orthonormal frame	Hamiltonian
β_α	n_α	$(\beta^0, \beta^\pm)$
k_α	Θ_α or $(H, \sigma_\pm)$	$(p_0, p_\pm)$

Connections between the three approaches

For ease of reference, we now summarize in Table 11.1 the basic variables used in Chapters 6, 10 and 11, and give the various connections between them.

Definitions:

$$k_\alpha = \ell^3 \frac{d\beta_\alpha}{dt}, \qquad (p_0, p_\pm) = 12e^{3\beta^0}\left(-\frac{d\beta^0}{dt}, \frac{d\beta^\pm}{dt}\right). \tag{11.33}$$

Connections:

$$\begin{cases} \beta_1 = \beta^0 - 2\beta^+, & \beta_{2,3} = \beta^0 + \beta^+ \pm \sqrt{3}\beta^-, \\[2mm] k_1 = \frac{1}{12}(-p_0 - 2p_+), & k_{2,3} = \frac{1}{12}(-p_0 + p_+ \pm \sqrt{3}p_-), \end{cases} \tag{11.34}$$

$$\ell = e^{\beta^0} = e^{\frac{1}{3}(\beta_1 + \beta_2 + \beta_3)}, \tag{11.35}$$

$$n_\alpha = \hat{n}_\alpha \ell^{-3} e^{2\beta_\alpha}, \qquad \Theta_\alpha = \frac{d\beta_\alpha}{dt}, \tag{11.36}$$

$$H = -\frac{1}{12}e^{-3\beta^0}p_0 = \frac{d\beta^0}{dt}, \qquad \sigma_\pm = \frac{1}{12}e^{-3\beta^0}p_\pm = \frac{d\beta^\pm}{dt}. \tag{11.37}$$

Dimensionless variables:

$$\Sigma_\pm = \frac{\sigma_\pm}{H} = \frac{p_\pm}{(-p_0)}, \qquad N_\alpha = \frac{n_\alpha}{H} = \frac{12\hat{n}_\alpha e^{2\beta_\alpha}}{(-p_0)}. \tag{11.38}$$

Time variables (t is clock time):

$$\frac{dt}{d\tau} = \frac{1}{H}, \quad \frac{dt}{dT} = \ell^3, \quad \frac{dt}{d\lambda} = 12e^{3\beta_0}, \quad \frac{dt}{d\eta} = \ell, \quad \frac{dt}{d\tilde{t}} = N \tag{11.39}$$

(dimensionless)   (BKL)   (Taub time)   (conformal)   (arbitrary)

BKL time and Taub time differ by only a factor of 12. Taub time is also called supertime (Misner 1972, page 450; Jantzen 1987, page 91). The dimensionless time variable τ corresponds, up to sign, to Misner's Ω–time

(Misner 1969a). In view of its use as the density parameter, the symbol Ω is unfortunately no longer appropriate for a time variable in cosmology. For completeness, we include conformal time η, which is often used in studying perturbations of FL models.

Part three

Inhomogeneous cosmologies

12

G_2 cosmologies

C.G. HEWITT
University of Waterloo

J. WAINWRIGHT
University of Waterloo

In this chapter we discuss the G_2 cosmologies, introduced in Section 1.6, from a dynamical systems point of view, following Hewitt & Wainwright 1990. In the interests of simplicity *we restrict our considerations to the subclass of diagonal models*, apart from giving a brief overview of the whole class in Section 12.4. As the source, we use a perfect fluid with equation of state $p = (\gamma - 1)\mu$. We introduce expansion–normalized variables, in analogy with Section 5.2, with the fluid 4–velocity chosen as the timelike frame vector, and write the evolution equations, given in Section 1.6.3, as a first–order system of quasi–linear partial differential equations (PDEs) of the form

$$\hat{\partial}_0 \mathbf{y} = \mathbf{B}(\mathbf{y})\hat{\partial}_1(\mathbf{y}) + \mathbf{C}(\mathbf{y}), \tag{12.1}$$

where $\mathbf{C} : \mathbb{R}^5 \to \mathbb{R}^5$ is a rational function, and $\mathbf{B}$ is a 5×5 matrix whose entries are rational functions of $\mathbb{R}^5$ into $\mathbb{R}$. The state vector $\mathbf{y}$ is an element of the infinite–dimensional space H^5, where H is an appropriate space of functions of one spatial variable. The operator $\hat{\partial}_0$ is a differential operator along the fluid flow lines and $\hat{\partial}_1$ is a spatial differential operator, both dimensionless.

The equilibrium points of the PDE (12.1) are defined by the condition

$$\hat{\partial}_0 \mathbf{y} = \mathbf{0}. \tag{12.2}$$

In analogy with the Bianchi case, we find that the corresponding cosmological models are self–similar, and are inhomogeneous provided that $\hat{\partial}_1 \mathbf{y} \neq \mathbf{0}$. In this case the models are not transitively self–similar; the orbits of the maximal similarity group are *timelike hypersurfaces*. When (12.2) is imposed, the structure of (12.1) leads to an autonomous DE of the form

$$\hat{\partial}_1 \mathbf{y} = \mathbf{G}(\mathbf{y}), \tag{12.3}$$

which describes the spatial dependence of the inhomogeneous self–similar models. A detailed qualitative analysis of (12.3) for the diagonal G_2 models has been given by Hewitt *et al.* 1988 and Hewitt *et al.* 1991. We summarize the properties of the models in Section 12.2. Note that in this application of dynamical systems, *the independent variable is a spatial variable, rather than time.*

We have seen in Part two, that for the Bianchi models the equilibrium points (i.e. self–similar models) play an important role in determining the asymptotic behaviour. The question naturally arises whether the equilibrium points play a similar role for the G_2 models. In Section 12.3 we investigate a special class of evolving (i.e. non–self–similar) G_2 cosmologies, the separable diagonal class, in order to shed light on this question.

12.1 Evolution equations

As in Section 1.6.2, we introduce a group–invariant orthonormal frame $\{e_a\}$, with $e_0 = u$, the fluid 4–velocity, and with e_2 and e_3 tangent to the orbits of the G_2. For the diagonal subclass the physical state of the models is described by the vector (H, x), where

$$x = (\sigma_+, \sigma_-, \dot{u}_1, a_1, n_\times) \tag{12.4}$$

(see (1.131)).

12.1.1 Dimensionless variables

In analogy with (5.10), we introduce dimensionless differential operators $\hat{\partial}_0$ and $\hat{\partial}_1$ according to

$$\hat{\partial}_0 = \frac{1}{H}\partial_0, \quad \hat{\partial}_1 = \frac{1}{H}\partial_1, \tag{12.5}$$

where H is the Hubble scalar, defined in terms of the fluid 4–velocity ($H = \frac{1}{3}\Theta = \frac{1}{3}u^a{}_{;a}$). For the Bianchi models, the deceleration parameter q satisfies (5.11). In the present case, it is convenient to define q by the analogue of this equation, namely

$$\hat{\partial}_0 H = -(1+q)H, \tag{12.6}$$

i.e. q determines the time evolution of H. We also introduce a dimensionless variable r according to

$$\hat{\partial}_1 H = -rH, \tag{12.7}$$

which determines *the spatial dependence of H.*

In analogy with (5.5), the dimensionless state **y** is defined by normalizing the physical state (12.4) with H,

$$\mathbf{y} = \frac{\mathbf{x}}{H}, \tag{12.8}$$

giving

$$\mathbf{y} = (\Sigma_+, \Sigma_-, \dot{U}, A, N_\times), \tag{12.9}$$

where

$$\Sigma_\pm = \frac{\sigma_\pm}{H}, \quad \dot{U} = \frac{\dot{u}_1}{H}, \quad A = \frac{a_1}{H}, \quad N_\times = \frac{n_\times}{H}. \tag{12.10}$$

In order to write the evolution equations in the dimensionless form (12.1), it is convenient to note the following identities

$$\frac{1}{H^2}\partial_0\,\mathbf{x} = [\hat{\partial}_0 - (1+q)]\mathbf{y},$$

$$\frac{1}{H^2}\partial_1\,\mathbf{x} = (\hat{\partial}_1 - r)\mathbf{y}, \tag{12.11}$$

which are a consequence of (12.5)–(12.8). In order to obtain (12.1), however, we need to express q and r in terms of **y** and $\hat{\partial}_1\,\mathbf{y}$. We accomplish this by writing the Raychaudhuri equation (1.141) (equivalently (1.42)), the generalized Friedmann equation (1.144) and the field equation (1.145) in dimensionless form. As in Section 5.2.2, we define the density parameter, shear parameter and curvature parameter according to (5.16)–(5.18):

$$\Omega = \frac{\mu}{3H^2}, \quad \Sigma^2 = \frac{\sigma^2}{3H^2}, \quad K = -\frac{{}^3R}{6H^2}. \tag{12.12}$$

Since the perfect fluid is irrotational the only difference in the Raychaudhuri equation is the term $\frac{1}{3}\dot{u}^a{}_{;a} = \frac{1}{3}(h^a{}_b\nabla_a\dot{u}^b + \dot{u}_a\dot{u}^a)$. To account for this, we define a new dimensionless variable B by

$$B = \frac{(h^a{}_b\nabla_a\dot{u}^b + \dot{u}_a\dot{u}^a)}{3H^2}. \tag{12.13}$$

As in Section 5.2.2 equation (1.57) now assumes the form (5.19),

$$\Omega + \Sigma^2 + K = 1, \tag{12.14}$$

and (1.42), in conjunction with (12.5), (12.6), (12.12), (12.13) and the equation of state gives

$$q = 2\Sigma^2 + \tfrac{1}{2}(3\gamma - 2)\Omega - B. \tag{12.15}$$

The variable B can be expressed in terms of **y** and $\hat{\partial}_1\,\mathbf{y}$ by comparing (1.141) and (1.42), and then making use of (12.8)–(12.11). The desired expression for r follows from (1.145) on using (1.125), (1.140) and (12.7)–(12.11).

There is one further obstacle to overcome before we can give the final evolution equation (12.1): the system of equations in Section 1.6.3 do not contain an evolution equation for $\dot{u}_1$, and hence for $\dot{U}$. We obtain an equation for $\hat{\partial}_0 \dot{U}$ by applying the commutator $[\hat{\partial}_0, \hat{\partial}_1]$ to Ω. Since we are using a perfect fluid with $p = (\gamma - 1)\mu$, the contracted Bianchi identities (1.152) and (1.153) give equations for $\hat{\partial}_0\mu$ and $\hat{\partial}_1\mu$ which can be converted to dimensionless form in the usual way, giving

$$\hat{\partial}_0\Omega = [2q - (3\gamma - 2)]\Omega, \quad \hat{\partial}_1\Omega = (2r - \tfrac{\gamma}{\gamma-1}\dot{U})\Omega. \tag{12.16}$$

Here we are assuming $\gamma \neq 1$, since otherwise $\dot{U} = 0$. The commutator (1.154), when translated to dimensionless form, reads

$$\left[\hat{\partial}_0, \hat{\partial}_1\right] = (\dot{U} - r)\hat{\partial}_0 + (q + 2\Sigma_+)\hat{\partial}_1, \tag{12.17}$$

which when applied to $\log\Omega$ gives the desired equation for $\hat{\partial}_0\dot{U}$.

We finally note that the anisotropic spatial curvature $^3S_{\alpha\beta}$, as defined by (1.146), (1.147) and (1.137), enters into the evolution equations. For the diagonal models the only non–zero components are $^3S_\pm$. We define their dimensionless counterparts in the usual way:

$$S_\pm = \frac{^3S_\pm}{H^2}. \tag{12.18}$$

We now summarize the full set of evolution equations, obtained by specializing the equations in Section 1.6.3 and transforming them to dimensionless form.

$$\hat{\partial}_0\Sigma_+ = -(2 - q)\Sigma_+ - B - \dot{U}A - S_+, \tag{12.19}$$
$$\hat{\partial}_0\Sigma_- = -(2 - q)\Sigma_- - \dot{U}N_\times - S_-, \tag{12.20}$$
$$\hat{\partial}_0\dot{U} = (q + 2\Sigma_+)\dot{U} + 3(\gamma - 1)(\dot{U} - r), \tag{12.21}$$
$$\hat{\partial}_0 A = (q + 2\Sigma_+)A - (\hat{\partial}_1 - r + \dot{U})(1 + \Sigma_+), \tag{12.22}$$
$$\hat{\partial}_0 N_\times = (q + 2\Sigma_+)N_\times - (\hat{\partial}_1 - r + \dot{U})\Sigma_-, \tag{12.23}$$

where q and r are given by

$$q = \tfrac{1}{2}(3\gamma - 2)(1 - K) + \tfrac{3}{2}(2 - \gamma)\Sigma^2 - B, \tag{12.24}$$

$$r(1 + \Sigma_+) = \hat{\partial}_1\Sigma_+ - 3(\Sigma_+ A + \Sigma_- N_\times), \tag{12.25}$$

and

$$B = \tfrac{1}{3}(\hat{\partial}_1 - r + \dot{U} - 2A)\dot{U}, \tag{12.26}$$

$$\Sigma^2 = \Sigma_+^2 + \Sigma_-^2 \,. \qquad (12.27)$$

The spatial curvature terms are given by

$$K = -\tfrac{2}{3}(\hat{\partial}_1 - r)A + A^2 + N_\times^2 \,, \qquad (12.28)$$

$$\mathcal{S}_+ = -\tfrac{1}{3}(\hat{\partial}_1 - r)A + 2N_\times^2 \,, \qquad \mathcal{S}_- = (\hat{\partial}_1 - r - 2A)N_\times \,. \qquad (12.29)$$

Comment: The evolution equation for $\dot{U}$ was defined under the assumptions $\gamma \neq 1$ and $\Omega > 0$. If $\gamma = 1$, then $\dot{U} = 0$, and the remaining equations (12.19)–(12.29) simplify somewhat and describe the dust models.

In assessing the structure of the evolution equation defined by (12.19)–(12.29), it should be noted that the spatial derivatives $\hat{\partial}_1 \mathbf{y}$ are contained in r, B, K (and hence in q) and $\mathcal{S}_\pm$. Since the derivatives $\hat{\partial}_1 \mathbf{y}$ appear linearly, *the system of evolution equations (12.19)–(12.29) is of the form (12.1)*. The matrix $\mathbf{B}(\mathbf{y})$ can be written down explicitly. Its explicit form is not important, only the fact that if $\gamma > 1$ it is a non–singular matrix. It should be stressed, however, that *(12.1) represents a formal system of PDEs*, since the differential operators are not simply partial derivative operators, but satisfy the commutator (12.17) (see the remarks following (1.77)).

For future reference we note that if the commutator is applied to $\log H$, then (12.6) and (12.7) lead to

$$\hat{\partial}_0 r - \hat{\partial}_1 q = (\dot{U} - r)(1 + q) + (q + 2\Sigma_+)r \,, \qquad (12.30)$$

which is a useful auxiliary equation.

12.1.2 The spatially homogeneous subclass

The diagonal G_2 cosmologies contain a subclass of non–tilted Bianchi cosmologies, which are determined by the conditions

$$\hat{\partial}_1 \mathbf{y} = 0, \quad \hat{\partial}_1 H = 0, \quad \dot{U} = 0. \qquad (12.31)$$

For this subclass equation (12.25), in conjunction with (12.7), leads to the restriction

$$\Sigma_+ A + \Sigma_- N_\times = 0.$$

The four possibilities† are listed in Table 12.1 (see Tables 6.1, 6.2, 7.1 and 7.2), together with the possible FL models.

---

† The connections between the state vectors (7.9) and (12.9) are $\tilde{A} = A^2, \tilde{\Sigma} = \Sigma_-^2$ and $\tilde{N} = N_\times^2$, with the restrictions $N_+ = \Delta = 0$, $3\Sigma_+^2 + \tilde{h}\tilde{\Sigma} = 0$, and $3\tilde{N} + \tilde{h}\tilde{A} = 0$. For Bianchi VI$_0$ $n_\alpha{}^\alpha = 0$, one can either use N_- (as in Figure 6.10) or $N_\times$ as a variable (see (1.125)).

Table 12.1. *Non–tilted Bianchi and FL cosmologies which are special cases of the diagonal G_2 cosmologies, and satisfy (12.31).*

Type	Restrictions	Phase portrait
Bianchi I	$A = N_\times = 0$	Figure 6.4
Bianchi VI$_0$ $(n_\alpha{}^\alpha = 0)$	$A = 0, N_\times \neq 0, \Sigma_- = 0$	Figure 6.10
Bianchi VI$_h$ $(n_\alpha{}^\alpha = 0)$	$A \neq 0, N_\times \neq 0,$	Figures 7.3(a,b)
Bianchi V	$A \neq 0, N_\times = 0, \Sigma_+ = 0$	Figure 7.3(c)
Open FL	$A \neq 0, N_\times = 0, \Sigma_\pm = 0$	Figure 7.3(c)
Flat FL	$A = N_\times = 0, \Sigma_\pm = 0$	All portraits

The diagonal G_2 cosmologies also contain some classes of spatially homogeneous models which do not satisfy all of the restrictions in (12.31). First, the closed FL universes arise when

$$\Sigma_\pm = 0, \quad \hat{\partial}_1 H = 0, \quad \dot{U} = 0, \quad K < 0.$$

It follows from (12.19), (12.20), (12.28) and (12.29) that

$$\hat{\partial}_1 A = 6N_\times^2, \quad \hat{\partial}_1 N_\times = 2AN_\times, \quad K = -3N_\times^2 + A^2.$$

Second, the Kantowski–Sachs models arise when

$$\hat{\partial}_1 \Sigma_\pm = 0, \quad \hat{\partial}_1 H = 0, \quad \dot{U} = 0,$$

and

$$N_\times^2 = \tfrac{1}{3}A^2, \quad \hat{\partial}_1(\hat{\partial}_1 A - 2A^2) = 0, \quad \hat{\partial}_1 A - 2A^2 > 0.$$

Finally, we note that the G_2 cosmologies also contain a subclass of tilted Bianchi cosmologies (see Table 12.4). The tilted models have $\hat{\partial}_1 y \neq 0$, since here we are using a fluid–aligned frame. They are more simply described using a frame that is aligned with the G_3.

12.2 Equilibrium points: self–similar models

As discussed in the introduction to this chapter, the equilibrium points of the PDE (12.1) are given by (12.2). Since the matrix $\mathbf{B}(\mathbf{y})$ is invertible (assuming $\gamma > 1$), (12.2) implies (12.3), and hence that

$$\hat{\partial}_0 \hat{\partial}_1 \mathbf{y} = 0.$$

It now follows from (12.24)–(12.29) and (12.14) that

$$\hat{\partial}_0 r = 0, \quad \hat{\partial}_0 q = 0, \quad \hat{\partial}_0 \Omega = 0.$$

In addition, the commutator (12.17) applied to **y** yields

$$(q + 2\Sigma_+)\hat{\partial}_1 \mathbf{y} = 0. \tag{12.32}$$

The spatially homogeneous equilibrium points, which satisfy (12.31), are contained in the Bianchi invariant sets listed in Table 12.1 and can be read off from the corresponding figures. Of particular importance are the Kasner equilibrium points, which are given by (12.31) and

$$A = N_\times = 0, \quad \Sigma_+^2 + \Sigma_-^2 = 1 \tag{12.33}$$

We now consider the spatially inhomogeneous equilibrium points for which

$$\hat{\partial}_1 \mathbf{y} \neq 0. \tag{12.34}$$

It is proved in Hewitt & Wainwright (1990, pages 2306–7) that *a diagonal G_2 cosmology corresponds to a spatially inhomogeneous equilibrium point if and only if the space–time is self–similar*, admitting a maximal similarity group H_3 acting on the timelike hypersurfaces generated by the KVFs of the G_2 and the fluid 4–velocity. This result is analogous to the characterization of the equilibrium points of the Bianchi models as self–similar space–times in Section 5.3.3. However, the equilibrium point solutions which satisfy (12.34) are *spatially inhomogeneous* since the dimensionless variables vary in the spacelike direction normal to the orbits of the H_3.

It is possible to give a complete qualitative analysis of these equilibrium point solutions (Hewitt *et al.* 1988 and Hewitt *et al.* 1991).† We now give a brief description of the procedure, and of the properties of the models. Equations (12.32) and (12.34) imply

$$q + 2\Sigma_+ = 0, \tag{12.35}$$

which, in conjunction with (12.21), (12.22) and (12.23) gives

$$r = \dot{U}, \quad \hat{\partial}_1 \Sigma_\pm = 0. \tag{12.36}$$

Thus q and $\Sigma_\pm$ are constants. It follows from (12.16) that

$$q = \tfrac{1}{2}(3\gamma - 2), \quad \Sigma_+ = -\tfrac{1}{4}(3\gamma - 2). \tag{12.37}$$

In addition, (12.36) and (12.25) relate $\dot{U}$ to A and $N_\times$:

$$(1 + \Sigma_+)\dot{U} = -3(\Sigma_+ A + \Sigma_- N_\times). \tag{12.38}$$

† These solutions have been discussed from a Hamiltonian point of view by Uggla (1992).

The only equations in (12.19)–(12.23) which are not identically satisfied at this stage are (12.19) and (12.20). These equations lead to an autonomous DE‡ for $(A, N_\times)$ and $\hat\partial_1$, involving two parameters, namely γ and Σ_-. Assuming $1 < \gamma < 2$, we write

$$\hat\partial_1 = \frac{3}{8(\gamma - 1)} \frac{d}{d\eta},$$

and introduce new variables U, V and a new parameter† b according to

$$A = \tfrac{3}{4}(2 - \gamma)U, \quad N_\times = \tfrac{\sqrt{3}}{4}(V + sbU),$$

$$\Sigma_- = \tfrac{\sqrt{3}}{4}bs, \qquad s^2 = (2 - \gamma)(3\gamma - 2). \tag{12.39}$$

The DE is

$$\begin{aligned} U' &= s^2(1 - b^2)(1 - U^2) - V^2, \\ V' &= -bs[k(1 - U^2) - V^2] + 4(\gamma - 1)(2 - \gamma)UV, \end{aligned} \tag{12.40}$$

where

$$k = s^2(1 - b^2) + 4(\gamma - 1)(2 - \gamma), \tag{12.41}$$

and $'$ denotes $d/d\eta$ (Hewitt *et al.* 1988, page 1318 and Hewitt & Wainwright 1990, page 2310). The density parameter Ω is given by

$$\Omega = \frac{3}{16(\gamma - 1)}[k(1 - U^2) - V^2]. \tag{12.42}$$

The DE (12.40) describes the 'spatial evolution' of the models, that is, the spatial variation of physical quantities. Keep in mind that *the evolution in time is determined by the existence of the similarity group, which implies that all dimensionless scalars are constant in time,* and that there is an initial singularity.

There are two cases $k > 0$ and $k < 0$, depending on whether the vacuum boundary $\Omega = 0$ is an ellipse or a hyperbola, respectively. The detailed analysis for the case $k > 0$ is given in Hewitt *et al.* 1988, and for the case $k < 0$ in Hewitt *et al.* 1991. We note that by (12.41) $k > 0$ is equivalent to $b < b_1(\gamma)$, where

$$b_1(\gamma) = \left[\frac{7\gamma - 6}{3\gamma - 2}\right]^{\frac{1}{2}}. \tag{12.43}$$

If $b < b_1(\gamma)$ it follows that the models are *spatially bounded,* i.e. all physical

---

‡ This DE is of the form (12.3), but on account of (12.36) and the fact that $\dot U$ is related to A and $N_\times$ by (12.38), it contains only two non–trivial equations.

† Since the evolution equations (12.19)–(12.23) are invariant under $(\Sigma_-, N_\times) \to (-\Sigma_-, -N_\times)$ we may, without loss of generality assume $b \geq 0$.

and geometrical scalars (e.g. μ, H) are bounded on each spacelike hyper-surface orthogonal to the fluid 4–velocity (Hewitt *et al.* 1988, pages 1319 and 1322). On the other hand, if $b > b_1(\gamma)$ (i.e. $k < 0$) a detailed analysis shows that the models are spatially bounded if and only if $b = b_2(\gamma)$, where

$$b_2(\gamma) = \left[\frac{3\gamma - 2}{2 - \gamma}\right]^{\frac{1}{2}} \tag{12.44}$$

(Hewitt *et al.* 1991, page 1509), and the spatial geometry has cylindrical symmetry (*ibid*, pages 1512–13).

In the case $b = b_2(\gamma)$ the orbits of the DE (12.40) are unbounded. In place of U and V it is convenient to use the variables W and Z, given by†

$$W = \dot{U}, \quad Z = n[b_2(\gamma)^2 A - \sqrt{3}N_x]\dot{U}, \tag{12.45}$$

where by (12.38),

$$\dot{U} = b_2(\gamma)^2(A - \sqrt{3}N_x), \tag{12.46}$$

and

$$n = \frac{2 - \gamma}{2(\gamma - 1)}.$$

Writing

$$\hat{\partial}_1 = \frac{d}{dx},$$

the resulting DE has the form

$$\begin{aligned}
W' &= -Z, \\[2mm]
Z' &= \frac{1}{b_2(\gamma)^2}\left[n^2W^2 - \frac{(5\gamma - 2)}{2(\gamma - 1)}Z - \tfrac{9}{4}(3\gamma - 2)^2\right],
\end{aligned} \tag{12.47}$$

where $'$ denotes d/dx. The density parameter is given by

$$\Omega = \frac{4(\gamma - 1)}{3(3\gamma - 2)^2}\left[n^2W^2 - \frac{(3\gamma - 2)}{(\gamma - 1)}Z - \tfrac{9}{4}(3\gamma - 2)^2\right]. \tag{12.48}$$

In summary, if $b < b_1(\gamma)$ (i.e. $k > 0$), the spatial structure of the self–similar models is determined by the DE (12.40), with Ω given by (12.42), in terms of the variables U, V. If $b > b_1(\gamma)$ (i.e. $k < 0$), the spatial structure is determined by the DE (12.47), with Ω given by (12.48), in terms of the variables W, Z, and b is given by $b = b_2(\gamma)$.

The phase portraits and density profiles (i.e. the graph of Ω as a function

---

† These variables are the variables $\tilde{W}$ and Z used in Hewitt *et al.* 1991, up to a constant scaling (see their equation (3.4)).

Table 12.2. *The self–similar diagonal G_2 cosmologies with μ and H*
bounded on each spacelike hypersurface orthogonal to **u**. *The equation of*
state parameter γ satisfies $1 < \gamma < 2$, and $b_1(\gamma) = [(7\gamma - 6)/(3\gamma - 2)]^{1/2}$,
$$b_2(\gamma) = [(3\gamma - 2)/(2 - \gamma)]^{1/2}.$$

Restriction	Description	Phase portrait and density profile
$b = 0$	Spatially periodic (except if $V \equiv 0$)	Figure 12.1(a)
$0 < b < 1$	Asymptotically spatially homogeneous and matter–dominated (except if $V \to 0$)	Figure 12.1(b)
$b = 1$	Asymptotically spatially homogeneous and vacuum–dominated	Figure 12.1(c)
$1 < b < b_1(\gamma)$	Asymptotically inhomogeneous and vacuum–dominated	Figure 12.1(c)
$b = b_2(\gamma) > b_1(\gamma)$	Cylindrically symmetric, asymptotically inhomogeneous and vacuum–dominated	Figure 12.2(a)

of the spatial variable) are shown in Figure 12.1 ($b < b_1(\gamma)$) and Figure 12.2(a) ($b = b_2(\gamma) > b_1(\gamma)$). The equilibrium points in the interior of the state space when $0 \le b < 1$ correspond to non–vacuum self–similar Bianchi models, the Collins type VI_0 model if $b = 0$ and the Collins type VI_h model if $0 < b < 1$ (see Tables 6.3 and 9.2). In each case the equilibrium points on the boundary of the state space correspond to vacuum models, with ($\dot{U} \ne 0$ and $B \ne 0$ (see (12.13)), except when $b = 1$. The names given to the various classes, summarized in Table 12.2, thus reflect the shape of the density profile and the asymptotic behaviour at large spatial distances ($\eta \to \pm\infty$ or $x \to \pm\infty$). Because the interior equilibrium points correspond to Bianchi models, *one can interpret the models with $0 \le b < 1$ as exact inhomogeneous perturbations of these Bianchi models.*

Comments:

(1) In the phase portrait in Figure 12.1(a) the orbit with $V = 0$ is exceptional. Since it joins two vacuum equilibrium points, the corresponding cosmological model is vacuum–dominated, and has a density profile as in Figure 12.1(c). This model plays an important role in Section 12.3.2.

(2) The phase portrait in Figure 12.2(a) requires comment. The points on the line $W = 0$ correspond to the axis in the cylindrically symmetric cosmological model (Hewitt *et al.* 1991, page 1513). Thus each half–orbit $W \geq 0$ describes the inhomogeneity in a particular model, with $W = 0$ giving the axis, and the equilibrium point with $W > 0$, $Z = 0$ corresponding to spatial infinity. The other half–orbit describes the same physical situation.

12.3 Evolving models: the separable class

In this section we give a qualitative analysis of the subclass of diagonal G_2 cosmologies which satisfy the conditions

$$r = \dot{U} \neq 0, \quad \hat{\partial}_1 \Sigma_{\pm} = 0. \tag{12.49}$$

It can be shown that these conditions are equivalent to the line–element being *separable* in the canonical coordinate system, i.e.

$$ds^2 = N^2(-dt^2 + dx^2) + R(fdy^2 + f^{-1}dz^2),$$

where N, R, f are functions of t and x which are separable (e.g. $N = N_1(t)N_2(x)$), and t is a co–moving coordinate:

$$\mathbf{u} = N^{-1}\frac{\partial}{\partial t}.$$

To the best of the authors' knowledge, all known exact inhomogeneous G_2 cosmologies with equation of state $p = (\gamma - 1)\mu$ and $0 < \gamma < 2$ have a line–element and fluid 4–velocity of this form.[†] It is thus of interest to give a unified analysis of this subclass, and to discuss their asymptotic behaviour. Although all of the self–similar diagonal models are separable (see (12.36)), the separable class also contains *non–self–similar models* (i.e. $\hat{\partial}_0 \mathbf{y} \neq \mathbf{0}$, so that the dimensionless state $\mathbf{y}$ evolves in time), which is our principal interest in this section.

12.3.1 Time evolution equation

Without loss of generality, we can locally write the differential operators $\hat{\partial}_0$ and $\hat{\partial}_1$ in the form

$$\hat{\partial}_0 = \alpha \frac{\partial}{\partial \tau}, \quad \hat{\partial}_1 = \beta \frac{\partial}{\partial \chi}, \tag{12.50}$$

[†] We would like to thank J. M. Senovilla and M. Mars for discussions about these solutions.

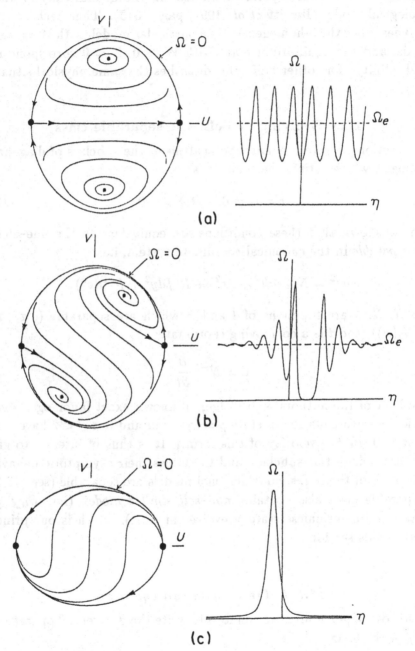

Fig. 12.1. Spatial phase portraits and density profiles for the self–similar models in Table 12.1, with $\Omega_e = \frac{3}{4}(1-\gamma)(1-b^2)$. Case (a) is $b = 0$, case (b) is $0 < b < 1$ and case (c) is $b = 1$ and $1 < b < b_1(\gamma)$.

where α and β are positive functions of τ and χ. It follows from (12.30), (12.49), (12.50) and (12.21) that

$$\frac{\partial q}{\partial \chi} = 0. \tag{12.51}$$

When the commutator (12.17) is evaluated using (12.49–(12.50) one finds that

$$\frac{\partial \alpha}{\partial \chi} = 0, \quad \frac{\partial^2 \log \beta}{\partial \chi \partial \tau} = 0.$$

Thus, by redefining χ and τ in (12.50), one can obtain

$$\hat{\partial}_0 = \frac{\partial}{\partial \tau}, \quad \hat{\partial}_1 = F(\tau)\frac{\partial}{\partial \chi}, \tag{12.52}$$

where

$$\frac{dF^2}{d\tau} = 2(q + 2\Sigma_+)F^2. \tag{12.53}$$

The function F can be rescaled by a constant multiple. Equations (12.21), (12.22) and (12.23) now imply that†

$$(\dot{U}, A, N_\times) = F(\tau)(\dot{U}^*(\chi), A^*(\chi), N_\times^*(\chi)), \tag{12.54}$$

where, on account of (12.25) and (12.49), the spatial functions satisfy

$$(1 + \Sigma_+)\dot{U}^* = -3(\Sigma_+ A^* + \Sigma_- N_\times^*). \tag{12.55}$$

By using (12.49), (12.26) and (12.28), equation (12.54) leads to

$$K = F^2(\tau)K^*(\chi), \quad B = F^2(\tau)B^*(\chi). \tag{12.56}$$

Equations (12.14) and (12.16), in conjunction with (12.49), (12.52), (12.53) and (12.56), now yield‡

$$\Sigma_+ = -\tfrac{1}{4}(3\gamma - 2), \quad 1 - \Sigma_+^2 - \Sigma_-^2 = mF^2(\tau), \tag{12.57}$$

where m is a constant, and with (12.24)

$$2 - q = \tfrac{3}{2}(2 - \gamma)cF^2(\tau), \tag{12.58}$$

where c is a constant, with the factor $\tfrac{3}{2}(2-\gamma)$ inserted for later convenience. When substituted in (12.53), equations (12.57) and (12.58) give the *time evolution equation*:

$$\frac{dF^2}{d\tau} = 3(2 - \gamma)(1 - cF^2)F^2. \tag{12.59}$$

---

† In the rest of this section, the notation ∗ is used to denote the spatially dependent part of a dimensionless variable that remains after the time dependence has been factored out.
‡ These conclusions depend on the assumption $\dot{U} \neq 0$ in (12.49).

We note that $dF^2/d\tau = 0$ implies $\hat{\partial}_0 y = 0$, on account of (12.33), (12.9), (12.54) and (12.57). Thus, equilibrium points of the DE (12.59) determine equilibrium points (12.2) of the PDE (12.1). The case $F^2 = 0$ leads to a Kasner solution, on account of (12.33), (12.54) and (12.57). Referring to (6.16) and (6.17), the Kasner exponents for this solution are given by

$$p_1 = \tfrac{1}{3}(1 - 2\cos\psi), \qquad p_{2,3} = \tfrac{1}{3}(1 + \cos\psi \pm \sqrt{3}\sin\psi),$$

$$\cos\psi = -\tfrac{1}{4}(3\gamma - 2). \tag{12.60}$$

The case $F^2 = 1/c$ leads to one of the spatially inhomogeneous solutions in Section 12.2, as will be explained in Sections 12.3.2 and 12.3.3.

At this stage, (12.55) can only be satisfied provided that

$$N_\times^* \frac{d\Sigma_-}{d\tau} = 0,$$

leading to two classes.

Class 1: $N_\times = 0$

Equation (12.20), with (12.57)–(12.59), leads to

$$m = \tfrac{3}{16}(3\gamma + 2)(2 - \gamma)c, \qquad \Sigma_-^2 = \tfrac{3}{16}(3\gamma + 2)(2 - \gamma)(1 - cF^2). \tag{12.61}$$

Class 2: $\dfrac{d\Sigma_-}{d\tau} = 0$

Since we are considering evolving (i.e. non–self–similar) solutions, we have $\dfrac{dF^2}{d\tau} \neq 0$. Thus (12.57) implies $m = 0$, i.e.

$$\Sigma_+^2 + \Sigma_-^2 = 1. \tag{12.62}$$

For both classes the time evolution is determined by (12.59), and depends on whether $c > 0$, $c = 0$ or $c < 0$.

12.3.2 Separable evolving models (class 1)

In class 1, on account of (12.55), there is only one independent spatial function, $A^*(\chi)$. Equation (12.19) leads to a differential equation for A^* which can be solved explicitly. The line–element for this class of solutions has been given by Wainwright & Goode (1980), and depends on two essential parameters, γ and a constant conformal factor.

The constant c in (12.59) is required to be positive, and on account of (12.61), $1 - cF^2 \geq 0$. Equation (12.59) thus implies that

$$\lim_{\tau \to -\infty} F^2 = 0, \qquad \lim_{\tau \to +\infty} F^2 = 1/c.$$

These limiting values determine equilibrium point solutions (12.2) of the PDE (12.1). The case $F^2 = 0$ gives a Kasner solution (12.61). The case $F^2 = 1/c$ gives a spatially inhomogeneous self-similar solution. This solution satisfies $N_\times = 0$ (class 1) and $\Sigma_- = 0$ (see (12.61)). Thus, in the DE (12.40), we have

$$b = 0, \quad V = 0.$$

Its orbit is shown in Figure 12.1(a), and joins the two vacuum equilibrium points on the U-axis. The corresponding model is vacuum-dominated at spatial infinity and its density parameter has a maximum value of

$$\Omega_{\max} = 3(2 - \gamma)(7\gamma - 6)/[16(\gamma - 1)]$$

as follows from (12.42) and (12.41). This solution was first given by Goode (1980; see Hewitt *et al.* 1988, page 1324).

The conclusion is that *the class 1 models are past asymptotic to the Kasner solution with exponents (12.60), and future asymptotic to the above inhomogeneous self-similar model.*

12.3.3 Separable evolving models (class 2)

In class 2 there are two independent spatial functions A^* and $N_\times^*$. Equations (12.19) and (12.20) lead to an autonomous DE for $(A^*, N_\times^*)$ which is similar to that obtained in Section 12.2 for the self-similar models (i.e. (12.40)), since the overall factors of $F(\tau)$ cancel. The only difference is the appearance of the constant c, first introduced in (12.58). Since the evolving models satisfy the restriction (12.62), it follows from (12.57), (12.39) and (12.41) that b and k are related to γ according to

$$b^2 = \frac{3\gamma + 2}{3\gamma - 2}, \quad k = -4(2 - \gamma)^2. \tag{12.63}$$

Since $k < 0$, the state space is unbounded. One finds that irrespective of the value of c, the models are spatially bounded if and only if $b = b_2(\gamma)$, where $b_2(\gamma)$ is given by (12.44). This condition, with (12.63), leads to

$$\gamma = \tfrac{4}{3}.$$

Table 12.3. *The class 2 separable models. All except 2A(ii) are evolving.*

Class	c	q	Values of τ	Values of F^2
2A (i)	$c > 0$	$q < 1$	$\tau_{min} < \tau < +\infty$	$+\infty \to 1/c$
(ii)	$c > 0$	$q = 1$	$-\infty < \tau < +\infty$	$1/c$
(iii)	$c > 0$	$1 < q < 2$	$-\infty < \tau < +\infty$	$0 \to 1/c$
2B	$c = 0$	$q = 2$	$-\infty < \tau < +\infty$	$0 \to +\infty$
2C	$c < 0$	$2 < q$	$-\infty < \tau < \tau_{max}$	$0 \to +\infty$

In analogy with Section 12.2, we introduce as variables W^* and Z^* defined by

$$W^* = \dot{U}^*, \quad Z^* = (3A^* - \sqrt{3}N_{\times}^*)\dot{U}^*, \tag{12.64}$$

where, by (12.55), (12.57) and (12.62),

$$\dot{U}^* = 3(A^* - \sqrt{3}N_{\times}^*).$$

The resulting DE† is

$$W^{*\prime} = -Z^*, \quad Z^{*\prime} = \tfrac{1}{3}(W^{*2} - 7Z^* - 9c) \tag{12.65}$$

where $'$ denotes $d/d\chi$ (see (12.52)). The density parameter is given by

$$\Omega = \tfrac{1}{9}(\dot{W}^{*2} - 6Z^* - 9c)F(\tau)^2. \tag{12.66}$$

Comment: The variables W^* and Z^* as given by (12.64) are closely related to the variables W and Z as defined by (12.45). Specifically, when $\gamma = \tfrac{4}{3}$ and $c = 1$ we have $W = W^*$, $Z = Z^*$, and the DE (12.47) specializes to (12.65).

In summary, the DE (12.65), with the restriction $\Omega \geq 0$, describes the inhomogeneities that arise in this class of models. There are three distinct subclasses, $c > 0$, $c = 0$ and $c < 0$, shown in Figure 12.2. The time evolution is again determined by (12.59) with the constant c determining q according to (12.58). Since $\gamma = \tfrac{4}{3}$, we have

$$\frac{dF^2}{d\tau} = 2(1 - cF^2)F^2, \quad 2 - q = cF^2. \tag{12.67}$$

The different possibilities are shown in Table 12.3.

The interpretation is as follows. The length scale function is defined as

---

† This DE follows from (12.19)–(12.23) using (12.59), (12.64) and the various restrictions that have arisen in this section.

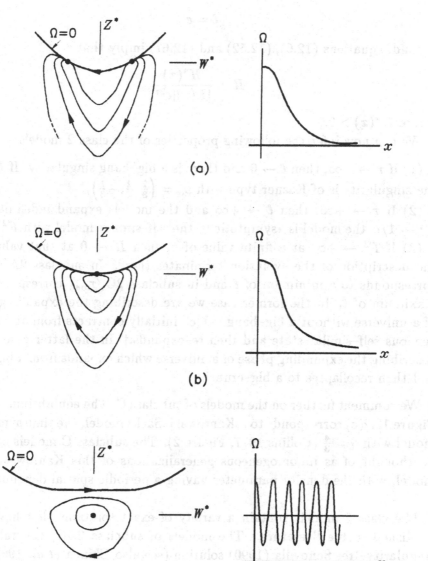

Fig. 12.2. Spatial phase portraits and density profiles for the class 2 evolving models in Table 12.2. Case (a) is subclass 2A, case (b) is subclass 2B and case (c) is subclass 2C.

usual (compare (5.6)) by

$$H = \frac{\partial_0 \ell}{\ell}. \tag{12.68}$$

It follows from (12.5) and (12.52) that

$$\ell = e^\tau.$$

Second, equations (12.6), (12.52) and (12.67) imply that

$$H = \frac{H^*(x)}{|F(\tau)|e^{2\tau}},$$

where $H^*(x) > 0$.

We can now infer the following properties of the class 2 models.

(1) If $\tau \to -\infty$, then $\ell \to 0$ and there is a big–bang singularity. If $F^2 \to 0$ the singularity is of Kasner type with $p_\alpha = \left(\frac{2}{3}, \frac{2}{3}, -\frac{1}{3}\right)$.

(2) If $\tau \to +\infty$, then $\ell \to +\infty$ and the models expand indefinitely. If $F^2 \to 1/c$, the model is asymptotic to the self–similar model with $F^2 = 1/c$.

(3) If $F^2 \to +\infty$ at a finite value of τ, then $H \to 0$ at that value, and the description of the evolution terminates there. In subclass 2A(i), $\tau_{\min}$ corresponds to a minimum of ℓ and in subclass 2C, $\tau_{\max}$ corresponds to a maximum of ℓ. In the former case we are describing the expanding phase of a universe without a big–bang, which initially contracts from an inhomogeneous self–similar state and then re–expands.† In the latter case we are describing the expanding phase of a universe which expands from a big–bang and then recollapses to a big–crunch.

We comment further on the models of subclass C. The equilibrium point in Figure 12.2(c) corresponds to a Kantowski–Sachs model, the time symmetric model with $\gamma = \frac{4}{3}$ (Collins 1977, Figure 2). The subclass C models can thus be thought of as inhomogeneous generalizations of this Kantowski–Sachs model, with the density parameter having a periodic spatial dependence.

The class 2 models contain a variety of exact solutions that have been published in the literature. The models of subclass 2A(i) generalize the singularity–free Senovilla (1990) solution (see also Chinea *et al.* 1992), and the models of subclass 2A(iii) generalize the Feinstein–Senovilla (1989) solutions. The models of subclass 2B correspond to the Davidson (1991) solution

---

† Note that these non–singular universes cannot be compatible with observations of the CBR, because if that radiation is close to isotropy back to the time of decoupling, and the usual energy conditions are satisfied, then a singularity will occur because of the Hawking–Penrose singularity theorems (Hawking & Ellis 1973).

and the models of subclass 2C correspond to solutions of Van den Bergh & Skea (1992, equations (8), (12) and (13)).

12.4 G_2 cosmologies and their specializations

As indicated in Section 1.6, for many models with high symmetry the isometry group G_r has an Abelian subgroup G_2. Hence these more special models are included in the general class of G_2 cosmologies. We now list the higher symmetry cases in Section 1.2.2 for which this is *not* true (i.e. the cases in which there is no Abelian subgroup G_2):

(i) Bianchi VIII (G_3, $s = 3$, $d = 0$),

(ii) Bianchi IX (G_3, $s = 3$, $d = 0$),

(iii) Inhomogeneous LRS, spherical orbits (G_3, $s = 2$, $d = 1$, $K > 0$),

(iv) Inhomogeneous LRS, hyperbolic orbits (G_3, $s = 2$, $d = 1$, $K < 0$).

Table 12.4 shows the five classes of perfect fluid G_2 cosmologies (see Figure 1.1) and their Bianchi specializations, that is *Bianchi models for which the isometry group G_3 admits an Abelian subgroup G_2*. The table does not include all G_2 models with higher symmetry, since there exist spatially homogeneous models with a G_4 or G_6 which has an Abelian subgroup G_2, but not an Abelian subgroup G_2 of a simply transitive subgroup G_3. The missing classes are:

(i) LRS Bianchi VIII† and IX,

(ii) Kantowski–Sachs,

(iii) closed FL ($k = +1$).

As indicated in Section 12.1.2, for the Kantowski–Sachs and closed FL universes the G_2 is diagonal, and so these models are described by the evolution equations (12.19)–(12.29).

In this chapter we have restricted our considerations to the two lowest levels in Table 12.4. We mention that the evolution equations have been formulated for the orthogonally transitive class by Hewitt & Wainwright (1990), and the associated self–similar solutions have been analyzed by Hewitt (1989).

---

† In the case of LRS Bianchi VIII, the metric also admits a one–parameter family of groups of type III (i.e. VI₋₁) with $n_\alpha{}^\alpha \neq 0$ (Ellis & MacCallum 1969, page 128).

Table 12.4. *The perfect fluid G_2 cosmologies and their specializations. The dimension of the group orbits is s and that of the isotropy subgroup is d. The columns of numbers indicate the number of arbitrary parameters in the respective class of solutions. The abbreviation HO is hypersurface–orthogonal.*

Spatially inhomogeneous $s=2$	Spatially homogeneous Anisotropic Bianchi $s=3$		LRS Bianchi $s=3$	
$d=0$	$d=0$		$d=1$	
Unrestricted	Tilted VI_0, VII_0	7		
	Tilted IV, VI_h, VII_h	7		
	Tilted II	5	None	
	Tilted V	5		
	Non–tilted $VI^*_{-1/9}$	5		
	$(n_\alpha^\alpha = 0)$			
One HO KVF	Tilted VI_0	4		
	Tilted V, VI_h	4	None	
	Non–tilted $VI^*_{-1/9}$	3		
Orthogonally transitive	Tilted VI_0, VII_0	5	–	
	Tilted IV, VI_h, VII_h	5	–	
	Tilted II	4	–	
	Non–tilted VI_0, VII_0	4	–	
	Non–tilted IV, VI_h, VII_h	4	Non–tilted VI_{-1}	3
	Non–tilted II	3	Non–tilted II	2
	$(n_\alpha^\alpha = 0)$		$(n_\alpha^\alpha = 0)$	
Diagonal	Tilted VI_0	3	–	
	Tilted V, VI_h	3	–	
	Non–tilted I, VI_0	2	–	
	Non–tilted V, VI_h	2	Non–tilted VI_{-1}	2
$d=1$	$d=1$		$d=3$	
Plane symmetric (LRS)	Tilted V LRS	2	FL ($k=-1$)	1
	Non–tilted I LRS	1	FL ($k=0$)	0

12.5 Summary

The diagonal G_2 cosmologies provide a convenient testbed for investigating the evolution of inhomogeneities, in particular density fluctuations, using both qualitative and numerical methods. We have seen that this class encompasses a wide variety of dynamical behaviour, and includes FL models, non–tilted and tilted Bianchi models, and Kantowski–Sachs models as special cases (see Sections 12.3 and 12.4), whose evolution can be described qualitatively (see Table 12.1 for the non–tilted and Section 8.5.3 for the tilted models).

In Section 13.2 we gave a qualitative description of the spatial structure that is permitted in self–similar models (the equilibrium point solutions). We have also given (Section 12.3) a complete qualitative description of the diagonal G_2 cosmologies whose line–element is separable. All separable models with an initial singularity are asymptotically self–similar (to a Kasner model) in the past, and, apart from the models of subclass 2B (which satisfy $q = 2$; see Table 12.3), all separable models which expand indefinitely are asymptotically self–similar (to an inhomogeneous model) in the future. It is thus natural to ask whether a typical diagonal G_2 cosmology which expands indefinitely from an initial singularity is asymptotically self–similar into the past and future. More specifically we ask

Q_1: is the Kasner circle $\mathcal{K}$ a past attractor, and

Q_2: is the class of inhomogeneous self–similar models a future attactor, in a suitably defined invariant set of the evolution equation (12.1) (i.e. (12.19)–(12.23))?

The first question is a special case of a long–standing conjecture of Belinskii and coworkers, that the singularity in a typical spatially inhomogeneous model will be locally spatially homogeneous (see for example Belinskii *et al.* 1982, Section 3; Zel'dovich & Novikov 1983, Section 23.3). We also note that the results of Isenberg & Moncrief (1990), on the initial singularity in the polarized Gowdy models (i.e. vacuum diagonal G_2 cosmologies with $S^1 \times S^1$ group orbits), support a positive answer to the first question.

If the second question is answered in the affirmative then the density inhomogeneities displayed by the self–similar models (see Figure 12.1) would be of considerable significance since they would represent typical inhomogeneities that are generated by the exact non–linear evolution of the EFE.

The important issue of intermediate isotropization has not arisen so far in this chapter. We have seen that the flat FL model is an equilibrium point of the evolution equations (12.19)–(12.23). Since this equilibrium point is a saddle in the Bianchi subclass it must be a saddle for the full evolution equations, and hence intermediate isotropization will occur for a subset of models. Lacking information about the stable manifold of this equilibrium point, however, we are unable to answer the following question,

Q_3: how 'large' is the subset of diagonal G_2 cosmologies which undergo intermediate isotropization?

Questions such as the above, which concern the system of PDEs (12.1), take us into the realm of dynamical systems on infinite–dimensional function spaces (e.g. Temam 1988a,b), which is well beyond the scope of this book. The natural starting point would be to investigate the local stability of the equilibrium points referred to in questions 1–3, within the infinite–dimensional state space.

In this chapter we have restricted our considerations to solutions with non–zero pressure (the case $p = \frac{1}{3}\mu$ being of particular interest). We pointed out in Section 12.1.1 that the case of dust ($p = 0, \gamma = 1$) is special as regards the structure of the evolution equation: the matrix $\mathbf{B}(\mathbf{y})$ in the general evolution equation (12.1) is singular. The evolution equations (12.19)–(12.29) simplify considerably. In particular the LRS subclass ($\Sigma_- = N_\times = 0$) has only two dependent functions Σ_+ and A. Some insight into the general evolution equations might be gained by analyzing this case, with reference to the corresponding exact solutions of the EFE, which are known explicitly (Eardley *et al.* 1972; they are also contained in the Szekeres 1975 class of solutions and hence are examples of silent universes, as considered in Chapter 13).

Further insight into the structure of the infinite–dimensional state space of the G_2 cosmologies might be obtained by systematically investigating all models which admit a simply transitive similarity group H_3.[†] The self–similar models that arose as equilibrium states of the evolution equations in Section 12.2 have the property that *the fluid 4–velocity is tangent to the orbits of the H_3*, which are thus timelike hypersurfaces. There are also self–similar G_2 cosmologies for which $\mathbf{u}$ is not tangent[‡] to the H_3 orbits. In this case the causal nature of the orbits can change over space–time. The special cases for which the H_3 is of type $_1\mathrm{I}$ or $_f\mathrm{V}$ in the Eardley (1974) classification, and the orbits are spacelike, have been analyzed qualitatively by Nilsson & Uggla (1996).

---

† In the classification of similarity groups H_3 by Eardley (1974; classes C, D and D_0 in Table 1), for all types except one the isometry subgroup G_2 is Abelian (*ibid*, page 294).
‡ Perfect fluid models for which u is orthogonal to the H_3 orbits are not of great interest, since in this case $p = \mu$ (McIntosh 1976, page 210).

13

Silent universes

M. BRUNI
Queen Mary & Westfield College

S. MATARRESE

O. PANTANO
Università di Padova, Italy

In this chapter we investigate the properties of a specific set of solutions of the Einstein field equations that describe dust ($p = 0$), with or without a cosmological constant. This class of solutions is specified by: *(i)* the assumption of zero pressure, which implies that the fluid world–lines are geodesic ($\dot{u}_a = 0$); *(ii)* the assumption that the dust is irrotational ($\omega_a = 0$); and *(iii)* the assumption that the magnetic Weyl curvature is zero ($H_{ab} = 0$; see (1.34)). In general, however, the solutions admit no symmetries. This class was initially studied in a cosmological setting by Matarrese *et al.* (1993); it contains the well–known Szekeres solutions as a special case (Szekeres 1975; see Goode & Wainwright 1982 for a unified treatment), as was shown in an earlier work by Barnes & Rowlingson (1989), who investigated a general perfect fluid subject to assumptions *(ii)* and *(iii)*. Related studies focusing on cosmological applications are those by Croudace *et al.* (1994) and Bertschinger & Jain (1994). The dynamical systems approach used here was first systematically applied to these models by Bruni *et al.* (1995a,b).

In analyzing these solutions we will make use of the evolution equations for the kinematic quantities and the Weyl curvature as given in Section 1.3. Considerable progress can be made because the assumptions imposed, namely

$$\dot{u}_a = 0, \quad \omega_a = 0, \quad H_{ab} = 0, \tag{13.1}$$

imply that the evolution equations for the shear σ_{ab}, the electric Weyl curvature E_{ab}, the matter density μ_m and the Hubble scalar H, decouple from the spatial derivatives, and hence can be written as an autonomous system of ordinary differential equations.

These solutions describe the evolution of an expanding or collapsing distribution of dust in which there is no exchange of information between different fluid elements, either by sound waves (since $p = 0$) or by gravitational waves

(since $H_{ab} = 0$). For this reason the models in this class have been called
silent universes (Matarrese *et al.* 1994a). Assuming a positive cosmological
constant Λ, taken to represent the vacuum energy density (e.g. Weinberg
1989), the study of silent universes may shed light on the conditions under
which inhomogeneous cosmologies undergo inflation. Furthermore, silent
universes with $\Lambda > 0$ will be relevant as regards the galactic epoch if the
observational data for the Hubble constant H_0 and the age of the universe
t_0 exclude FL models with $\Lambda = 0$ (see Section 3.5.1). Finally, it is worth
pointing out that originally the study of these solutions in a cosmological
setting was motivated by the hope that the condition $H_{ab} = 0$, which holds
true for first–order scalar perturbations (Goode 1989, page 2885), would
remain valid at higher order (Matarrese *et al.* 1993). Despite the fact that
this has been shown not to be the case (Matarrese *et al.* 1994a,b, and Kofman & Pogosyan 1995), these solutions, corresponding to both expanding
and collapsing distributions, may still be relevant for the study of the nonlinear stages of the evolution of inhomogeneities in the matter–dominated
epoch. These solutions are also related to the Zel'dovich (1970) approximation that is widely used to study gravitational collapse in Newtonian theory
(e.g. Matarrese *et al.* 1993).

13.1 Basic equations for silent universes

13.1.1 Decoupling of the time dependence

We introduce an orthonormal frame $\{u, e_\alpha\}$, $\alpha = 1, 2, 3$, where u is the 4–
velocity of the dust. Since u is geodesic and irrotational we can introduce a
time variable t such that

$$u = \frac{\partial}{\partial t},$$
(13.2)

and t is constant on the hypersurfaces orthogonal to u. The spatial frame
components of the shear tensor and the electric Weyl curvature are symmet-
ric trace–free matrices $\sigma_{\alpha\beta}$ and $E_{\alpha\beta}$. The constraint (13.47), given in the
Appendix, implies that $\sigma_{\alpha\beta}$ and $E_{\alpha\beta}$ commute, and hence we can choose
e_α to be a common eigenframe of these matrices. Since $\sigma_{\alpha\beta}$ and $E_{\alpha\beta}$ are
diagonal, equation (13.45), with $\alpha \neq \beta$, reduces to

$$(\sigma_{22} - \sigma_{33})\Omega_1 = 0, \quad (\sigma_{33} - \sigma_{11})\Omega_2 = 0, \quad (\sigma_{11} - \sigma_{22})\Omega_3 = 0.$$

It follows† that

$$\Omega_\alpha = 0, \tag{13.3}$$

i.e. the eigenframe is Fermi–propagated (see (1.60)).

The diagonal trace–free matrices $\sigma_{\alpha\beta}$ and $E_{\alpha\beta}$ have only two independent components, which we label using†

$$\sigma_+ = \tfrac{1}{2}(\sigma_{22} + \sigma_{33}), \quad \sigma_- = \tfrac{1}{2\sqrt{3}}(\sigma_{22} - \sigma_{33})$$

$$E_+ = \tfrac{1}{2}(E_{22} + E_{33}), \quad E_- = \tfrac{1}{2\sqrt{3}}(E_{22} - E_{33}). \tag{13.4}$$

The physical state of the silent universes is thus described by a vector $(H, \mathbf{x})$, where

$$\mathbf{x} = (\sigma_\pm, \ E_\pm, \ \mu_m, \ \Lambda), \tag{13.5}$$

where μ_m is the density of the dust and Λ is the cosmological constant.

The evolution equations for $\sigma_\pm$ and $E_\pm$ follow from (13.45) and (13.46), using (13.2)–(13.4):

$$\begin{aligned}
\dot\sigma_+ &= -2H\sigma_+ + \sigma_+^2 - \sigma_-^2 - E_+, \\
\dot\sigma_- &= -2H\sigma_- - 2\sigma_+\sigma_- - E_-, \\
\dot E_+ &= -3HE_+ - 3(\sigma_+E_+ - \sigma_-E_-) - \tfrac{1}{2}\mu_m\sigma_+, \\
\dot E_- &= -3HE_- + 3(\sigma_+E_- + \sigma_-E_+) - \tfrac{1}{2}\mu_m\sigma_-,
\end{aligned} \tag{13.6}$$

where an overdot denotes d/dt. These equations are augmented by

$$\dot\mu_m = -3H\mu_m, \quad \dot\Lambda = 0, \tag{13.7}$$

and

$$\dot H = -H^2 - \tfrac{2}{3}\sigma^2 - \tfrac{1}{6}\mu_m + \tfrac{1}{3}\Lambda, \tag{13.8}$$

as follows from (1.42) and (1.50) with

$$\mu = \mu_m + \Lambda, \quad p = -\Lambda. \tag{13.9}$$

Equations (13.6)–(13.8) govern the local dynamics of silent universes, i.e. their evolution along individual fundamental world–lines. The spatial inhomogeneity in the solutions is governed by the constraint equations which are not identically satisfied, namely (1.45), (1.47), (1.52) and (1.53). It has been shown that if these equations are satisfied on an initial hypersurface, then

† If two of the σ's are equal, say $\sigma_{22} = \sigma_{33}$, then $\Omega_2 = \Omega_3 = 0$ but Ω_1 is unrestricted. In this case, however, (13.45) implies $E_{22} = E_{33}$, and the eigenframe is not uniquely determined. We can thus set $\Omega_1 = 0$ by rotating the frame in the 23–plane using (1.121) and (1.124).

† This notation is compatible with the notation (6.2) and (6.45) introduced in Chapter 6 for the class A Bianchi universes.

they are satisfied for all time (Lesame *et al.* 1995). We note that equations
equivalent to (13.6) have been derived by Matarrese *et al.* (1993, page 1317),
and by Croudace *et al.* (1994, page 24). These authors use $\sigma_{11}, \sigma_{22}, E_{11}$ and
E_{22} as variables instead of our $\sigma_{\pm}$ and $E_{\pm}$; the two descriptions are, however,
completely equivalent.

13.1.2 Dimensionless evolution equations

We now write the evolution equations (13.6)–(13.8) in terms of expansion–
normalized variables, following Bruni *et al.* (1995a). We make some changes
in notation, however, in order to be compatible with Section 5.2 and subsequent chapters (in particular Chapters 6, 7 and 12).

The dimensionless state vector is denoted by

$$y = (\Sigma_{\pm}, \mathcal{E}_{\pm}, \Omega_m, \Omega_\Lambda), \tag{13.10}$$

where

$$\Sigma_{\pm} = \frac{\sigma_{\pm}}{H}, \quad \mathcal{E}_{\pm} = \frac{E_{\pm}}{H^2}, \tag{13.11}$$

$$\Omega_m = \frac{\mu_m}{3H^2}, \quad \Omega_\Lambda = \frac{\Lambda}{3H^2}. \tag{13.12}$$

We also introduce a dimensionless time variable τ by

$$\frac{dt}{d\tau} = \frac{1}{H}, \tag{13.13}$$

(see (5.10)), and write the evolution equation for H in terms of the deceleration parameter q, as in (5.11):

$$\frac{dH}{d\tau} = -(1+q)H. \tag{13.14}$$

Comparison of (13.8) with (13.14), using (13.13), yields

$$q = 2\Sigma^2 + \tfrac{1}{2}\Omega_m - \Omega_\Lambda, \tag{13.15}$$

where

$$\Sigma^2 = \frac{\sigma^2}{3H^2}. \tag{13.16}$$

It follows, as in Section 6.1.1, that

$$\Sigma^2 = \Sigma_+^2 + \Sigma_-^2, \tag{13.17}$$

(see (6.13)).

The evolution equation

$$\frac{d\mathbf{y}}{d\tau} = \mathbf{f}(\mathbf{y})$$

for the dimensionless state $\mathbf{y}$ is derived from (13.6) and (13.7), in conjunction with (13.11)–(13.14). We obtain

$$\Sigma'_+ = (q-1)\Sigma_+ + \Sigma_+^2 - \Sigma_-^2 - \mathcal{E}_+,$$

$$\Sigma'_- = (q-1)\Sigma_- - 2\Sigma_+\Sigma_- - \mathcal{E}_-,$$

(13.18)

$$\mathcal{E}'_+ = (2q-1)\mathcal{E}_+ - 3(\Sigma_+\mathcal{E}_+ - \Sigma_-\mathcal{E}_-) - \tfrac{3}{2}\Omega_m\Sigma_+,$$

$$\mathcal{E}'_- = (2q-1)\mathcal{E}_- + 3(\Sigma_+\mathcal{E}_- + \Sigma_-\mathcal{E}_+) - \tfrac{3}{2}\Omega_m\Sigma_-,$$

(13.19)

$$\Omega'_m = (2q-1)\Omega_m,$$

(13.20)

$$\Omega'_\Lambda = 2(q+1)\Omega_\Lambda,$$

(13.21)

where $'$ denotes $d/d\tau$, and q is given by (13.15) and (13.17). The scalars Ω_m and Ω_Λ are the density parameters associated with the dust and the cosmological constant. The (total) density parameter Ω is defined as usual by

$$\Omega = \frac{\mu}{3H^2},$$

(13.22)

(see (5.16) and (12.12)). It follows from (13.9) and (13.12) that

$$\Omega = \Omega_m + \Omega_\Lambda.$$

(13.23)

It should be noted at this point that *equations (13.18)–(13.21) describe the local dynamics of silent universes directly in terms of physically relevant expansion–normalized variables.* We note that, because of our choice of sign in (13.13), these equations describe an expanding distribution of dust.† A collapsing distribution is described by replacing the time variable τ by $-\tau$, leading to an overall change in sign in (13.18)–(13.21).

13.1.3 Spatial curvature variables

For the purpose of the dynamical systems analysis, to be given in the next section, it is desirable to introduce new variables in the evolution equations (13.18)–(13.21). These equations assume a more transparent form if the

---

† It is assumed here that $dt/d\tau > 0$, so that (13.13) holds during an expanding phase ($H > 0$), while in a collapsing phase ($H < 0$), (13.13) would be replaced by $dt/d\tau = -1/H$.

Weyl curvature and energy density are expressed in terms of the spatial curvature $^3S_{\alpha\beta}$ and ^3R and the shear, as given by (1.80) and (1.57):

$$E_{\alpha\beta} = {}^3S_{\alpha\beta} + H\sigma_{\alpha\beta} - \sigma_\alpha{}^\mu \sigma_{\mu\beta} + \tfrac{2}{3}\sigma^2 \delta_{\alpha\beta} \qquad (13.24)$$

$$\mu = \tfrac{1}{2}\,{}^3R + 3H^2 - \sigma^2. \qquad (13.25)$$

Since $\sigma_{\alpha\beta}$ and $E_{\alpha\beta}$ are diagonal, so is $^3S_{\alpha\beta}$; we label its two independent components according to

$$^3S_+ = \tfrac{1}{2}({}^3S_{22} + {}^3S_{33}), \quad {}^3S_- = \tfrac{1}{2\sqrt{3}}({}^3S_{22} - {}^3S_{33}). \qquad (13.26)$$

The dimensionless spatial curvature variables are defined by

$$S_\pm = \frac{{}^3S_\pm}{H^2}, \quad K = -\frac{{}^3R}{6H^2}. \qquad (13.27)$$

It follows from (13.4), (13.11), (13.16), (13.17) and (13.24)–(13.27) that

$$\begin{aligned}\mathcal{E}_+ &= S_+ + \Sigma_+ + (\Sigma_+^2 - \Sigma_-^2), \\ \mathcal{E}_- &= S_- + \Sigma_- - 2\Sigma_+\Sigma_-, \end{aligned} \qquad (13.28)$$

$$\Omega = 1 - \Sigma^2 - K. \qquad (13.29)$$

At this stage one can derive the evolution equations for $S_\pm$ and K, using (13.18)–(13.21), (13.23), (13.28) and (13.29). One finds that a further simplification occurs, namely that the equations partly decouple, if one introduces the following linear combinations of $S_\pm$ and K:

$$M_1 = \tfrac{1}{3}(K - 2S_+), \quad M_{2,3} = \tfrac{1}{3}(K + S_+ \pm \sqrt{3}S_-). \qquad (13.30)$$

We now give the resulting evolution equations for the variables $(\Sigma_\pm, M_\alpha, \Omega_\Lambda)$:

$$\Sigma'_\pm = -(2 - q)\Sigma_\pm - S_\pm, \qquad (13.31)$$

$$M'_1 = 2(q - \Sigma_+)M_1, \qquad (13.32)$$

$$M'_{2,3} = (2q + \Sigma_+ \pm \sqrt{3}\Sigma_-)M_{2,3}, \qquad (13.33)$$

$$\Omega'_\Lambda = 2(1 + q)\Omega_\Lambda, \qquad (13.34)$$

where

$$q = \tfrac{1}{2}(1 + 3\Sigma_+^2 + 3\Sigma_-^2 - K - 3\Omega_\Lambda), \qquad (13.35)$$

and

$$K = M_1 + M_2 + M_3,$$

$$S_+ = \tfrac{1}{2}(M_2 + M_3 - 2M_1), \qquad (13.36)$$

$$S_- = \tfrac{\sqrt{3}}{2}(M_2 - M_3).$$

The equation for q follows from (13.15), (13.17), (13.23) and (13.29), and the equations for K and $S_\pm$ are obtained by inverting (13.30). In this formulation we regard the Weyl curvature variables $\mathcal{E}_\pm$, as given by (13.28) and (13.36), and the matter density parameter Ω_m, as given by (13.38) below, as auxiliary variables.

Equations (13.31)–(13.36) describe the local dynamics of silent universes in terms of expansion-normalized spatial curvature variables; again, these equations hold during an expanding phase, because of our choice of sign in (13.13).

Before continuing, we make some remarks concerning the deceleration parameter q, as given by (13.15) or (13.35). It follows from (13.15) that if $\Omega_\Lambda > 0$ one can have $q < 0$, in which case the model is regarded as *inflationary* (the length scale satisfies $\ddot{\ell} > 0$). By using (13.23) one can also write q in the form

$$q = -1 + 3\Sigma^2 + K + \tfrac{3}{2}\Omega_m.$$

Thus if $K > 0$ (i.e. $^3R < 0$) then $q > -1$, but if $K < 0$ one can have $q < -1$ in which case one says that *superinflation* occurs i.e. $\ddot{\ell} > 0$ and $\dot{H} > 0$ (Lucchin & Matarrese 1985).

13.2 Dynamical systems analysis

In this section we apply dynamical systems methods to the evolution equations (13.31) in order to describe the local dynamics of the silent universes.

13.2.1 Invariant sets

The form of the evolution equations (13.31) implies the existence of a variety of invariant sets (see Proposition 4.1), which we now describe. The dimensionless state of a silent universe is given by a point

$$(\Sigma_\pm, M_\alpha, \Omega_\Lambda) \in \mathbb{R}^6. \qquad (13.37)$$

The density parameter Ω_m, which is now an auxiliary variable, is given by

$$\Omega_m = 1 - \Sigma_+^2 - \Sigma_-^2 - M_1 - M_2 - M_3 - \Omega_\Lambda, \qquad (13.38)$$

as follows from (13.17), (13.23), (13.29) and (13.36). Equation (13.20) implies that the subset of $\mathbb{R}^6$ defined by

$$\Omega_m \geq 0, \tag{13.39}$$

which is the physical region of state space, is an invariant set, as is the boundary $\Omega_m = 0$. Since we consider a non–negative cosmological constant, we have the restriction

$$\Omega_\Lambda \geq 0. \tag{13.40}$$

Equation (13.34) implies that this inequality defines an invariant set, whose boundary $\Omega_\Lambda = 0$ is also invariant.

Secondly, it follows from (13.32) and (13.33) that any combination of the conditions $M_\alpha > 0$, $M_\alpha = 0$ or $M_\alpha < 0$ defines an invariant set; this explicitly shows the usefulness of this choice of variables. It is of interest that *any invariant set which satisfies*

$$M_\alpha \geq 0, \quad \text{for} \quad \alpha = 1, 2, 3, \tag{13.41}$$

is bounded, as follows from (13.38)–(13.40). Finally, restrictions on the shear lead to invariant sets:

$$\Sigma_- = 0, \quad M_2 = M_3, \tag{13.42}$$

and

$$\Sigma_\pm = 0, \quad M_1 = M_2 = M_3, \tag{13.43}$$

as follows from (13.31)–(13.33). We note that the restrictions† (13.42) describe the Szekeres (1975) solutions ($\Omega_\Lambda = 0$), and their generalizations with $\Omega_\Lambda > 0$ (Barrow & Stein–Schabes 1984).

13.2.2 Spatially homogeneous models

It is important to keep in mind that equations (13.31)–(13.34) describe the evolution of silent universes *along individual fundamental world–lines*; the evolution, in particular the asymptotic behaviour and whether or not recollapse occurs, will depend on the specific world–line in general. The special cases, in which the evolution is the same for all fundamental world–lines, are spatially homogeneous models. We list these models and the corresponding invariant sets in Table 13.1. One can view the spatially inhomogeneous silent universes as 'exact non–linear perturbations' of these spatially homogeneous universes.

† There are two other invariant sets which are equivalent to (13.42) under interchange of spatial axes, namely $\Sigma_- = \sqrt{3}\Sigma_+, M_3 = M_1$ and $\Sigma_- = -\sqrt{3}\Sigma_+, M_1 = M_2$.

Table 13.1. *Spatially homogeneous silent universes.*

Class of universes	Invariant set
FL	$\Sigma_\pm = 0$, $M_1 = M_2 = M_3$
Bianchi I	$M_1 = M_2 = M_3 = 0$
Bianchi VI$_{-1}(n_\alpha{}^\alpha = 0)$	$\Sigma_- = 0$, $M_2 = M_3$, $M_1 > 0$
Kantowski–Sachs	$\Sigma_- = 0$, $M_2 = M_3$, $M_1 < 0$.

Table 13.2. *Isolated equilibrium points of the DE that describes the evolution of the silent universes. The notation is that of Bruni et al. (1995a); the letters in brackets correspond to the notation in Chapters 6, 7 and 9. All points have $\Sigma_- = \mathcal{E}_- = \mathcal{S}_- = 0$. Points marked with a ? have not been identified with known exact solutions.*

Notation	Ω_m	Ω_Λ	Σ_+	$\mathcal{E}_+$	M_1	M_2	M_3	Exact solution
LI	0	1	0	0	0	0	0	de Sitter
LII	0	3	−1	4	−3	0	0	?
LIII	0	9	2	0	0	−6	−6	?
DI (F)	1	0	0	0	0	0	0	flat FL
DII (M)	0	0	0	0	$\frac{1}{3}$	$\frac{1}{3}$	$\frac{1}{3}$	Milne
DIII (D)	0	0	$\frac{1}{2}$	0	$\frac{3}{4}$	0	0	flat Bianchi III
DVI	0	0	$-\frac{1}{4}$	$\frac{9}{32}$	0	$\frac{15}{32}$	$\frac{15}{32}$	?

13.2.3 Equilibrium points

The equilibrium points of the DE (13.31)–(13.34) can be found by successively considering the four possibilities $M_1 M_2 M_3 \neq 0$, one M_α zero, two M_α zero and all M_α zero. As canonical choices we use $M_1 = 0$, $M_2 M_3 \neq 0$ and $M_1 \neq 0$, $M_2 = M_3 = 0$. We obtain a number of isolated equilibrium points, which are listed‡ in Table 13.2, and a circle of equilibrium points which correspond to the Kasner solutions. For all the isolated points the shear is restricted: $\Sigma_- = 0$ for the above canonical choices, which implies $\mathcal{E}_- = 0$. We also give the values of the auxiliary variables Ω_m and $\mathcal{E}_+$. All equilibrium points with $\Omega_\Lambda \neq 0$ have $q = -1$.

‡ Not listed in the table are some equilibrium points which are unphysical, corresponding to $\Omega < 0$.

Kasner circle $\mathcal{K}$:

The Kasner equilibrium points form a circle $\mathcal{K}$ given by

$$\Omega_\Lambda = 0, \ M_\alpha = 0, \ \Sigma_+^2 + \Sigma_-^2 = 1, \ \Sigma_\pm = \text{constant},$$

and satisfy $\Omega_m = 0$, $q = 2$. The Kasner exponents p_α are given by

$$p_1 = \tfrac{1}{3}(1 - 2\Sigma_+), \quad p_{2,3} = \tfrac{1}{3}(1 + \Sigma_+ \pm \sqrt{3}\Sigma_-).$$

The cases $(p_1, p_2, p_3) = (1, 0, 0)$ (and cycle on 1, 2, 3) give the Taub form of flat space–time. These points are denoted by T_α in Chapter 6. The other points with two equal exponents are denoted Q_α and correspond to LRS Kasner solutions. We refer to Section 6.2.2 and Figure 6.2 for further details.

Comment: We briefly discuss the equilibrium points from the point of view of Section 5.3.3. The equilibrium points F, M, D and $\mathcal{K}$ arise as equilibrium points in the analysis of the Bianchi models, and correspond to self–similar spatially–homogeneous solutions of the EFE (see Tables 6.3, 7.3, 9.1 and 9.2). The equilibrium point LI also arises as an equilibrium point for the Bianchi models, and corresponds to the de Sitter solution. It is an exceptional point in that the corresponding solution is not self–similar (see Section 6.2.1). We have been unable to identify the equilibrium points LII, LIII and DVI with known exact solutions. However, LII is a model in which an initially spherical distribution of test particles acquires a prolate shape, while in LIII the same distribution becomes oblate. DVI was found to be the limit of a subset of the Szekeres (1975) models (see Section 13.2.5) by Bonnor & Tomimura (1976).

13.2.4 Stability of the equilibrium points

It is straightforward to show that the isolated equilibrium points are hyperbolic (Re $(\lambda) \neq 0$ for all eigenvalues λ; see Section 4.3.2). In addition, for the Kasner circle $\mathcal{K}$, all points have one zero eigenvalue and the rest positive. Thus $\mathcal{K}$ is a normally hyperbolic equilibrium set which is a local source (see Section 4.3.4) . Information about the isolated equilibrium points and their stable manifolds (see Section 4.3.3) is summarized in Table 13.3.

One can use the DE (13.31)–(13.34), or equivalently (13.18)–(13.21), to determine the asymptotic form of general solutions in the neighbourhood of the equilibrium points. We refer to Bruni *et al.* (1995b, page 445) for details.

13.2.5 Evolution of the Szekeres solutions

The Szekeres (1975) solutions are described by the three–dimensional invariant set (13.42) with $\Omega_\Lambda = 0$. The DE (13.31)–(13.34) simplifies to

$$
\begin{aligned}
\Sigma'_+ &= -(2-q)\Sigma_+ - M_2 + M_1, \\
M'_1 &= 2(q - \Sigma_+)M_1, \\
M'_2 &= (2q + \Sigma_+)M_2,
\end{aligned}
$$

where

$$
q = \tfrac{1}{2}(1 + 3\Sigma_+^2 - M_1 - 2M_2).
$$

The density parameter (13.38) reduces to

$$
\Omega_m = 1 - \Sigma_+^2 - M_1 - 2M_2.
$$

The physically significant equilibrium points† in this invariant set are F (flat FL), M (Milne), D (flat Bianchi III), DVI, and two Kasner points, T (the Taub form of flat space–time) and Q (LRS Kasner).

A clear picture of the dynamics can be obtained from the phase portraits of the invariant sets $M_1 = 0$, $M_2 = 0$ and $\Omega_m = 0$, which are shown in Figures 13.1 and 13.2.‡ The conclusion is that the future attractor is the Milne point M, and its domain of attraction is the subset with $M_1 > 0$ and $M_2 > 0$. All other orbits escape to infinity. The past attractor is $\{T, Q\}$. The equilibrium point DVI is the future attractor in the invariant set $M_1 = 0$ and D is the future attractor in the invariant set $M_2 = 0$. Both points, as well as F, are saddles in the three–dimensional state space.

13.3 Physical implications

We have shown that the de Sitter equilibrium point LI is a local sink in the invariant set $\Omega_\Lambda > 0$, the Milne equilibrium point M is a local sink in the invariant set $\Omega_\Lambda = 0$, and that the Kasner circle $\mathcal{K}$ is a local source in the invariant set $\Omega_\Lambda \geq 0$ (see Table 13.3). In this section we use these local stability results to make conjectures (which are also supported by our numerical experiments) about the evolution of the whole family of silent universes, and we then discuss their physical implications.

---

† Again, we exclude points with $\Omega_m < 0$.
‡ The vacuum Szekeres solutions corresponding to the invariant set $\Omega_m = 0$ are alternatively described by a planar DE for Σ_+ and $\mathcal{E}_+$; see Bruni *et al.* (1995b) for a discussion and phase portraits.

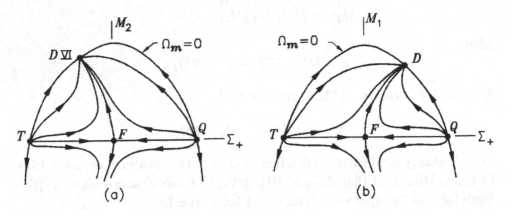

Fig. 13.1. Phase portraits for the invariant sets $M_1 = 0$ (a) and $M_2 = 0$ (b), showing the evolution of the Szekeres solutions.

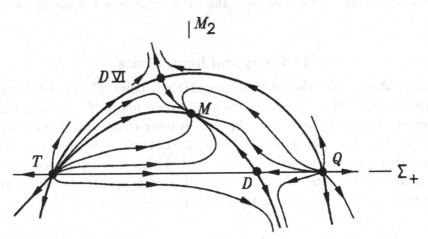

Fig. 13.2. Phase portrait for the invariant set $\Omega_m = 0$, showing the evolution of the vacuum Szekeres solutions. We have projected this invariant set, which is the surface $\Sigma_+^2 + M_1 + 2M_2 = 1$ in (Σ_+, M_1, M_2)–space into the $\Sigma_+ M_2$–plane.

Table 13.3. *Local stability of the equilibrium points for the case of*
expanding universes $(H > 0)$. *The invariant sets marked §are bounded. We*
have not been able to describe the invariant sets marked ?.

Equilibrium points	Dimension of the stable manifold	Invariant set in which the equilibrium point is a local sink
LI (de Sitter)	6	$\Omega_\Lambda > 0$
LII	5	?
LIII	4	?
DI (flat FL)	2	$\{M_1 = M_2 = M_3 = 0, \Omega_\Lambda = 0\}^\S$
DII (Milne)	5	$\{M_\alpha > 0, \Omega_\Lambda = 0\}^\S$
DIII (flat Bianchi III)	3	$\{M_{2,3} = 0, M_1 > 0, \Omega_\Lambda = 0\}^\S$
DVI	4	$\{M_1 = 0, M_{2,3} > 0, \Omega_\Lambda = 0\}^\S$
$\mathcal{K}$ (Kasner circle)	0	(Local source)

13.3.1 Asymptotic evolution near a singularity

All of the equilibrium points, apart from LI (de Sitter) and the Kasner circle $\mathcal{K}$, have an unstable manifold whose dimension n satisfies $1 \leq n \leq 5$. It follows that some orbits (a set of measure zero) will be past asymptotic to these equilibrium points. We conjecture that all other orbits are past asymptotic to $\mathcal{K}$ i.e. $\mathcal{K}$ *is the past attractor in the invariant set* $\Omega_\Lambda \geq 0$, *for expanding models* (i.e. $H > 0$). Similarly, by reversing the direction of time, it follows that $\mathcal{K}$ *is the future attractor in the invariant set* $\Omega_\Lambda \geq 0$, *for collapsing models* ($H < 0$).

The singularity type of an equilibrium point on $\mathcal{K}$ (a Kasner solution) is a cigar (also known as a spindle) except for the three points T_α (the Taub form of flat space–time) which describe a pancake singularity (see Section 6.2.2). The validity of the above conjecture means that *silent universes generically have cigar singularities* (past or future). This is surprising, in that use of the Zel'dovich (1970) approximation in Newtonian theory has led to the conjecture that the generic case is collapse to a pancake. This issue is still under debate. In General Relativity, only the axisymmetric silent universes (i.e. Szekeres solutions) can have pancake singularities.†

It is of interest that the stability of the Kasner circle $\mathcal{K}$ for silent universes differs significantly from its stability for non–tilted Bianchi universes. For the simplest type, namely Bianchi I, for which the magnetic Weyl curvature

† This possibility was pointed out by Goode & Wainwright (1982, page 3319).

is zero, the stability of $\mathcal{K}$ is necessarily the same. However, when the Bianchi type becomes more general (leading to $H_{ab} \neq 0$), only a subset of $\mathcal{K}$ is a local source and for the most general types (i.e. VIII and IX) no subset of $\mathcal{K}$ is a local source, which gives rise to the oscillatory approach to the singularity (see Sections 6.2.2, 6.4 and 7.2.3).

13.3.2 Future evolution of expanding models with $\Lambda > 0$

The de Sitter equilibrium point LI is the only local sink in the bounded invariant set

$$\Omega_\Lambda > 0, \quad M_1 \geq 0, \quad M_2 \geq 0, \quad M_3 \geq 0. \tag{13.44}$$

It is thus plausible to conjecture that all orbits in this invariant set are future asymptotic to LI. This result can be proved using the LaSalle Invariance Principle (see Section 4.4.3) and the fact that Ω_Λ is a monotone function on this invariant set. On account of (13.27) and (13.36) orbits in this invariant set describe models with $^3R \leq 0$. We also conjecture that the orbits in an open set with one or more of the M_α negative are future asymptotic to LI. This conjecture is supported by our numerical experiments and the fact that some closed† FL models with $\Lambda > 0$ expand indefinitely and approach the de Sitter universe. The physical interpretation is that *there is an open set of initial conditions with $H > 0$ for which silent universes undergo inflation as $t \to +\infty$*. This can occur even if $^3R > 0$ initially. However, there is also the implication that *there is an open set of initial conditions with $H > 0$ for which silent universes do not undergo inflation as $t \to +\infty$*, i.e. an inflationary epoch (characterized by $q < 0$), if it occurs, is followed by recollapse. This can occur even if $^3R < 0$ initially. The results of some numerical experiments which illustrate these results are given by Bruni *et al.* (1995a, Figure 1). If recollapse occurs, the effect of Λ is soon negligible, and the results are the same as for the $\Lambda = 0$ case

13.3.3 Future evolution of expanding models with $\Lambda = 0$

The Milne equilibrium point M is the only local sink in the invariant set $\Omega_\Lambda = 0$ (see Table 13.3), and it lies in the interior of the bounded subset $M_1 > 0, M_2 > 0, M_3 > 0$. It is thus plausible to conjecture that all orbits in this invariant set are future asymptotic to M. This result can be proved using the LaSalle Invariance Principle (see Section 4.4.3) and the fact that

---

† These models satisfy $M_1 = M_2 = M_3 < 0$ and $^3R > 0$, as follows from (13.27), (13.36) and Table 13.1.

$\Delta = M_1 M_2 M_3$ is a monotone function on this invariant set. For initial states, apart from a set of measure zero, that satisfy $\Omega_\Lambda = 0$ but are not in this invariant set, the orbit will escape to infinity, which means that the cosmological model reaches a state of maximum expansion ($H = 0$) and then recollapses. The exceptional initial states are those with one or more of the M_α equal to zero, whose orbits are future asymptotic to the flat FL equilibrium point F, or to D or DVI (see Tables 13.2 and 13.3).

The physical interpretation is that *there is an open set of initial conditions with $\Lambda = 0$ and $H > 0$ for which silent universes expand indefinitely and approach a Milne model, representing a spherical void* (since $\Omega_m = 0$ and $\Sigma_\pm = 0$). However, any initial condition with $\Lambda = 0$, $H > 0$ and $^3R > 0$, and an open set with $^3R \leq 0$, will lead to a state of maximum expansion and subsequent recollapse, with the orbit approaching one of the points of the Kasner circle $\mathcal{K}$, i.e. towards a cigar singularity, in general. In the axisymmetric case, these results are illustrated by the analysis of the Szekeres models given by Goode & Wainwright (1982, page 1321), which is confirmed by our analysis of Section 13.2.5.

13.4 Summary

In this chapter we have given a detailed qualitative analysis of the local dynamics of silent universes, and have briefly discussed the physical significance of the results. We have not been able to establish all of the conclusions rigorously, and in some ways our understanding is incomplete. In particular, we have not been able to give a complete description of the set of initial points in the state space for silent universes whose orbits are attracted to the de Sitter equilibrium point, and hence cannot describe the full set of initial conditions that lead to inflation as $t \rightarrow +\infty$ (see Section 13.3.2). The analysis is also incomplete in that we have not discussed the constraint equations for silent universes, which arise from (1.52)–(1.54) when the restrictions (13.1) are imposed. The analysis of Lesame *et al.* (1995) shows that these constraint equations are compatible (i.e. are preserved by the evolution equations) but we do not know, in some intuitive sense, how 'large' is the class of silent universes, except in the axisymmetric case (13.42), for which the full solutions are known explicitly (the Szekeres solutions). Nevertheless these solutions are of interest, both in terms of their applications as inhomogeneous distributions of dust, and in that their dynamics can be described by using a *finite–dimensional* invariant set of the infinite–dimensional state space of inhomogeneous models.

Appendix to Chapter 13

Equations (1.44) and (1.51) govern the evolution of silent universes. When simplified using (13.1) and expanded relative to an orthonormal frame, these equations assume the form

$$\partial_0 \sigma_{\alpha\beta} - 2\varepsilon^{\mu\nu}{}_{(\alpha}\sigma_{\beta)\mu}\Omega_\nu = -2H\sigma_{\alpha\beta} - \sigma_\alpha{}^\mu\sigma_{\mu\beta} + \tfrac{2}{3}\sigma^2\delta_{\alpha\beta} - E_{\alpha\beta}, \quad (13.45)$$

$$\partial_0 E_{\alpha\beta} - 2\varepsilon^{\mu\nu}{}_{(\alpha}E_{\beta)\mu}\Omega_\nu = -3H E_{\alpha\beta} + 3\sigma_{(\alpha}{}^\mu E_{\beta)\mu} - \sigma_\mu{}^\nu E_\nu{}^\mu\delta_{\alpha\beta} - \tfrac{1}{2}\mu\sigma_{\alpha\beta},$$

$$(13.46)$$

where

$$\partial_0 = \frac{\partial}{\partial t},$$

(see (13.2)). The constraint equation (1.54) yields

$$\varepsilon_\alpha{}^{\mu\nu}\sigma_\mu{}^\lambda E_{\nu\lambda} = 0. \tag{13.47}$$

Note added in proof:

It has recently been pointed out by Bonilla *et al* (1996) and by Maartens (private communication) that there is an error in the paper Lesame *et al.* (1995), which invalidates the conclusion that the silent universe constraint equations are preserved by the evolution equations (see Section 13.4). This consistency question is currently being investigated by Maartens, and by van Elst and Uggla (private communication). It is clear that if the restrictions (13.42), which characterize the Szekeres solutions and their generalizations with $\Lambda \neq 0$, are imposed then the silent universe constraint equations are preserved, since these solutions are known explicitly. At present, however, it is not known whether there are any other inhomogeneous silent universes. Indeed, it is possible that the silent universe assumptions (13.1) and the inhomogeneity condition $h_a{}^b\nabla_b\mu \neq 0$ uniquely determine the generalized Szekeres solutions, thereby restricting one to the invariant set (13.42).

14

Cosmological density perturbations

P. K. S. DUNSBY

University of Cape Town

Over the past four decades cosmological perturbation theory has played an important role in our attempts to understand the formation of large–scale structures in the universe. So far, most of the work done in this field has been concerned with linear perturbations of the FL cosmologies, the underlying assumption being that on a sufficiently large scale the universe can be described by a homogeneous and isotropic model. A number of approaches to this problem have been presented in the literature since the pioneering work of Lifshitz, notably the gauge–invariant formulation of Bardeen (1980). Although this approach has been widely used to describe both the origin and evolution of small perturbations from the quantum era through to the time when the linear approximation breaks down, it has three shortcomings. First, the variables are non–local, depending on unobservable boundary conditions at infinity. Second, many of the key variables have a clear physical meaning only in a particular gauge. Finally, the approach is inherently limited to linear perturbations of FL models.

Recently, Ellis & Bruni (1989), building on Hawking (1966), developed a geometrical method for studying cosmological density perturbations.† This approach, which is based on the spatial gradients of the energy density μ and Hubble scalar H, is both coordinate–independent and gauge–invariant, and the variables have an unambiguous physical interpretation. In addition their approach is of a general nature, because it starts from exact non–linear equations that can in principle be linearized about any FL or non–tilted Bianchi model. The background model is described by n dimensionless variables, denoted by $\mathbf{y} \in \mathbb{R}^n$, that satisfy a DE

$$\frac{d\mathbf{y}}{d\tau} = \mathbf{f}(\mathbf{y}), \tag{14.1}$$

---

† Lyth & Mukherjee (1988) have developed a related approach that also focuses on the fluid kinematical quantities.

where τ is the dimensionless time variable (5.9). The evolution equation for the dimensionless perturbation variables, denoted by $\mathbf{U} \in \mathbb{R}^m$, are written in the form

$$\frac{d\mathbf{U}}{d\tau} = \mathbf{g}(\mathbf{U}, \mathbf{y}), \qquad (14.2)$$

so that (14.1) and (14.2) form a coupled system of DEs with state vector $(\mathbf{y}, \mathbf{U}) \in \mathbb{R}^{n+m}$. One can then use dynamical systems techniques to investigate the stability of the perturbations, without having to solve explicitly for the background variables. This approach has so far been applied to the flat and open FL models (Woszczyna 1992, Bruni 1993, and Bruni & Piotrkowska 1993) and to the LRS Bianchi I models (Dunsby 1993).

Our goal in this chapter is to give an introduction to the geometrical perturbation method of Ellis and Bruni, and to illustrate its dynamical systems formulation (14.1)–(14.2) for flat and open FL models, and for Bianchi I models.

14.1 The geometrical approach to density perturbations

14.1.1 Preliminaries

We consider a cosmological model $(\mathcal{M}, \mathbf{g}, \mathbf{u})$, where the matter content is a perfect fluid with 4–velocity $\mathbf{u}$ and barotropic equation of state $p = p(\mu)$. The Hubble scalar is defined by

$$H = \tfrac{1}{3} u^a{}_{;a},$$

(see (1.27) with (1.31)), and determines a scale factor ℓ, up to a constant multiple, by

$$H = \frac{\dot{\ell}}{\ell}, \qquad (14.3)$$

where $\dot{\ell} = \ell_{,a} u^a$ (compare (2.10), (5.6) and (12.68)). In general, the derivative of any tensor along the fundamental congruence $\mathbf{u}$ is defined, for example, by

$$\dot{T}_{ab} = u^c \nabla_c T_{ab} = T_{ab;c} u^c. \qquad (14.4)$$

We shall make use of the concepts introduced in Section 1.1.3 for characterizing cosmological models, namely the kinematic quantities σ_{ab}, ω_{ab} and $\dot{u}_a$, and the electric (E_{ab}) and magnetic (H_{ab}) parts of the Weyl tensor. The projection tensor h_{ab}, defined by

$$h_{ab} = g_{ab} + u_a u_b,$$

(see (1.25)), will play an important role. For example, the *orthogonal co-variant derivative* $\hat{\nabla}_a$ is defined for scalars by

$$\hat{\nabla}_a f = h_a{}^b \nabla_b f, \tag{14.5}$$

and for tensors orthogonal to **u** by formulas such as

$$\hat{\nabla}_a T_{cd} = h_a{}^b h_c{}^e h_d{}^f \nabla_b T_{ef}, \tag{14.6}$$

where $u^c T_{cd} = 0$, and all free indices are projected orthogonal to u^a (see Ellis *et al.* 1990, page 1044).

14.1.2 Gauge–invariant variables

To define perturbations it is necessary to choose a one–to–one map between the given background space–time and the real inhomogeneous universe, so that at any point P the perturbation $\delta\mu$ is defined as $\delta\mu = \mu - \overline{\mu}$, where μ is the actual value and $\overline{\mu}$ the background value at P. A change in this map, keeping the background space–time fixed, is called a *gauge transformation*. In describing perturbations, in particular density perturbations, it is desirable to use gauge–invariant variables i.e. quantities that are independent of the choice of the map. We refer to Ellis & Bruni (1989, pages 1804–8) for a detailed discussion of these matters.

A fundamental result in perturbation theory is the Stewart–Walker lemma (Stewart & Walker 1974, lemma 2.2), which states that for a geometrically defined quantity to be gauge–invariant, it has to be zero in the background space–time.† In order to give a gauge–invariant description of density perturbations in a cosmological model $(\mathcal{M}, \mathbf{g}, \mathbf{u})$ we focus on the spatial density gradient

$$\hat{\nabla}_a \mu = h_a{}^b \nabla_b \mu, \tag{14.7}$$

defined as in (14.5). This quantity gives a covariant description of density inhomogeneities, and is in principle observable (Ellis & Bruni 1989, pages 1809–10). We also wish to compare the density gradient to the density itself, and so we form the fractional density gradient

$$\frac{\hat{\nabla}_a \mu}{\mu}. \tag{14.8}$$

The final consideration is that we would like to work with a dimensionless

† Or be a constant scalar or a tensor that is a constant linear combination of Kronecker deltas; these cases are not of consequence to us here.

quantity, and so we multiply (14.8) by the scale factor ℓ, and define the *comoving fractional density gradient* $\mathcal{D}_a$ by

$$\mathcal{D}_a = \frac{\ell \hat{\nabla}_a \mu}{\mu}, \tag{14.9}$$

(see Ellis & Bruni 1989, page 1810). The essential point is that, by the Stewart–Walker lemma, $\mathcal{D}_a$ is a *gauge–invariant quantity for perturbations of FL universes and of non–tilted Bianchi universes.* We note that the magnitude

$$\mathcal{D} = (\mathcal{D}_a \mathcal{D}^a)^{\frac{1}{2}} \tag{14.10}$$

is the gauge–invariant scalar that most closely corresponds to the intention of the usual gauge–dependent quantity $\delta\mu/\mu$.

When one calculates $\dot{\mathcal{D}}_a$ (Section 14.2) one finds that it depends on the spatial gradient of H. It is convenient to represent this by the gauge–invariant comoving gradient:

$$\mathcal{Z}_a = 3\ell \; \hat{\nabla}_a H. \tag{14.11}$$

In addition to the gauge–invariant quantities (14.10) and (14.11), which are directly related to density perturbations, we note that the kinematic quantities σ_{ab}, ω_{ab} and $\dot{u}_a$, and the Weyl tensor variables E_{ab} and H_{ab}, are gauge–invariant variables for FL models (on account of the Stewart–Walker lemma; see (2.55) and (2.56)). Of these only ω_{ab} and $\dot{u}_a$ are, in general, gauge–invariant for non–tilted Bianchi models.

14.1.3 Exact non–linear evolution equations

We now derive evolution equations for the gauge–invariant vectors $\mathcal{D}_a$ and $\mathcal{Z}_a$. In doing so we make use of the contracted Bianchi identities (1.48) and (1.49), which for a perfect fluid reduce to

$$\dot{\mu} + 3(\mu + p)H = 0 \tag{14.12}$$

$$(\mu + p)\dot{u}_a + \hat{\nabla}_a p = 0, \tag{14.13}$$

in terms of the notation (14.5). For later use, we note that we are assuming a barotropic equation of state $p = p(\mu)$, and introduce the notation

$$w = \frac{p}{\mu}, \quad c_s^2 = \frac{dp}{d\mu}, \tag{14.14}$$

where c_s is the speed of sound. It follows from (14.12) that

$$\dot{w} = -3H(1 + w)(c_s^2 - w). \tag{14.15}$$

In order to derive the evolution equations we need an identity that permits the interchange of the two derivative operators, ˙ and $\hat{\nabla}_a$. A straightforward calculation using (1.26) yields

$$h_a{}^b(\ell\hat{\nabla}_b f)^{\cdot} = \ell(\hat{\nabla}_a + \dot{u}_a)\dot{f} - \ell(\sigma_a{}^b + \omega_a{}^b)\hat{\nabla}_b f, \qquad (14.16)$$

for any scalar f (see Ellis *et al.* 1990, equation (92)). We choose $f = \log\mu$ in (14.16), and after using (14.9) and (14.11)–(14.13), obtain

$$h_a{}^b(\mathcal{D}_a)^{\cdot} = 3wH\mathcal{D}_a - (\sigma_a{}^b + \omega_a{}^b)\mathcal{D}_b - (1+w)\mathcal{Z}_a. \qquad (14.17)$$

In deriving the second equation it is convenient to introduce the notation

$$\mathcal{R} = -3H^2 + \mu + A - 2\sigma^2 + 2\omega^2, \qquad (14.18)$$

where

$$A = \dot{u}^a{}_{;a}. \qquad (14.19)$$

The Raychaudhuri equation (1.42) then assumes the form

$$\dot{H} = \tfrac{1}{3}\mathcal{R} - \tfrac{1}{2}(\mu + p). \qquad (14.20)$$

We now choose $f = H$ in (14.16) and after using (14.9), (14.11)–(14.13), (14.18) and (14.20), obtain

$$\begin{aligned} h_a{}^b(\mathcal{Z}_b)^{\cdot} = {} & -2H\mathcal{Z}_a - (\sigma_a{}^b + \omega_a{}^b)\mathcal{Z}_b - \tfrac{1}{2}\mu\mathcal{D}_a \\ & + \ell\mathcal{R}\dot{u}_a + \ell\,\hat{\nabla}_a(A - 2\sigma^2 + 2\omega^2). \end{aligned} \qquad (14.21)$$

The exact evolution equations (14.17) and (14.21) play a central role in the geometrical approach to density perturbations.

14.2 Perturbations of FL universes

In this section we apply the fundamental evolution equations (14.17) and (14.21) to the FL universes. We shall make use of the Friedmann equation (1.57) which, for an FL model, reads

$$\mu - 3H^2 = \tfrac{1}{2}\,{}^3R = \frac{3k}{\ell^2}, \qquad (14.22)$$

(see (2.4)).

14.2.1 The linearized evolution equations

In order to linearize (14.17) and (14.21) about a FL background, we treat the quantities that are not zero in the background, namely μ, p and H, as *zero order*, and those that are zero in the background (and hence gauge invariant) as *first-order*.

The linearization is performed by dropping all products of first–order quantities in (14.17) and (14.21). We also use (14.13)–(14.15) to express $\dot{u}_a$ in terms of $\mathcal{D}_a$. The result is

$$\dot{\mathcal{D}}_{\perp a} = 3wH\mathcal{D}_a - (1+w)\mathcal{Z}_a, \tag{14.23}$$

$$\dot{\mathcal{Z}}_{\perp a} = -2H\mathcal{Z}_a - \left[\tfrac{1}{2}\mu + c_s^2\frac{(\mu - 3H^2)}{1+w}\right]\mathcal{D}_a + \ell\,\hat{\nabla}_a A, \tag{14.24}$$

where $\dot{\mathcal{D}}_{\perp a} = h_a{}^b(\mathcal{D}_b)\dot{}$ etc.

To proceed with the analysis in the general case $p \neq 0$ we have to rewrite the term $\hat{\nabla}_a A$ (which is zero for pressure–free matter). In order to simplify matters we now restrict our considerations to an irrotational fluid (i.e. $\omega_{ab} = 0$). To first order, it then follows, using (14.13)–(14.15) and the properties of $\hat{\nabla}_a$, that

$$\ell\hat{\nabla}_a A = -\frac{c_s^2}{(1+w)}\left(\hat{\nabla}^2 - \frac{2k}{\ell^2}\right)\mathcal{D}_a, \tag{14.25}$$

where $\hat{\nabla}^2 = \hat{\nabla}_a\hat{\nabla}^a$ (Ellis *et al.* 1990, equations (17) and (23)).

Using (14.25), we can now combine (14.23) and (14.24) to give a second–order DE in $\mathcal{D}_a$ alone:

$$\ddot{\mathcal{D}}_{\perp a} + \mathcal{A}(t)H\dot{\mathcal{D}}_{\perp a} + \mathcal{B}(t)H^2\mathcal{D}_a - c_s^2\left(\hat{\nabla}^2 - \frac{2k}{\ell^2}\right)\mathcal{D}_a = 0, \tag{14.26}$$

where

$$\mathcal{A}(t) = 2 - 3w - 3(w - c_s^2),$$

$$\mathcal{B}(t) = -\tfrac{3}{2}\left[(1-w)(1+3w) + 6(w - c_s^2)\right]\Omega + \frac{12k}{H^2\ell^2}(w - c_s^2), \tag{14.27}$$

(Ellis *et al.* 1990, equation (26) with $\omega_{ab} = 0$, $\Lambda = 0$ and $\Theta = 3H$; we have also introduced the density parameter Ω as defined by (2.12)).

Equation (14.26) is a linearised equation for structure growth in a FL background. Because the perturbation quantity is a vector, in general the time derivatives conceal connection terms; however as this is a covariant equation one can evaluate it in any desired basis, and a convenient choice is a frame that is parallel–propagated along the fluid flow lines. Relative to this basis, the covariant derivatives become ordinary derivatives.

14.2.2 The pressure–free case

In the case of pressure–free matter ($p = 0$), it follows from (14.14) that

$$w = 0, \quad c_s^2 = 0,$$

and the coefficients (14.27) simplify significantly yielding $\mathcal{A}(t) = 2$, $\mathcal{B}(t) = -\frac{3}{2}$. Using a parallel–propagated frame, so that we can drop the projections in the covariant derivatives, (14.26) becomes

$$\ddot{\mathcal{D}}_a + 2H\dot{\mathcal{D}}_a - \tfrac{3}{2}\Omega H^2 \mathcal{D}_a = 0.$$

For a flat FL model ($\Omega = 1$, $H = 2/(3t)$), the general solution of this DE is (14.34).

14.2.3 The harmonic decomposition

In order to proceed further when $p \neq 0$, it is necessary to introduce *scalar harmonics*, which are defined as comoving eigenfunctions of the covariant Laplace–Beltrami operator (Hawking 1966):

$$\hat{\nabla}^2 Q^{(n)} = -\frac{n^2}{\ell^2} Q^{(n)}, \qquad \dot{Q}^{(n)} = 0.$$

These scalar harmonics can be used to expand gauge–invariant first–order scalars:†

$$f = \Sigma f_{(n)} Q^{(n)}.$$

In order to expand vectors we define (Bruni *et al.* 1992, page 51)

$$Q_a^{(n)} = -\frac{\ell}{n}\hat{\nabla}_a Q^{(n)},$$

which implies $\dot{Q}_a^{(n)} = 0$. It follows (*ibid*, page 52) that

$$\left(\hat{\nabla}^2 - \frac{2k}{\ell^2}\right) Q_a^{(n)} = -\frac{n^2}{\ell^2} Q_a^{(n)}. \tag{14.28}$$

The vector $\mathcal{D}_a$ can thus be expanded according to

$$\mathcal{D}_a = \sum_n \mathcal{D}_{(n)} Q_a^{(n)}. \tag{14.29}$$

---

† We use Σ as a symbol for a sum that could be a summation over a discrete set of values or an integral over a continuously varying index, depending on the eigenvalues n (Bruni *et al.* 1992, page 51).

On substituting (14.29) into (14.26) and using (14.28), the harmonics decouple, yielding

$$\ddot{\mathcal{D}}_{(n)} + \mathcal{A}(t)H\dot{\mathcal{D}}_{(n)} + \left(\mathcal{B}(t) + \frac{c_s^2 n^2}{H^2 \ell^2}\right) H^2 \mathcal{D}_{(n)} = 0, \qquad (14.30)$$

where $\mathcal{A}(t)$ and $\mathcal{B}(t)$ are given by (14.27).

Thus, the density perturbations of a FL model are described by a second–order linear DE with coefficients that depend on the given background and equation of state, through H, Ω, ℓ, w and c_s^2 (see (14.27)). These coefficients are thus written as functions of t in (14.30). This DE corresponds to the basic DE (4.9) in Bardeen (1980).† Specifically, the variable $\mathcal{D}_{(n)}$ satisfies the same evolution equation as Bardeen's density perturbation variable ε_m. Bardeen's DE (4.9) has different coefficients because he uses $\mu \ell^3 \varepsilon_m$ as the dependent variable and conformal time η, defined by $dt/d\eta = \ell$, as the independent variable. We also note that for the flat FL model ($k = 0, \Omega = 1$), (14.30) coincides with the linearized DE of Lyth & Stewart (1990, page 346, equation (38) with $\eta = 0$; see also Padmanabhan 1993, page 145, equation (4.88)).

We now restrict our considerations to an equation of state $p = (\gamma - 1)\mu$, where γ is a constant. It follows from (14.14) that

$$w = \gamma - 1, \quad c_s^2 = \gamma - 1. \qquad (14.31)$$

The DE (14.30) assumes the form

$$\ddot{\mathcal{D}}_{(n)} + (5 - 3\gamma)H\dot{\mathcal{D}}_{(n)} + \left[-\tfrac{3}{2}(2 - \gamma)(3\gamma - 2)\Omega + \frac{(\gamma - 1)n^2}{H^2 \ell^2}\right] H^2 \mathcal{D}_{(n)} = 0. \qquad (14.32)$$

14.2.4 Perturbations of flat FL

We now consider the case of the flat FL background, described by setting

$$\Omega = 1, \qquad H = \tfrac{2}{3\gamma}t^{-1}, \qquad \ell = t^{\frac{2}{3\gamma}}, \qquad (14.33)$$

in (14.32). For dust ($\gamma = 1$), we observe that the solution of (14.32) is independent of n and so we drop the subscript (n). The general solution is

$$\mathcal{D} = c_+ t^{\frac{2}{3}} + c_- t^{-1}, \qquad (14.34)$$

---

† The terms on the right side of Bardeen's equation are zero here because of our assumptions about the equation of state.

giving the usual growing and decaying models. Similar results have been obtained by many other workers, but the present derivation has the great advantage of being both covariant and gauge–invariant (many papers that use specific gauges include unphysical gauge modes, and the physical meaning of the other modes is gauge–dependent).

For other values of γ, the general solution of (14.32) can be written in terms of Bessel functions. We refer to Goode (1989, page 2889, equation (6.1)), who gives the solution for Bardeen's DE in this situation. The use of harmonics, however, enables one to define a 'long–wavelength limit', i.e. to consider the solutions for wavelengths $\lambda = 2\pi\ell/n$ larger than the Hubble distance: $H^{-1}/\lambda \ll 1$, or equivalently

$$\frac{n}{H\ell} \ll 1, \tag{14.35}$$

(Bruni *et al.* 1992, page 44). For a flat background, in the long–wavelength limit (14.30) reduces to

$$\ddot{\mathcal{D}} + (5 - 3\gamma)H\dot{\mathcal{D}} - \tfrac{3}{2}(2 - \gamma)(3\gamma - 2)H^2\mathcal{D} = 0,$$

independently of n. Introducing the dimensionless time variable τ defined by $\tau = \log(\ell/\ell_0)$ (see (2.43)), we obtain a DE with constant coefficients:†

$$\mathcal{D}'' + \tfrac{1}{2}(10 - 9\gamma)\mathcal{D}' - \tfrac{3}{2}(2 - \gamma)(3\gamma - 2)\mathcal{D} = 0,$$

where $'$ denotes $d/d\tau$. The general solution, in terms of t, is

$$\mathcal{D} = c_+ t^{\frac{2(3\gamma-2)}{3\gamma}} + c_- t^{-\frac{(2-\gamma)}{\gamma}}. \tag{14.36}$$

Thus for radiation ($\gamma = \tfrac{4}{3}$), in the long–wavelength limit, we have

$$\mathcal{D} = c_+ t + c_- t^{-\frac{1}{2}}.$$

We refer to Ellis *et al.* (1989, page 1852) for a discussion of this result.

14.2.5 The autonomous DE

We now consider the case of the open FL background. In this case it is not possible to treat the coefficients of (14.32) as functions of t and obtain explicit solutions. An alternative approach to analyzing the perturbations, suggested by Woszcznya (1992) (see also Bruni 1993), is to regard the coefficients in the DE (14.30) as functions of the state variables that describe the background space–time, and augment the DE with evolution equations for

† Use (14.33), or the general equations (2.44)–(2.46).

the background variables. One can then determine the nature of the perturbations by giving a qualitative analysis of the resulting autonomous system
of DEs. The idea is to use expansion–normalized dimensionless variables to
describe the state of the background space–time. As in Section 2.3, we use
the density parameter Ω, whose evolution equation is (2.48) in terms of the
time variable $\tau = \log(\ell/\ell_0)$ (see (2.43)). We write the basic DE (14.30) in
terms of τ, using (2.44) and (2.46), and express the coefficients in terms of
Ω using (2.12) and (2.13) with $k = -1$. The result is

$$\mathcal{D}''_{(n)} + \psi(\Omega)\mathcal{D}'_{(n)} + \xi(\Omega)\mathcal{D}_{(n)} = 0, \tag{14.37}$$

where

$$\psi(\Omega) \;=\; (4 - 3\gamma) - \tfrac{1}{2}(3\gamma - 2)\Omega,$$

$$\xi(\Omega) \;=\; -\tfrac{3}{2}(2 - \gamma)(3\gamma - 2)\Omega + n^2(\gamma - 1)(1 - \Omega). \tag{14.38}$$

We now write (14.37) as a first–order DE, by introducing

$$\mathcal{U}_{(n)} = \frac{\mathcal{D}'_{(n)}}{\mathcal{D}_{(n)}}. \tag{14.39}$$

This leads to

$$\mathcal{U}'_{(n)} = -\mathcal{U}^2_{(n)} - \psi(\Omega)\mathcal{U}_{(n)} - \xi(\Omega). \tag{14.40}$$

Finally, we recall that Ω satisfies (2.48), i.e.

$$\Omega' = -(3\gamma - 2)(1 - \Omega)\Omega, \tag{14.41}$$

Equations (14.40) and (14.41), which correspond to (14.2) and (14.1), re-
spectively, form an autonomous DE for $(\Omega, \mathcal{U}_{(n)})$ which describes the evolu-
tion of density perturbations in an open $(\Omega < 1)$ FL universe.

The definition (14.39) requires comment. The usual way to write a second–
order DE such as (14.37) in first–order form is to use $(X, Y) = (\mathcal{D}_{(n)}, \mathcal{D}'_{(n)})$
as variables. We should thus regard $\mathcal{U}_{(n)}$ as $\tan\theta_{(n)}$, where $\theta_{(n)}$ is the usual
polar angle in the $(\mathcal{D}_{(n)}, \mathcal{D}'_{(n)})$–plane, with $0 \leq \theta_{(n)} < 2\pi$. Consider the
subset D of $\mathbb{R}^2$ defined by

$$0 \leq \Omega \leq 1, \quad -\infty < \mathcal{U}_{(n)} < +\infty, \tag{14.42}$$

and the *cylinder* $[0, 1] \times S^1$ defined by

$$0 \leq \Omega \leq 1, \quad -\tfrac{\pi}{2} \leq \theta_{(n)} < \tfrac{3\pi}{2}, \tag{14.43}$$

with $-\frac{\pi}{2}$ and $\frac{3\pi}{2}$ identified. The set D is the image, under the $2-1$ map

$$(\Omega, \theta) \mapsto (\Omega, \tan \theta),$$

of the two halves of the above cylinder that are given by $-\frac{\pi}{2} < \theta < \frac{\pi}{2}$ and $\frac{\pi}{2} < \theta < \frac{3\pi}{2}$. It follows that as a point makes one revolution on the cylinder, the image point traverses the strip D twice. In principle we should use $\theta_{(n)}$ as the variable and derive the DE for $\theta'_{(n)}$ that corresponds to (14.45). In practice, however, it is more convenient to use the DE (14.40), rather than the corresponding one for $\theta'_{(n)}$, and to draw the phase portraits in the subset of $\mathbb{R}^2$ defined by (14.42). From a conceptual point of view, however, we shall regard the state space as the cylinder $[0, 1] \times S^1$ given by (14.43). Since this set is compact, there will be a future attractor for all values of γ and n, which will determine the nature of the perturbations. We also note that the invariant sets $\Omega = 0$ and $\Omega = 1$ (see (14.41) are *circles* on the cylinder.

Because of the above interpretation one can think of $\mathcal{U}_{(n)}$ as the *phase of the perturbation*. If $\mathcal{U}_{(n)} > 0$ during the evolution, it follows from (14.39) that the perturbation variables $\mathcal{D}_{(n)}$ are growing at that time (since $\mathcal{D}_{(n)} > 0$ and $\mathcal{D}'_{(n)} > 0$, or $\mathcal{D}_{(n)} < 0$ and $\mathcal{D}'_{(n)} < 0$), while if $\mathcal{U}_{(n)} < 0$, the perturbation is decaying. If an orbit of the DE (14.40) is asymptotic to an equilibrium point, then the perturbation approaches a stationary state, decaying to zero if $\mathcal{U}_{(n)} < 0$ and growing if $\mathcal{U}_{(n)} > 0$. If an orbit is asymptotic to a periodic orbit (on the cylinder), then the perturbation oscillates indefinitely, i.e. the perturbation propagates as sound waves.

14.2.6 Stability of the open FL models

The qualitative analysis of the DE (14.40)–(14.41) on the invariant set (14.42) is simple, since Ω is a monotone function for $0 < \Omega < 1$. The equilibrium points are given by

$$\Omega = 1, \quad \mathcal{U}_{(n)} = -\tfrac{3}{2}(2 - \gamma), \quad (3\gamma - 2),$$

$$\Omega = 0, \quad \mathcal{U}_{(n)} = -\tfrac{1}{2}(4 - 3\gamma)\left[1 \pm \sqrt{1 - (\tfrac{n}{n_{\text{crit}}})^2}\right],$$

(14.44)

where

$$n_{\text{crit}}^2 = \frac{(4 - 3\gamma)^2}{4(\gamma - 1)}. \tag{14.45}$$

The parameter n satisfies $n^2 > 1$ (Bruni 1993, page 740). In order for there to be equilibrium points with $\Omega = 0$, we must have $n^2 < n_{\text{crit}}^2$, which entails

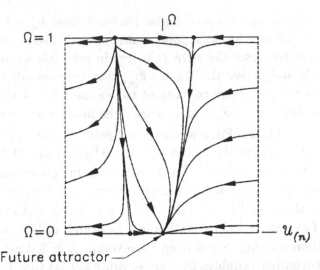

Fig. 14.1. Density perturbations in the open FL models, described by orbits in the $(\mathcal{U}_{(n)}, \Omega)$ state space, showing the cases in which there are two equilibrium points with $\Omega = 0$, one of which is the future attractor. The physical interpretation depends on whether the future attractor has $\mathcal{U}_{(n)} = 0$ or $\mathcal{U}_{(n)} < 0$ (see Table 14.1).

$n_{\text{crit}}^2 > 1$. By (14.44), this condition is equivalent to

$$1 < \gamma < \tfrac{10}{9} \quad \text{or} \quad \gamma > 2.$$

If $\gamma = 1$, the equilibrium points with $\Omega = 0$ are

$$\Omega = 0, \quad \mathcal{U}_{(n)} = 0 \quad \text{or} \quad -1,$$

independently of the value of n.

The phase portraits can be drawn by elementary means. There are two topologically inequivalent cases, shown in Figures 14.1 and 14.2. In accordance with the remarks at the end of Section 14.2.3, we again interpret the state space as the cylinder (14.43). In the case in which there are no equilibrium points with $\Omega = 0$ (Figure 14.2), the invariant set $\Omega = 0$ is a periodic orbit, which is the future attractor. In the case in which there are two equilibrium points with $\Omega = 0$ (Figure 14.1), one is a saddle and the other is the future attractor. The physical interpretation is described in Table 14.1.

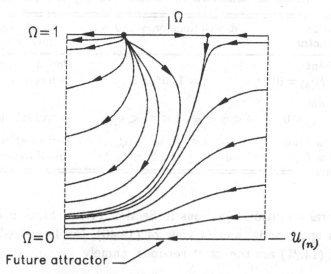

Fig. 14.2. Density perturbations in the open FL models, described by orbits in the $(\mathcal{U}_{(n)}, \Omega)$ state space, showing the cases in which there are no equilibrium points with $\Omega = 0$, and the future attractor is the periodic orbit $\Omega = 0$ (see Table 14.1).

14.3 Perturbations of Bianchi I universes

In this section we show that the dynamical systems approach to density perturbations can be applied to models more general than FL by considering the simplest class of anisotropic models, namely the Bianchi I solutions (see Section 6.3.1 and Figure 6.4). These solutions are given explicitly in Section 9.3.1, but we will use the qualitative analysis of the solutions given in Section 6.3.1 instead. We give a brief discussion, referring to Dunsby (1993) for details. Density perturbations in Bianchi I models have been studied, using a traditional approach, by Perko *et al.* (1972) and Doroshkevich *et al.* (1971).

14.3.1 The linearized evolution equations

Since our goal is to illustrate the dynamical systems approach in a simple context, we restrict our considerations to perturbed solutions for dust with zero vorticity, thereby imposing the restrictions

$$\dot{u}_a = 0, \quad \omega_{ab} = 0, \quad w = 0, \quad c_s^2 = 0, \tag{14.46}$$

Table 14.1. *The different behaviours of density perturbations in the open FL model, as described by the DE (14.40) and (14.41).*

Future attractor	Restrictions on γ and n	Physical interpretation
Point $\Omega = 0$, $U_{(n)} = 0$	$\gamma = 1$, $n^2 > 1$	Perturbation evolves to a constant value
Point $\Omega = 0$, $U_{(n)} < 0$	$1 < \gamma < \frac{10}{9}$, $1 < n^2 < n_{\text{crit}}^2$	Perturbation decays
Periodic orbit $\Omega = 0$	$1 < \gamma < \frac{10}{9}$, $n^2 > n_{\text{crit}}^2$, or $\frac{10}{9} \leq \gamma < 2$, $n^2 > 1$	Perturbation is a sound wave

on the general evolution equations in Section 14.1.3. Since $\sigma_{ab} \neq 0$ in the background space–time, and in view of (14.46), the linearized versions of (14.17) and (14.21) are the exact versions, namely

$$\dot{\mathcal{D}}_{\perp a} = -\sigma_a{}^b \mathcal{D}_b - \mathcal{Z}_a, \tag{14.47}$$

$$\dot{\mathcal{Z}}_{\perp a} = -2H\mathcal{Z}_a - \sigma_a{}^b \mathcal{Z}_b - \tfrac{1}{2}\mu\mathcal{D}_a - 2\mathcal{S}_a, \tag{14.48}$$

where

$$\mathcal{S}_a = \ell\hat{\nabla}_a \sigma^2. \tag{14.49}$$

One can derive a second–order DE for $\mathcal{D}_a$ by eliminating $\mathcal{Z}_a$:

$$\ddot{\mathcal{D}}_{\perp a} + 2H\dot{\mathcal{D}}_{\perp a} + 2\sigma_a{}^b\dot{\mathcal{D}}_b - \tfrac{1}{2}\mu\mathcal{D}_a - H\sigma_a{}^b\mathcal{D}_b + \sigma_a{}^b\sigma_b{}^c\mathcal{D}_c - 2\mathcal{S}_a = 0, \tag{14.50}$$

(a special case of equation (44) in Dunsby 1993).

Due to the presence of $\mathcal{S}_a$, the DE (14.50) does not, as it stands, determine the evolution of $\mathcal{D}_a$. Various assumptions can be made to deal with this term. The simplest is to consider *isocurvature perturbations*, defined by the condition

$$\hat{\nabla}_a{}^3R = 0. \tag{14.51}$$

This restriction leads to the elimination of $\mathcal{S}_a$, as follows. The generalized Friedmann equation (1.57) reads

$$^3R = 2(-3H^2 + \sigma^2 + \mu). \tag{14.52}$$

This equation, in conjunction with (14.9), (14.11) and (14.49) implies that

$$\mathcal{S}_a = 2H\mathcal{Z}_a - \mu\mathcal{D}_a. \tag{14.53}$$

By using (14.47) we can thus express $\mathcal{S}_a$ in terms of $\mathcal{D}_a$ and $\dot{\mathcal{D}}_{\perp a}$. We substitute the resulting expression in (14.50), obtaining

$$\ddot{\mathcal{D}}_{\perp a} + 6H\dot{\mathcal{D}}_{\perp a} + 2\sigma_a{}^b\dot{\mathcal{D}}_{\perp b} + \tfrac{3}{2}\mu\mathcal{D}_a + 3H\sigma_a{}^b\mathcal{D}_b + \sigma_a{}^b\sigma_b{}^c\mathcal{D}_c = 0, \quad (14.54)$$

(Dunsby 1993, equation (76)).

14.3.2 The autonomous DE

In Section 6.3.1, the Bianchi I models are described using an orthonormal frame $\{\mathbf{u}, \mathbf{e}_\alpha\}$ that is an eigenframe of the shear tensor, and is Fermi-propagated ($\Omega_\alpha = 0$; see (1.60)). The two independent frame components of the shear tensor are denoted by $\sigma_\pm$, where

$$\sigma_+ = \tfrac{1}{2}(\sigma_{22} + \sigma_{33}), \quad \sigma_- = \tfrac{1}{2\sqrt{3}}(\sigma_{22} - \sigma_{33}), \quad (14.55)$$

(see (6.2)). The expansion–normalized counterparts are defined by

$$\Sigma_\pm = \frac{\sigma_\pm}{H}, \quad (14.56)$$

(see (6.8)).

We now expand the DE (14.54) relative to an orthonormal frame $\{\mathbf{u}, \mathbf{e}_\alpha\}$ that coincides with the above frame when $\mathcal{D}_a = 0$. We can use the freedom in the choice of the $\mathbf{e}_\alpha$ (a time–dependent rotation) to ensure that the frame is Fermi–propagated ($\Omega_\alpha = 0$). The perturbed shear tensor $\sigma_{\alpha\beta}$ relative to this frame will not be diagonal in general. However, we can regard $\sigma_{\alpha\beta}$ as being effectively diagonal in (14.54): since this is a linearized DE, the terms involving non–diagonal components of $\sigma_{\alpha\beta}$ will be products of first–order quantities and hence will be dropped. Since $\Omega_\alpha = 0$, the covariant derivatives will become ordinary derivatives with respect to clock time t. In this way, (14.54) yields three uncoupled second–order DEs for the $\mathcal{D}_\alpha$, $\alpha = 1, 2, 3$, with the coefficients depending on H and $\Sigma_\pm$. Note that we use (14.52) to eliminate μ, the terms involving ^3R, which is zero in the background, being dropped. The shear scalar σ^2 is expressed in terms of the dimensionless shear scalar $\Sigma^2 = \sigma^2/(3H^2)$ (see (5.17)), which is given by

$$\Sigma^2 = \Sigma_+^2 + \Sigma_-^2, \quad (14.57)$$

(see (6.13)).

The next step is to introduce the dimensionless time variable $\tau = \log(\ell/\ell_0)$, as in Section 14.2.4. It follows from (5.10) and (5.11) that

$$\dot{\mathcal{D}}_\alpha = H\mathcal{D}'_\alpha, \quad \ddot{\mathcal{D}}_\alpha = H^2[\mathcal{D}''_\alpha - (1+q)\mathcal{D}'_\alpha], \quad (14.58)$$

where ' denotes differentiation with respect to τ. Equation (6.11) with $\gamma = 1$ gives

$$1 + q = \tfrac{3}{2}(1 + \Sigma^2) - \tfrac{1}{2}K, \qquad (14.59)$$

where we note that $K = 0$ in the background. The result is the following three uncoupled second-order DEs:

$$\mathcal{D}_\alpha'' + \psi_\alpha(\Sigma_\pm)\mathcal{D}_\alpha' + \xi_\alpha(\Sigma_\pm)\mathcal{D}_\alpha = 0, \qquad (14.60)$$

where

$$\psi_\alpha(\Sigma_\pm) = 3 + 2s_\alpha + \tfrac{3}{2}(1 - \Sigma^2),$$

$$\xi_\alpha(\Sigma_\pm) = s_\alpha(3 + s_\alpha) + \tfrac{9}{2}(1 - \Sigma^2), \qquad (14.61)$$

and

$$s_1 = -2\Sigma_+, \quad s_{2,3} = \Sigma_+ \pm \sqrt{3}\Sigma_-. \qquad (14.62)$$

The final step is to write (14.60) in first-order form. As in Section 14.2.5, we let

$$\mathcal{U}_\alpha = \frac{\mathcal{D}_\alpha'}{\mathcal{D}_\alpha}, \qquad (14.63)$$

for $\alpha = 1, 2, 3$. The DEs (14.60) assume the form

$$\mathcal{U}_\alpha' = -\mathcal{U}_\alpha^2 - \psi_\alpha(\Sigma_\pm)\mathcal{U}_\alpha - \xi_\alpha(\Sigma_\pm). \qquad (14.64)$$

The evolution equation for the shear variables $\Sigma_\pm$ that describe the background is given by (6.19) and (14.59) with $K = 0$:

$$\Sigma_\pm' = -\tfrac{3}{2}(1 - \Sigma^2)\Sigma_\pm, \qquad (14.65)$$

where Σ^2 is given by (14.57). The *DE in* $\mathbf{R}^5$ *given by (14.64)–(14.65), with (14.61)–(14.63), describes the evolution of isocurvature density perturbations for pressure-free irrotational matter in a Bianchi I background.* The evolution can in fact be represented by phase portraits in the plane: equations (14.65) imply that $\Sigma_- = C\Sigma_+$, where C is an arbitrary constant, and the $\mathcal{U}_\alpha$ can be treated separately since the DEs (14.64) are decoupled.

The equilibrium points of the combined DE (14.64)–(14.65) can be written down. There are two cases, corresponding to the equilibrium points in the Bianchi I background, which are the flat FL equilibrium point F,

$$\Sigma_\pm = 0,$$

and the equilibrium points forming the Kasner circle $\mathcal{K}$,

$$\Sigma_+^2 + \Sigma_-^2 = 1, \quad \Sigma_\pm = \text{constant},$$

(Section 6.3.1). The associated Kasner exponents are given by (6.16)

$$p_1 = \tfrac{1}{3}(1 - 2\Sigma_+), \quad p_{2,3} = \tfrac{1}{3}(1 + \Sigma_+ \pm \sqrt{3}\Sigma_-).$$

By (14.61) and (14.62) the combined equilibrium points are

$$\Sigma_+^2 + \Sigma_-^2 = 1, \quad \mathcal{U}_\alpha = 1 - 3p_\alpha \text{ or } -(2 + 3p_\alpha), \tag{14.66}$$

and

$$\Sigma_\pm = 0, \quad \mathcal{U}_1 = \mathcal{U}_2 = \mathcal{U}_3 = -\tfrac{3}{2} \text{ or } -3. \tag{14.67}$$

By (14.63) the corresponding isocurvature density perturbations are

$$\mathcal{D}_\alpha = a_\alpha e^{(1-3p_\alpha)\tau} + b_\alpha e^{-(2+3p_\alpha)\tau}, \tag{14.68}$$

(no sum on α) for the Kasner solution, and

$$\mathcal{D}_\alpha = a_\alpha e^{-\frac{3}{2}\tau} + b_\alpha e^{-3\tau}, \tag{14.69}$$

for the flat FL solution, where a_α and b_α are constants. Since $\ell^3 = t$ and $\ell = e^\tau$ after rescaling ℓ, we obtain

$$\mathcal{D}_\alpha = a_\alpha t^{\frac{1}{3}-p_\alpha} + b_\alpha t^{-\frac{2}{3}-p_\alpha}, \tag{14.70}$$

and

$$\mathcal{D}_\alpha = a_\alpha t^{-\frac{1}{2}} + b_\alpha t^{-1}, \tag{14.71}$$

respectively.

To illustrate the evolution of the perturbations in the general Bianchi I background, we restrict our considerations to the LRS case ($\Sigma_- = 0, \mathcal{U}_2 = \mathcal{U}_3$). The phase plane for $(\mathcal{U}_1, \Sigma_+)$ and $(\mathcal{U}_2, \Sigma_+)$, which are topologically equivalent, are shown in Figure 14.3.

14.4 Summary

In this chapter we have given an introduction to the Ellis–Bruni geometrical approach to cosmological density perturbations and, following Woszczyna (1992), have given a qualitative analysis of the perturbations using dynamical systems methods, assuming that the background space–time is a FL universe or a Bianchi I universe. As regards the Bianchi I perturbations, the analysis simply serves to illustrate the method for an anisotropic background, since we restricted our considerations to the special case of isocurvature perturbations. A preliminary analysis shows that in order to discuss arbitrary perturbations, one will have to augment the basic second–order DE (14.54) with additional equations to describe the evolution of the shear

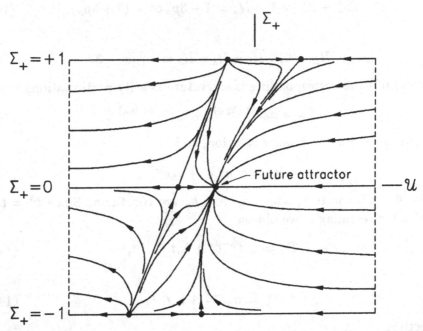

Fig. 14.3. Isocurvature density perturbations in an LRS Bianchi I model, described by orbits in the $(\mathcal{U}_1, \Sigma_+)$ and $(\mathcal{U}_2, \Sigma_+)$ state spaces. Here $\mathcal{U}$ represents $\mathcal{U}_1$ or $\mathcal{U}_2$, since the two phase portraits are topologically equivalent. The equilibrium points on the line $\Sigma_+ = 0$ has $\mathcal{U} = -3$ or $-\frac{3}{2}$ in both cases. The equilibrium points on the line $\Sigma_+ = +1$ have $\mathcal{U}_1 = -1$ or 2 and $\mathcal{U}_2 = -4$ or -1, and on the line $\Sigma_- = -1$ have $\mathcal{U}_1 = -5$ or -2 and $\mathcal{U}_2 = -2$ or 1. In both cases the future attractor is $(\mathcal{U}, \Sigma_+) = (-\frac{3}{2}, 0)$.

gradient $\mathcal{S}_a$. In principle, the exact evolution equations (14.17) and (14.21) can be linearized about any non–tilted Bianchi universe. It would be of interest to apply the method to other background universes. Two cases which may be tractable are the Bianchi V models and the Bianchi VI$_h$ models with $n_\alpha{}^\alpha = 0$ (see Table 7.2), both of which, like the Bianchi I models, have a two–dimensional state space.

Part four

Conclusion

15

Overview

G. F. R. ELLIS
University of Cape Town

J. WAINWRIGHT
University of Waterloo

The first goal of theoretical cosmology is to find a model of the universe, the simplest model, that is in agreement with observational data. The second goal is to explore the range of models that are compatible with observational data, in order to understand whether the simplest model is highly probable, and to understand the full range of cosmological possibilities in epochs that are not constrained by observations. This book describes results and techniques of analysis that pertain to the second goal.

The FL models are widely accepted as meeting the first goal (e.g. Peebles *et al.* 1991), although some uncertainties remain. First, insufficient evidence is available from redshift and peculiar velocity surveys to convincingly establish the averaging scale over which the universe can be regarded as isotropic and homogeneous. Second, a fully satisfactory theory of the formation of structure (i.e. of galaxies and their distribution in space) in a FL model has not yet been found. Third, the fact that the FL models (with $\Lambda = 0$), in particular the flat model, are unstable makes it implausible that the real universe can be approximated by a FL model over its entire evolution up to the present and into the future. Fourth, inflation is motivated by the desire to make a flat FL universe in the present epoch inevitable, or at least highly probable. In attempting to reach this goal one has to work with models more general than FL in the pre–inflation epoch. All of these considerations underline the importance of studies relating to the above second goal.

The second goal is clearly too broad in scope to reach in the short term. In this book we have described some of the efforts that have been made to date. The results described in Parts II and III range from a detailed (though still incomplete) qualitative analysis of the dynamics of non–tilted Bianchi cosmologies, and a unified description of the known exact solutions, to an inevitably tentative and preliminary discussion of methods for gaining qualitative insight into the dynamics of inhomogeneous cosmologies. In this chapter we discuss some of the implications of these results, particularly as regards cosmological modelling, comment on the advantages of the methods used here, and give some indication of possible future avenues of research.

15.1 Cosmological dynamical systems

In this section we discuss various aspects of the use of dynamical systems methods in cosmology.

15.1.1 Overview of Bianchi universes

Chapters 6 and 7 contain a detailed qualitative analysis of the dynamics of Bianchi universes whose matter content is a non–tilted perfect fluid. The use of dynamical systems methods in conjunction with expansion–normalized variables leads to a number of insights. First, it highlights the significance of self–similar solutions, which are represented as equilibrium points of the cosmological DEs. Although they are very special solutions in that they admit additional symmetry (i.e. a homothetic vector field) the analysis shows that they play a fundamental role in determining the evolution of generic models, either as part of an attractor, or as part of a heteroclinic sequence. Second, it permits one to analyze the dynamics near the initial singularity, both analytically and numerically, since the evolution equations are 'regularized'. Third, the stability properties of special solutions are determined as an integral part of the analysis.

   The analysis is incomplete in two respects. First, the evolution equations break down at the moment of maximum expansion in a type IX model. Second the description of the late–time behaviour of models of types VII_0 and VIII is incomplete, since the state space is unbounded. We do know, however, that the models are not asymptotically self–similar at late times, and that physical and geometrical quantities such as the normalized Weyl curvature and spatial curvature are unbounded on the state space, in contrast to all other ever–expanding non–tilted models. The first gap is filled by the reformulation of the type IX equations in Section 8.5.2 (although the resulting DE (8.13) has not been studied in detail). The second gap remains as an unsolved problem.

15:1.2 The role of Bianchi I universes

The Bianchi type I models with a perfect fluid and equation of state $p = (\gamma-1)\mu$, are the simplest and most studied anisotropic cosmologies. In terms of dimensionless variables they have a two–dimensional state space $B(\mathrm{I})$ that contains the flat FL equilibrium point F, and a circle of equilibrium points, the Kasner circle $\mathcal{K}$ (Section 6.3 and Figure 6.4). The non–vacuum models are asymptotic at early times to Kasner vacuum solutions and at late times to the flat FL model. The Bianchi I DE admits monotone functions: the dimensionless shear Σ is monotone decreasing (from 1 to 0) and the density parameter Ω is monotone increasing (from 0 to 1). The spatial geometry is flat ($^{3}R = 0$, $^{3}S_{\alpha\beta} = 0$) and the magnetic Weyl curvature is zero ($H_{\alpha\beta} = 0$). The Bianchi I models and their equilibrium points $\mathcal{K}$ and F are universal, in that the state space $B(\mathrm{I})$ is an invariant subset of the state space of all other Bianchi types.† The dynamics in $B(\mathrm{I})$ thus influence the dynamics in the higher–dimensional Bianchi sets, but, as we have seen, the overall behaviour changes significantly as additional degrees of freedom are included. One lesson, however, can be learned from the type I models: if $\Omega \approx 1$ now it does not follow, unless the model is an exact FL model, that Ω must have been closer to 1 in the past.

The DEs that we have encountered in Parts II and III exhibit a number of special features that to a large extent have their origins in the Bianchi I models. We now discuss in turn these special features.

15.1.3 Hierarchies of invariant sets

For the Bianchi cosmologies, the variables N_{α}, which determine the Bianchi type, define a hierarchy of invariant sets (see Table 6.1) depending on how many are zero. The lower–dimensional invariant sets serve as building blocks for the higher–dimensional sets, and thereby simplify the analysis of the dynamics. For tilted models, the hierarchy is extended as velocity variables or additional shear variables are incorporated, eventually leading to an eight–dimensional state space (Table 9.5). In the state space $B(\mathrm{I})$, $\mathcal{K}$ is a local source and F is a local sink. As additional degrees of freedom are incorporated, the local stability properties of $\mathcal{K}$ and F change. This process has not been fully investigated for tilted models, but for class A models $\mathcal{K}$ gains one negative and two positive eigenvalues, thereby becoming a saddle set, while F gains three positive eigenvalues, thereby becoming a saddle point.

The DE (13.31)–(13.36) that determines the dynamics of the silent universes also admits a hierarchy of invariant sets, determined by the spatial curvature variables M_{α}, and by Ω_{Λ}. The Bianchi I invariant set $B(\mathrm{I})$ is again the basic building block, corresponding to $M_1 = M_2 = M_3 = 0$, $\Omega_{\Lambda} = 0$, but

---

† $B(\mathrm{I})$ is a subset of the Bianchi V state space for tilted models, but the Bianchi V state space for non–tilted models only contains the LRS Bianchi I models as a subset.

in this case, as the additional degrees of freedom are added, $\mathcal{K}$ gains only positive eigenvalues, thus remaining a source, while as before F gains positive eigenvalues, becoming a saddle point. The fact that F gains positive eigenvalues in both cases is a manifestation of the instability of the flat FL model. This instability is further exemplified by the density perturbations of the model, which have one growing mode (see (14.34)).

The G_2 cosmologies admit a complicated hierarchy of invariant sets, involving both inhomogeneous and homogeneous models, in particular the Bianchi I set $B(\mathrm{I})$, as can be seen from Table 13.3. Relatively little progress has been made in investigating the dynamics of this hierarchy of models. In Chapter 12, only (parts of) the two lowest levels in this hierarchy were considered, and the local stability of $\mathcal{K}$ and F in the infinite–dimensional state space was not investigated. However, the results of Isenberg & Moncrief (1990) suggest that $\mathcal{K}$ is a local source in the infinite–dimensional state space of diagonal G_2 cosmologies.

It would be of interest to perform a systematic investigation of the local stability of $\mathcal{K}$ and F in the full Bianchi state space (including tilt) and in the infinite–dimensional state space of the G_2 cosmologies.

15.1.4 Monotone functions

An important feature of the DEs that describe the non–tilted Bianchi cosmologies is the existence of monotone functions in the various invariant sets (see the Appendices to Chapters 6 and 7). This feature simplifies the dynamics dramatically: there are no periodic orbits, recurrent orbits or homoclinic orbits† in the dimensionless state space. As a result the dynamical behaviour is dominated by equilibrium points and heteroclinic sequences. In this way self–similar models, which correspond to equilibrium points, play a dominant role in the dynamics. In addition the LaSalle Invariance Principle and the Monotonicity Principle (Section 4.4.3) can be used to determine the past and future attractors. One type of quasi–periodic behaviour is permitted, however, in that the DE for class A cosmologies admits heteroclinic cycles.‡ These cycles are contained in the Mixmaster attractor and occur because the Kasner map, as defined in Section 6.4.1, has periodic points of period 3 and greater (Creighton & Hobill 1994, pages 445–7).

The origin of the monotone functions, in mathematical terms, is not fully understood. Some of them can be found by inspection of the DEs, while others are well disguised; some of these have been derived from the Hamil-

---

† See Definitions 4.4–4.6 for this terminology.
‡ This type of invariant set is illustrated in Figure 4.2(b).

tonian approach (see the Appendix to Chapter 10). There is a conspicuous gap in the library of monotone functions for non–tilted models, namely as regards the full state space for the non–exceptional class B models.

There is an interplay between the monotone functions and the hierarchy of invariant sets in that often the monotone functions cause the orbits to be attracted (into the past and into the future) to lower–dimensional invariant sets in the hierarchy, essentially because the functions attain their maximum/minimum on these sets.

15.1.5 Equilibrium sets

An important feature of the DEs that describe the Bianchi cosmologies is the existence of equilibrium sets, in addition to isolated equilibrium points. The Kasner set $\mathcal{K}$ is ubiquitous, occurring for both class A and class B models and also in the presence of tilt (e.g. Hewitt & Wainwright 1992). It also occurs in the finite–dimensional state space of the silent universes and in the infinite–dimensional state space of the G_2 cosmologies. The second most important case is the plane wave set† $\mathcal{L}(\tilde{h})$ which occurs in the non–exceptional class B state space and also in the infinite–dimensional state space of the G_2 cosmologies, as can be inferred from Table 13.3.

The sets $\mathcal{K}$ and $\mathcal{L}(\tilde{h})$, since they describe vacuum models, exist independently of the equation of state parameter γ. Other equilibrium sets, corresponding to non–vacuum models, exist for specific values of γ. The most important are the FL arcs $\mathcal{F}_\alpha^\pm$ and $\mathcal{F}^\pm(\text{IX})$ (Table 6.3), and their analogues in the class B state space (Hewitt & Wainwright 1993, the lines $\mathcal{F}_0$ and F_k in Table 4). These lines of equilibrium points, which exist only for $\gamma = \frac{2}{3}$, emanate from the flat FL point F. They are partly responsible for the loss of stability that F experiences as γ increases through $\frac{2}{3}$. At this value of γ, the flat FL model loses its inflationary nature (i.e. $q < 0$) and changes from a local sink to a saddle. Other equilibrium sets occur for tilted Bianchi models (e.g. Hewitt & Wainwright 1992 for type V) and for magnetic Bianchi models (LeBlanc *et al.* 1995), and are associated with a change of stability.

An implication of the existence of the equilibrium set $\mathcal{K}$ (for all values of γ) is that *the cosmological DEs are structurally unstable‡* in the sense that a small change in the vector field which defines the DE gives rise to qualitatively different dynamics (i.e. the flows are not topologically equivalent). This structural instability is not an artefact of simplifying the EFEs

---

† The lines $\mathcal{L}_\alpha^\pm$ in the VII$_0$ state space are essentially limits of $\mathcal{L}(\tilde{h})$ as $\tilde{h} \to +\infty$ ($h \to 0$).

‡ Any DE which admits non–hyperbolic equilibrium points is not structurally stable, since a small change in the DE can change a zero eigenvalue to a non–zero one.

by imposing symmetry conditions, but is a property of the full EFEs (since the Kasner solutions are solutions of the full equations).

15.2 Bianchi universes and observations

In this section we turn to the problem of assessing which of the non–tilted Bianchi models are potentially compatible with observational data. Most current research in cosmology is based on the assumption that on a sufficiently large scale the universe is isotropic about each point, and hence spatially homogeneous, leading to the FL models. As described in Section 3.4, various observations, in conjunction with a weak Copernican Principle provide some direct support for the assumption of large–scale isotropy. At best, however, *observations can only provide upper bounds on the deviations of the universe from a FL model in the past, up to a finite value of the redshift parameter z* ($0 \leq z \lesssim 4$ for observations of discrete sources, and $0 \leq z \lesssim 1000$ for observations of the CBR). These ranges correspond to a *finite* interval in terms of the dimensionless time variable τ. Thus the Bianchi models which are potentially compatible with observations are those which are *close to FL*, in the sense of Definition 2.1, *at least during an epoch of finite duration.*

15.2.1 Flat quasi–isotropic epochs

In Section 5.3.3 we introduced the notion of a flat quasi–isotropic epoch, namely an epoch during which the dimensionless state lies in an ε–neighbourhood of the flat FL equilibrium point F (which implies that conditions (2.58) are satisfied). Since F is universal in that it is contained in the state space of each Bianchi type, *flat quasi–isotropic epochs will occur in universes of all Bianchi types.* Moreover, since an open set of orbits intersects any ε–neighbourhood of F, *these epochs will occur for an open set of initial conditions.* If an orbit passes sufficiently close to F the approximation property of orbits (Section 4.4.4) ensures that the state vector **y** will remain close to F for an arbitrarily long time.† Thus a *flat quasi–isotropic epoch can have any pre–assigned duration.* However, increasing the duration will further restrict the possible initial conditions. The detailed evolution of models of this type will be described by orbits which shadow a heteroclinic sequence, one of whose saddle points is the equilibrium point F (e.g. Figures 6.9 and 6.12). We note that these conclusions apply equally to tilted and non–tilted

---

† Both dimensionless time τ and clock time t. The relation between $\Delta\tau$ and Δt is given by (5.31).

models, the reason being that since F is a saddle point in the state space for non–tilted models, it will be a saddle point in the state space for tilted models. During a flat quasi–isotropic epoch, however, we have $\Omega \approx 1$ (since $\Omega = 1$ at F), which has not been confirmed by observational data concerning Ω_0; however, this could be correct if sufficient (so far undetected) dark matter is present (see Section 3.5.2). It is thus important to investigate whether there are other possibilities for Bianchi models to be close to FL.

15.2.2 Quasi–Milne epochs

A quasi–Milne epoch occurs when the dimensionless state lies in an ε–neighbourhood of a Milne equilibrium point M or $M^\pm$ (see Table 7.3). These epochs can only occur in non–exceptional class B models, and their occurrence is due to the fact that the subset $\mathcal{A}(\tilde{h})$ of the plane wave curve of equilibrium points is a future attractor, and that M and $M^\pm$ are end points of this subset. It follows from Figure 7.2 that

(i) for any fixed $\tilde{h} \geq 0$, there is an open set of orbits in the Bianchi VII$_h$ state space (Bianchi IV if $\tilde{h} = 0$) that are future asymptotic to a point in an ε–neighbourhood of the Milne equilibrium points $M^\pm$ (or M if $\tilde{h} = 0$),

(ii) for $\tilde{h} < 0$ and sufficiently close to zero, there is an open set of orbits in the Bianchi VI$_h$ state space that are future asymptotic to a point in an ε–neighourhood of M (note that $D(\tilde{h})$ approaches M as $\tilde{h} \to 0^-$).

As with a flat quasi–isotropic epoch, a universe in a quasi–Milne epoch is highly isotropic. An important difference is that a quasi–Milne epoch lasts indefinitely and $\Omega \to 0$ as $t \to +\infty$ (since $\Omega = 0$ on the attractor $\mathcal{A}(\hat{h})$). This property may prevent a universe with a quasi–Milne epoch from being compatible with observations, i.e. Ω may be too close to zero by the time the shear Σ gets close enough to zero. This fact may exclude Bianchi VI$_h$ models, but there will be an open set of type IV and type VII$_h$ models whose orbits shadow the open FL orbit ($F \to M$ or $F \to M^\pm$ in Figures 7.3(c) and (d)) and hence have Σ sufficiently small.

15.2.3 Quasi–de Sitter epochs

Wald's theorem (Section 8.3) states that any Bianchi model with $\Lambda > 0$ and with physically reasonable matter content is future asymptotic to the de Sitter model. Since the de Sitter model has $\Sigma = 0$, this theorem provides a third possibility for a Bianchi model to be close to FL, namely a quasi–de

Sitter epoch. More specifically, a non–tilted Bianchi model with $\Lambda > 0$ can be described as a two–fluid model as in Section 8.2 with $\gamma_2 = 0$ and $\gamma_1 = 1$ (dust), and the Corollary to Theorem 8.2 applies (Section 8.3). The density parameter Ω is a sum of two parts

$$\Omega = \Omega_m + \Omega_\Lambda.$$

The future attractor, for all Bianchi types, is the flat equilibrium point F with $\gamma_2 = 0$ (i.e. the de Sitter model) and $\Omega_m \to 0$ and $\Omega_\Lambda \to 1$ as $t \to +\infty$. This implies that any Bianchi model with $\Lambda > 0$ will have a quasi–de Sitter epoch. In principle, a quasi–de Sitter epoch lasts indefinitely, but compatibility with observations requires that Ω_m not become too close to zero.

15.2.4 *Isotropization: historical perspective*

The question of isotropization of Bianchi models has a lengthy history, dating back to the work of Misner, Collins and Hawking, and Novikov in the 1960s and 1970s. Misner (1967, 1968) posed the question as to whether a Bianchi universe with arbitrary initial conditions at the Planck time would inevitably isotropize, in an attempt to explain why the universe since last–scattering should be close to isotropy (see also Barrow 1982a). Collins & Hawking (1973b) showed that Misner's conjecture was false by proving that for physically reasonable matter, the set of homogeneous initial data that gives rise to models that isotropize† as $t \to +\infty$ is of measure zero in the space of all homogeneous initial data (*ibid*, page 300, Theorem 2). Collins and Hawking also proved that for type VII_0 models containing dust there is a neighbourhood N of the Einstein–de Sitter initial data such that all initial data in N give rise to models that approach isotropy at late times (*ibid*, page 333, Theorem 3). It is of interest to understand how this Collins & Hawking isotropization is described in terms of the dimensionless state space in Chapter 6. The orbits of type VII_0 models cannot be asymptotic to the flat FL point F since only type I orbits have this property (Conclusion C1 in Section 6.2.1). The answer lies in the fact that the flat FL models are also described by unbounded ordinary orbits $F_\alpha^\pm$ in the invariant sets $S_\alpha^\pm$ (VII_0), which satisfy $\Sigma_\pm = 0$ (see Table 6.2 and Section 6.1.4). Universes which undergo Collins & Hawking isotropization are described by unbounded orbits in $B(VII_0)$ which are attracted to (and in fact spiral around) the flat FL orbits $F_\alpha^\pm$, so that $\sigma/H \to 0$ as $t \to \infty$.

† In the sense that $\sigma/H \to 0$ and $\int_0^t \sigma dt$ approaches a constant as $t \to +\infty$ (Barrow 1982a, page 348).

The Collins & Hawking (1973b) investigations were based on a definition of isotropization as $t \to +\infty$. They proved that this type of isotropization can occur only in the Bianchi types that are compatible with the FL symmetry‡ (see Table 1.1). Novikov and coworkers (e.g. Novikov 1974, Zel'dovich & Novikov 1983, page 548–50) pointed out that agreement with observations does not require that a model isotropize as $t \to +\infty$, and that 'close to FL behaviour' over an extended time interval was possible in other Bianchi types (see also Barrow 1982a, page 347). They thus introduced the notion of 'observational isotropy', which means that the expansion is close to isotropic and proceeds at a rate that is the same as for a FL model for a sufficiently long time (Zel'dovich & Novikov 1983, page 549). In terms of the notation introduced in Chapter 5, this presumably means that the shear scalar Σ satisfies $\Sigma \ll 1$, and that $\Omega \approx 1$ and $q \approx \frac{1}{2}(3\gamma - 2)$, with $\gamma = 1$ or $\frac{4}{3}$ (see (5.20)). They specifically did not impose an assumption on the anisotropic spatial curvature ${}^3S_{\alpha\beta}$ (or equivalently, on the Weyl curvature), on the grounds that this curvature is not restricted by observations§ (*ibid*, page 549). In this respect, their notion of observational isotropy differs from the notion of flat quasi–isotropic epoch, discussed in Section 15.2.1. In particular, it does not correspond to the state of the universe being close to an equilibrium point in the dimensionless state space. Zel'dovich & Novikov (1983) argue that isotropization in the above sense is typical in models of type IX and VII$_h$: 'The conclusion to be drawn from the analysis is that for a wide range in the choice of parameters of the models (with respect to generality, coinciding with a complete selection of the arbitrary parameters determining a model), a given model isotropizes according to the character of deformation¶ and, in this sense, approximates a Friedmann model' (*ibid*, page 555). The analysis of the Bianchi models in Chapters 6 and 7 does not confirm this expectation.

15.2.5 Quasi–equilibrium epochs

Any saddle equilibrium point of a DE determines a quasi–equilibrium epoch for the corresponding physical system (Section 4.4.4). We have seen that the flat FL equilibrium point F determines such an epoch, with $\Omega \approx 1$ and $\Sigma \approx 0$. The other matter–dominated equilibrium points (i.e. self–similar models, see Tables 6.3 and 7.3), while being anisotropic ($\Sigma \neq 0$), have a

---

‡ The Collins & Hawking (1973b) paper had a significant impact and had the effect of influencing other workers to exclude models of other Bianchi types from the class of models that are potentially compatible with observations.

§ The analysis of Maartens *et al.* (1995a,b) does lead to such restrictions (see (3.10)).

¶ i.e. shear.

number of properties in common with F, which make them of potential interest as models of the early universe.† These properties stem from the fact that they have the same value for the deceleration parameter q, namely (5.33),

$$q^* = \tfrac{1}{2}(3\gamma - 2),$$

which implies

$$H = \frac{2}{3\gamma}t^{-1}. \tag{15.1}$$

(see (5.34)). In the case of radiation ($\gamma = \tfrac{4}{3}$), the matter density μ is related to the temperature T by

$$\mu = aT^4$$

where a is the black–body constant. It follows from (5.16) and (15.1) with $\gamma = \tfrac{4}{3}$ that

$$T = \left(\frac{3\Omega^*}{4a}\right)^{\frac{1}{4}} t^{-\frac{1}{2}}$$

where Ω^* is the value of Ω at the equilibrium point. The FL relation arises when $\Omega^* = 1$ (e.g. Narlikar 1993, page 137, equation (5.7)), but in all cases $T \propto t^{-1/2}$, i.e. (3.12) holds. The Collins–Stewart type II equilibrium point $P(II)$ (Tables 6.3 and 7.3) is the most general of the anisotropic matter–dominated equilibrium points in the sense that it can give rise to a quasi–equilibrium epoch in all models of more general Bianchi type (i.e. all except types I, II and V). It is of interest that of all the self–similar models considered by Barrow (1984), this model (for which $\Omega^* = \tfrac{7}{8}, \Sigma_* = \tfrac{1}{4}$) gave the smallest increase in helium production compared to the flat FL universe (Barrow 1984). It would be of interest to know whether a universe with a Collins–Stewart epoch could subsequently become close to FL.†

15.2.6 Observational constraints on state space

One can think of flat quasi–isotropic, quasi–Milne and quasi–de Sitter epochs as defining observationally acceptable domains in the dimensionless state space, *based on anisotropy constraints*. The orbit of an observationally acceptable cosmological model must pass through one of these domains. As

† These are the models for which Barrow (1984) estimated the helium production using timescale arguments (see Section 3.3.3).

† This is not possible in a type VI_0 universe with $\Lambda = 0$, since the unstable manifold of $P(II)$ is future asymptotic to the attractor $P(VI_0)$, but by Wald's theorem (Section 8.3) if $\Lambda > 0$ a quasi–de Sitter epoch will occur for all Bianchi types.

indicated in Sections 15.2.1–15.2.3 observational bounds on the density parameter Ω_0 provide an additional constraint that may reduce the size of, or eliminate, the observationally acceptable domain. In addition the age parameter $\alpha_0 = t_0 H_0$ (see (5.23)) provides a third constraint.‡ For the flat FL model $\alpha = \frac{2}{3}$ (since $\ell = t^{2/3}$) and for the Milne model $\alpha = 1$ (since $\ell = t$); for the de Sitter model we write '$\alpha = +\infty$' to indicate that the age is not finite. If observational data for t_0 and H_0 exclude $\alpha_0 = \frac{2}{3}$ then models with a flat quasi–isotropic epoch are excluded.§ Likewise if the data exclude $\alpha_0 = 1$, then models with a quasi–Milne epoch are excluded.

It should be noted that if observational data for H_0 and t_0 turn out to be incompatible with dust FL models with $\Lambda = 0$, by violating the age inequality $t_0 H_0 \leq 1$ (see (3.16)), then, on account of (5.28), these data will also be incompatible with Bianchi models with dust and zero cosmological constant. Indeed, since the inequality (5.28) depends only on the Raychaudhuri equation (5.14), which, for an arbitrary irrotational dust source, has the same form as for non–tilted Bianchi models, *the data will also be incompatible with any model with irrotational dust and $\Lambda = 0$, irrespective of symmetry properties* (Wainwright 1996, Section 4.1).

15.3 Inhomogeneous cosmologies

Regarded as evolution equations for inhomogeneous cosmologies, the EFE are partial differential equations, and the state space is thus an infinite dimensional function space. The investigations in Part III while differing significantly in scope and applicability, have a common mathematical theme, namely the reduction of evolution equations (partial differential equations) in an infinite–dimensional space to evolution equations (ordinary differential equations) in a finite–dimensional invariant subset.

For the diagonal G_2 cosmologies the process of reduction in Chapter 12 is performed by considering equilibrium states of the evolution equations in the infinite dimensional state space. The resulting DEs and phase planes describe the spatial structure of the models, their evolution in time being determined explicitly by the fact that the models are self–similar. This class of solutions do not provide realistic models of the universe since they do not admit an epoch in which they are close to FL. However, the density profiles of the models give a glimpse of the density inhomogeneities that could develop

‡ In FL models with $\Lambda = 0$, $t_0 H_0$ is determined by Ω_0 (see (2.32)). In Bianchi models, $\alpha = tH$ is a well–defined function on the dimensionless state space (see (5.27)), but α_0 and Ω_0 are, in general, not related.

§ In making this statement we are assuming that $tH = \alpha(y)$ is a continuous function on state space. This claim is plausible, if one considers the definition (5.27), but has not been proved.

in a fully non–linear regime in a radiation–dominated universe ($\gamma = \frac{4}{3}$). It should be noted that this process of reduction is possible only when there is one independent spatial variable.

In the class of silent universes the reduction occurs because the time evolution decouples from the space evolution, giving a DE in $\mathbb{R}^6$ and a set of spatial constraints. The analysis deals only with the DE, which describes the evolution along individual world–lines of the dust. The variations between world–lines are governed by the constraints. The phase planes for the silent universes thus do not give information about the growth of density inhomogeneities, and are analogous to the phase planes for the Bianchi universes.

In perturbation analyses of inhomogeneous cosmologies the reduction occurs as a result of a harmonic expansion,† leading to a one–parameter family of DEs, parameterized by the wave number n, which is coupled to the DE that describes the evolution of the background universe. The phase planes display the evolution of the linear density perturbations, coupled with the non–linear evolution of the background. They thus differ significantly, as regards the information they convey, from the phase portraits of the silent universes.

The relation between infinite– and finite–dimensional dynamical systems, established by some process of reduction, is a topic of current research in the theory of dynamical systems (e.g. Temam 1988a,b). The concept of an *inertial manifold* is of importance as regards the asymptotic behaviour of infinite–dimensional dynamical systems. Loosely speaking, an inertial manifold is a finite dimensional invariant set that attracts orbits at an exponential rate and whose orbits approximate the orbits of the infinite–dimensional system as $t \to +\infty$. Thus, if an inertial manifold exists, the late–time behaviour can be obtained by analyzing the DE (the 'inertial system') that is obtained by restricting the partial differential equation to the inertial manifold (e.g. Temam 1988b, in particular pages 3–4).

The challenge in this whole area is to gain further insight into the structure of the infinite–dimensional state space of the inhomogeneous models, and to determine to what extent this structure is determined by finite–dimensional invariant sets. We expect that a combination of analytical and numerical work will be needed.

---

† The case of dust is special in that, with the condition $\dot{u}_a = 0$, the linearized equations lead directly to an DE for the density gradient $\mathcal{D}_a$.

References

Adams, P.J., Hellings, R.W., Zimmerman, R.L., Farhoosh, H., Levine, D.I. & Zeldich, S. (1982). Inhomogeneous cosmology: gravitational radiation in Bianchi backgrounds, *Astrophys. J.* **253**, 1–18. [Chapter 1]

Amann, H. (1990). *Ordinary Differential Equations.* De Gruyter. [Chapter 4]

Anninos, P., Centrella, J. & Matzner, R.A. (1991). Numerical methods for solving the planar vacuum Einstein equations, *Phys. Rev.* D **43**, 1808–24. [Introduction, Chapter 1]

Arrowsmith, D.K. & Place, C.M. (1990). *An Introduction to Dynamical Systems.* Cambridge University Press. [Chapter 4] .

Aulbach, B. (1984). *Continuous and Discrete Dynamics near Manifolds of Equilibria,* Lecture Notes in Mathematics, No. 1058. Springer–Verlag. [Chapter 4]

Auslander, J., Bhatia, N.P. & Seibert, P. (1964). Attractors in dynamical systems, *Bol. Soc. Mat. Mexicana* **9**, 55–66. [Chapter 4]

Bahcall, N.A., Lubin, L.M. & Dorman, V. (1995). Where is the dark matter?, *Astrophys. J.* **447**, L81–5. [Chapter 3]

Bajtlik, R. Justkiewicz, M. Proszynski, & Amsterdamski, P. (1986). 2.7 K Radiation and the isotropy of the universe, *Astrophys. J.* **300**, 463–73. [Chapter 3]

Bardeen, J. M. (1980). Gauge invariant cosmological perturbations, *Phys. Rev.* D **22**, 1882–905. [Chapters 3, 14]

Barnes, A. & Rowlingson, R.A. (1989). Irrotational perfect fluids with a purely electric Weyl tensor, *Class. Quantum Grav.* **6**, 949–60. [Chapter 13]

Barrow, J.D. (1976). Light elements and the isotropy of the universe, *Mon. Not. R. Astron. Soc.* **175**, 359–70. [Chapter 3]

Barrow, J.D. (1982a). The isotropy of the universe, *Q. Jl. R. Astr. Soc.* **23**, 344–57. [Introduction, Chapter 15]

319

Barrow, J.D. (1982b). Chaotic behaviour in general relativity, *Phys. Rep.* **85**, 1–49. [Chapter 11]

Barrow, J.D. (1984). Helium formation in cosmologies with anisotropic curvature, *Mon. Not. R. Astron. Soc.* **211**, 221–7. [Chapters 3, 15]

Barrow, J.D. (1987a). Some topics in relativistic cosmology. In *Gravitation in Astrophysics, Cargèse 1986*, eds. B. Carter and J.B. Hartle. Plenum Press. [Chapter 3]

Barrow, J.D. (1987b). Cosmic no–hair theorems and inflation, *Phys. Lett.* **B 187**, 12–16. [Chapter 8]

Barrow, J.D., Galloway, G.J. & Tipler, F.J., 1986, The closed–universe recollapse conjecture, *Mon. Not. R. Astron. Soc.* **223**, 835–44. [Chapter 8]

Barrow, J.D. Juskiewicz, R. & Sonoda, D.H. (1983). Structure of the cosmic microwave background, *Nature* **309**, 397–402. [Chapter 3]

Barrow, J.D., Juszkiewicz, R. & Sonoda, D.H. (1985). Universal rotation: how large can it be?, *Mon. Not. R. Astron. Soc.* **213**, 917–43. [Chapter 3]

Barrow, J.D. & Sonoda D.H. (1986). Asymptotic stability of Bianchi universes, *Phys. Rep.* **139**, 1–49. [Chapter 7]

Barrow, J.D. & Stein–Schabes, J. (1984). Inhomogeneous cosmologies with cosmological constant, *Phys. Lett. A* **103**, 315–17. [Chapter 13]

Belinskii, V.A., Grishchuk, L.P., Zel'dovich, Ya.B. & Khalatnikov, I.M. (1985). Inflationary stages in cosmological models with a scalar field, *Sov. Phys. JETP* **62**, 195–203. [Chapter 8]

Belinskii, V.A., Ishihara, H., Khalatnikov, I.M. & Sato, H. (1988). On the degree of generality of inflation in Friedmann cosmological models with a massive scalar field, *Prog. Theoret. Phys.* **79**, 676–84. [Chapter 8]

Belinskii, V.A. & Khalatnikov, I.M. (1976). Influence of viscosity on the character of cosmological evolution, *Sov. Phys. JETP* **42**, 205–10. [Chapter 8]

Belinskii, V.A. & Khalatnikov, I.M. (1989). On the degree of generality of inflationary solutions in cosmological models with a scalar field. In *Proceedings of the Fifth Marcel Grossmann meeting on General Relativity*, eds. D.G. Blair and M.J. Buckingham. World Scientific. [Chapter 8]

Belinskii, V.A., Khalatnikov, I.M. & Lifshitz, E.M. (1970). Oscillatory approach to a singular point in the relativistic cosmology, *Adv. Phys.* **19**, 525–73. [Introduction, Chapters 5, 6, 11]

Belinskii, V.A., Khalatnikov, I.M. & Lifschitz, E.M. (1982). A general solution of the Einstein equations with a time singularity, *Adv. Phys.* **31**, 639–67. [Chapters 8, 12]

Belinskii, V.A., Nikomarov, E.S. & Khalatnikov, I.M. (1979). Investigation of the cosmological evolution of viscoelastic matter with causal thermodynamics, *Sov. Phys. JETP* **50**, 213–21. [Chapter 8]

Benettin, G., Galgani, L., Giorgilli, A., & Strelcyn J.-M. (1980). Lyapunov characteristic exponents for smooth dynamical systems and for Hamiltonian systems; a method for computing all of them I, II, *Meccanica* **15**, 9–20, 21–30. [Chapter 11]

Berger, B.K. (1991). Comments on the computation of Liapunov exponents for the Mixmaster universe, *Gen. Rel. Grav.* **23**, 1385–402. [Chapter 11]

Berger, B.K. & Moncrief, V. (1993). Numerical investigation of cosmological singularities, *Phys. Rev. D* **48**, 4676–87. [Introduction, Chapter 1]

Bertschinger, E. & Dekel, A. (1989). Recovering the full velocity and density fields from large–scale redshift–distance samples, *Astrophys. J.* **336**, L5–8. [Chapter 3]

Bertschinger, E. & Jain, B. (1994). Gravitational instability of cold matter, *Astrophys. J.* **432**, 486–94. [Chapter 13]

Birkhoff, G.D. (1927). *Dynamical Systems.* American Mathematical Society Colloquium Publications, volume 9. [Chapter 4]

Bogoyavlensky, O.I. (1985). *Methods in the Qualitative Theory of Dynamical Systems in Astrophysics and Gas Dynamics.* Springer–Verlag. [Introduction, Chapters 4, 5, 6, 7]

Bogoyavlensky, O.I. & Novikov, S.P. (1973). Singularities of the cosmological model of the Bianchi IX type according to the qualitative theory of differential equations, *Sov. Phys. JETP* **37**, 747–55. [Chapter 9]

Bonilla, M.A.C., Mars, M., Senovilla, J.M.M., Sopuerta, C.F. & Vera, R. (1996). Comment on "Integrability conditions for irrotational dust with a purely electric Weyl tensor: a tetrad analysis", preprint. [Chapter 13]

Bonnor, W. B. & Tomimura, N. (1976). Evolution of Szekeres's cosmological models, *Mon. Not. R. Astron. Soc.* **175**, 85–93. [Chapter 13]

Bradley, M. (1988). Dust EPL cosmologies, *Class. Quantum Grav.* **5**, L15–19. [Chapter 8]

Bruni, M. (1993). On the stability of open universes, *Phys. Rev. D* **47**, 738–42. [Chapter 14]

Bruni, M., Dunsby, P.K.S. & Ellis, G.F.R. (1992). Cosmological perturbations and the physical meaning of gauge–invariant variables, *Astrophys. J.* **395**, 34–53. [Chapter 14]

Bruni, M., Matarrese, S., & Pantano, O. (1995a). A local view of the observable universe, *Phys. Rev. Lett.* **74**, 1916–9. [Chapter 13]

Bruni, M., Matarrese, S. & Pantano, O. (1995b). Dynamics of silent universes, *Astrophys. J.* **445**, 958–77. [Chapter 13]

Bruni, M. & Piotrkowska, K. (1994). Dust–radiation universes: stability analysis, *Mon. Not. R. Astron. Soc.* **270**, 630–40. [Chapter 14]

Burd, A.B., Buric, N. & Ellis, G.F.R. (1990). A numerical analysis of chaotic behaviour in Bianchi IX models, *Gen. Rel. Grav.* **23**, 349–63. [Chapters 6, 11]

Burd, A. & Tavakol, R. (1993). Invariant Lyapunov exponents and chaos in cosmology, *Phys. Rev. D* **47**, 5336–41. [Chapter 11]

Cahill, M.E. & Taub, A.H. (1971). Spherically symmetric similarity solutions of the Einstein field equations for a perfect fluid, *Commun. Math. Phys.* **21**, 1–40. [Chapter 1]

Carmeli, M., Charach, Ch. & Malin, S. (1981). Survey of cosmological models with gravitational, scalar and electromagnetic waves, *Phys. Rep.* **76**, 79–156. [Chapter 1]

Carr, B. (1994). Baryonic dark matter, *Annual Rev. Astron. and Astrophys.* **32**, 531–90. [Chapter 3]

Carr, B.J. & Verdaguer, E. (1983). Soliton solutions and cosmological gravitational waves, *Phys. Rev. D* **28**, 2995–3006. [Chapter 1]

Carr, J. (1981). *Applications of Center Manifold Theory*. Springer–Verlag. [Chapter 4]

Carter, B. (1973). Blackhole equilibrium states. In *Black Holes*, eds. C. DeWitt and B.S. DeWitt. Gordon and Breach. [Chapter 1]

Carter, B. & Henriksen, R.N. (1989). A covariant characterization of kinematic self–similarity, *Ann. Physique Supp. No. 6*, **14**, 47–53. [Chapter 1]

Centrella, J. & Wilson, J. (1983). Planar numerical cosmology I. The differential equations, *Astrophys. J.* **273**, 428–35. [Chapter 3]

Centrella, J. & Wilson, J. (1984). Planar numerical cosmology. II. The difference equations and numerical tests, *Astrophys. J. Suppl.* **54**, 229–49. [Chapter 3]

Chaboyer, B., Demarque, P., Kernan, P.J. & Krauss, L.M. (1996). A lower limit on the age of the universe, *Science* **271**, 957–61. [Chapter 3]

Chinea, F.J., Fernández–Jambrina, L. & Senovilla, J.M.M. (1992). Singularity–free space–time, *Phys. Rev. D* **45**, 481–86. [Chapter 12]

Chmielowski, P. & Page, D.N. (1988). Probability of Bianchi type I inflation, *Phys. Rev. D* **38**, 2392–8. [Chapter 8]

Cohen, J.M. (1967). Friedmann cosmological models with both radiation and matter, *Nature* **246**, 249. [Chapter 2]

Coles, P. & Ellis, G.F.R. (1994). The case for an open universe, *Nature* **370**, 609–15. [Chapter 3]

Coles, P. & Lucchin, F. (1995). *Cosmology: the Origin and Evolution of Cosmic Structure.* John Wiley and Sons. [Chapters 3, 11]

Coley, A.A. & Tupper, B.O.J. (1989). Special conformal Killing vector space–times and symmetry inheritance, *J. Math. Phys.* **30**, 2616–25. [Chapter 1]

Coley, A.A. & van den Hoogen, R.J. (1995). Causal viscous fluid FRW models, *Class. Quantum Grav.* **12**, 1977–94. [Chapter 8]

Coley, A.A., van den Hoogen, R.J. & Maartens, R. (1996). Qualitative viscous cosmology, *Phys. Rev. D*, to appear. [Chapter 8]

Coley, A.A. & Wainwright, J. (1992). Qualitative analysis of two–fluid Bianchi cosmologies, *Class. Quantum Grav.* **9**, 651–65. [Chapters 2, 8]

Collins, C.B. (1971). More qualitative cosmology, *Commun. Math. Phys.* **23**, 137–56. [Introduction, Chapters 5, 8, 9]

Collins, C.B. (1977). Global structure of Kantowski–Sachs cosmological models, *J. Math. Phys.* **18**, 2116–24. [Chapters 8, 12]

Collins, C.B. & Ellis, G.F.R. (1979). Singularities in Bianchi cosmologies, *Phys. Rep.* **56**, 65–105. [Chapters 1, 8]

Collins, C.B. & Hawking, S.W. (1973a). The rotation and distortion of the universe, *Mon. Not. R. Astron. Soc.* **162**, 307–20. [Chapters 3, 5]

Collins, C.B. & Hawking, S.W. (1973b). Why is the universe isotropic?, *Astrophys. J.* **180**, 317–34. [Chapters 5, 6, 15]

Collins, C.B. & Stewart, J.M. (1971). Qualitative cosmology, *Mon. Not. R. Astron. Soc.* **153**, 419–34. [Chapter 9]

Collinson, C.D. & French, D.C. (1967). Null tetrad approach to motions in empty space–time, *J. Math. Phys.* **8**, 701–8. [Chapter 9]

Copi, C.J., Schramm, D.N. & Turner, M.S. (1995). Big–bang nucleosynthesis and the baryon density of the universe, *Science* **267**, 192–9. [Chapter 3]

Creighton, T.D. & Hobill, D.W. (1994). Continuous time dynamics and iterative maps of Ellis–MacCallum–Wainwright variables. In *Deterministic Chaos in General Relativity*, eds. D. Hobill, A. Burd and A. Coley. Plenum Press. [Chapters 6, 11, 15]

Croudace, K.M., Parry, J., Salopek, D.S. & Stewart, J.M. 1994). Applying the Zel'dovich approximation to general relativity, *Astrophys. J.* **423**, 22–32. [Chapter 13]

Davidson, W. (1991). A big–bang cylindrically symmetric radiation universe, *J. Math. Phys.* **32**, 1560–1. [Chapter 12]

Dekel, A. (1994). Dynamics of cosmic flows, *Annual Rev. Astron. and Astrophys.* **32**, 371–418. [Chapter 3]

de Lapparent, V., Geller, M.J. & Huchra, J.P. (1986). A slice of the universe, *Astrophys. J.* **302**, L1–5. [Chapter 3]

Doroshkevich, A.G., Lukash, V.N. & Novikov, I.D. (1973). The isotropization of homogeneous cosmological models, *Sov. Phys. JETP* **37**, 739–46. [Introduction, Chapters 3, 5, 9]

Doroshkevich, A.G., Lukash, V.N. & Novikov, I.D. (1975). Primordial radiation in a homogeneous anisotropic universe, *Sov. Astr.* **18**, 554–60. [Chapters 3, 5]

Doroshkevich, A.G., Zel'dovich, Ya.B. & Novikov, I.D. (1971). Perturbations in an anisotropic homogeneous universe, *Sov. Phys. JETP* **33**, 1–4 [Chapter 14].

Dowker, Y.N. & Friedlander, F.G. (1954). On limit sets in dynamical systems, *Proc. London Math. Soc.* **4**, 168–76. [Chapter 4]

Dunsby, P.K.S. (1993). Covariant perturbations of anisotropic cosmological models. *Phys. Rev. D* **48**, 3562–76. [Chapter 14]

Dyer, C.C. (1976). The gravitational perturbation of the cosmic background radiation by density concentrations, *Mon. Not. R. Astron. Soc.* **175**, 429–47. [Chapter 3]

Eardley, D. (1974). Self–similar spacetimes: geometry and dynamics, *Commun. Math. Phys.* **37**, 287–309. [Chapters 1, 5]

Eardley, D., Liang, E. & Sachs, R. (1972). Velocity–dominated singularities in irrotational dust cosmologies, *J. Math. Phys.* **13**, 99–107. [Chapter 12]

Eckhart, C. (1940). The thermodynamics of irreversible processes, *Phys. Rev.* **58**, 919–24. [Chapter 8]

Ehlers, J. (1961). Beiträge zur relativistischen Mechanik kontinuierlicher Medien, *Akad. Wiss. Lit. Mainz, Abhandl. Math.-Nat. Kl.* **11**. [Chapter 1]

Ehlers, J. (1993). Contributions to the relativistic mechanics of continuous media, *Gen. Rel. Grav.* **25**, 1225–66. [Chapter 1]

Ehlers, J. Gerens, P. & Sachs, R.K. (1968). Isotropic solutions of the Einstein–Liouville equations, *J. Math. Phys.* **9**, 1344–9. [Chapter 3]

Ehlers, J. & Rindler, W. (1989). A phase–space representation of FL universes containing both dust and radiation and the inevitability of a big–bang, *Mon. Not. R. Astron. Soc.* **238**, 503–21. [Chapter 2]

Einstein, A. & de Sitter, W. (1932). On the relation between the expansion and the mean density of the universe, *Proc. Natl. Acad. Sci. USA* **18**, 213–14. [Chapters 2, 9, 12]

Ellis, G.F.R. (1967). Dynamics of pressure–free matter in general relativity, *J. Math. Phys.* **8**, 1171–94. [Chapters 1,5]

Ellis, G.F.R. (1971). Relativistic cosmology. In *General Relativity and Cosmology*, Proceedings of XLVII Enrico Fermi Summer School, ed. R. K. Sachs. Academic Press. [Introduction, Chapter 1]

Ellis, G. F. R. (1973). Relativistic cosmology. In *Cargèse Lectures in Physics, Volume 6*, ed. E. Schatzmann. Gordon and Breach. [Introduction, Chapter 1]

Ellis G.F.R. (1975). Cosmology and verifiability, *Q. Jl. R. Astr. Soc.* **16**, 245–64.

Ellis, G.F.R. (1991). Standard and inflationary cosmologies. In *Gravitation*, eds. R. Mann and P. Wesson. World Scientific. [Chapter 3]

Ellis, G.F.R. & Baldwin, J. (1984). On the expected anisotropy of radio source counts, *Mon. Not. R. Astron. Soc.* **206**, 377–81. [Chapter 3]

Ellis, G.F.R. & Bruni, M. (1989). A covariant and gauge–invariant approach to cosmological density fluctuations, *Phys. Rev. D*, **40**, 1804–18. [Introduction, Chapters 3, 4, 14, 15]

Ellis, G.F.R., Bruni, M. & Hwang, J. (1990). Density gradient–vorticity relation in perfect fluid Robertson-Walker perturbations, *Phys. Rev. D*, **42**, 1035–46. [Chapter 14]

Ellis, G.F.R., Ehlers, J., van den Bergh, S., Kirshner, R.P., Thielemann, F.-K., Börner, G., Press, W.H., Roffelt, G., Buchert, T. & Hogan, C. (1996). What do we know about global properties of the universe? In *Proceedings of the Dahlem Workshop ES19*, eds. S. Gottlober and G. Borner. [Chapter 3]

Ellis, G.F.R., Hwang, J. & Bruni, M. (1989). Covariant and gauge–independent perfect fluid Robertson-Walker perturbations, *Phys. Rev. D*, **40**, 1819–26. [Chapter 14]

Ellis, G.F.R. & King, A.R. (1974). Was the big bang a whimper?, *Comm. Math. Phys.* **38**, 119–56. [Chapter 1]

Ellis, G.F.R. & MacCallum, M.A.H. (1969). A class of homogeneous cosmological models, *Commun. Math. Phys.* **12**, 108–41. [Chapters 1, 5, 8, 9]

Ellis, G.F.R., Matravers, D.R. & Treciokas, R. (1983). An exact anisotropic solution of the Einstein–Liouville equations, *Gen. Rel. Grav.* **15**, 931–44. [Chapter 1]

Estabrook, F.B., Wahlquist, H.D. & Behr, C.G. (1968). Dyadic analysis of spatially homogeneous world models, *J. Math. Phys.* **9**, 497–504. [Chapter 1]

Evans, A.B. (1978). Correlation of Newtonian and relativistic cosmologies, *Mon. Not. R. Astron. Soc.* **183**, 727–48. [Chapter 9]

Feinstein, A. & Senovilla, J.M.M. (1989). A new inhomogeneous cosmological perfect fluid solution with $p = \rho/3$, *Class. Quantum Grav.* **6**, L89–91. [Chapter 12]

Francisco, G. & Matsas, G.E.A. (1988). Qualitative and numerical study of Bianchi IX models, *Gen. Rel. Grav.* **20**, 1047–54. [Chapter 11]

Freedman, W. *et al.* (1994). Distance to the Virgo cluster galaxy M1000 from Hubble space telescope observations of Cepheids, *Nature* **371**, 757–62. [Chapter 3]

Fukugita, M., Hogan, C.J. & Peebles, P.J.E. (1993). The cosmic distance scale and the Hubble constant, *Nature* **366**, 309–12. [Chapter 3]

Gelfand, I.M. & Fomin, S.V. (1963). *Calculus of Variations*. Prentice Hall. [Chapter 10]

Glendinning, P. (1994). *Stability, Instability and Chaos*. Cambridge University Press. [Chapter 11]

Goode, S.W. (1980). Some Aspects of Spatially Inhomogeneous Cosmologies. M.Math. thesis, University of Waterloo. [Chapter 12]

Goode, S.W. (1989). Analysis of spatially inhomogeneous perturbations of the FRW cosmologies, *Phys. Rev.* D **39**, 2882–92. [Chapters 13, 14]

Goode, S.W. & Wainwright, J. (1982). Singularities and evolution of the Szekeres cosmological models, *Phys. Rev.* D **26**, 3315–26. [Chapter 13]

Goode, S.W. & Wainwright, J. (1985). Isotropic singularities in cosmological models, *Class. Quantum Grav.* **2**, 99–115. [Chapter 6]

Gott, J.R. Gunn, J.E. Schramm, D.N. & Tinsley, B.M. (1974). An unbounded universe?, *Astrophys. J.* **194**, 543–53. [Chapter 3]

Guckenheimer, J. & Holmes, P. (1983). *Nonlinear Oscillations, Dynamical Systems, and Bifurcations of Vector Fields*. Springer–Verlag. [Chapter 4]

Hale, J.K. (1969), *Ordinary Differential Equations*. John Wiley and Sons. [Chapter 4]

Hale, J.K. (1988). *Asymptotic Behaviour of Dissipative Systems*, Mathematical Surveys and Monographs, No. 25. American Mathematical Society. [Introduction, Chapter 4]

Hall, G.S. & Steele, J.D. (1990). Homothety groups in space-time, *Gen. Rel. Grav.* **22**, 457–68. [Chapter 1]

Halliwell, J.J. (1987). Scalar fields in cosmology with an exponential potential, *Phys. Lett.* B **185**, 341–4. [Chapter 8]

Hancock, S., Davies, R.D., Lasenby, A.N., Gutierrez, de la Cruz, C.M., Watson, R.A., Rebolo, R. & Beckman, J.E. (1994). Direct observation of structure in the cosmic microwave background, *Nature* **367**, 333–7. [Chapter 3]

Harrison, E.R. (1967). Classification of uniform cosmological models, *Mon. Not. R. Astron. Soc.* **137**, 69–79. [Chapters 2, 9]

Hartman, P. (1982). *Ordinary Differential Equations*, 2nd edition. Birkhäuser. [Chapter 4]

Harvey, A. & Tsoubelis, D. (1977). Exact Bianchi IV cosmological model, *Phys. Rev. D* **15**, 2734–7. [Chapter 9]

Hata, N., Scherrer, R.J., Steigman, G., Thomas, D. & Walker, T.P. (1995). Big–bang nucleosynthesis in crisis? *Phys. Rev. Lett.* **75**, 3977–80. [Chapter 3]

Hawking, S.W. (1966). Perturbations of an expanding universe, *Astrophys. J.* **145**, 544–54. [Chapters 1, 14]

Hawking, S.W. & Ellis, G.F.R. (1973). *The Large Scale Structure of Space–Time*. Cambridge University Press. [Introduction, Chapter 12]

Hawking, S.W. & Tayler, R.J. (1966). Helium production is an anisotropic big–bang cosmology, *Nature* **209**, 1278–9. [Chapter 3]

Heckmann, O. & Schücking, E. (1962). Relativistic cosmology. In *Gravitation: an Introduction to Current Research*, ed. L. Witten. John Wiley and Sons. [Chapter 9]

Hewitt, C.G. (1989). Asymptotic States of G_2 Cosmologies. Ph.D. thesis, University of Waterloo. [Chapter 12]

Hewitt, C.G. (1991a). An investigation of the dynamical evolution of a class of Bianchi $VI_{-1/9}$ cosmological models, *Gen. Rel. Grav.* **23**, 691–712. [Chapter 8]

Hewitt, C.G. (1991b). An exact tilted Bianchi II cosmology, *Class. Quantum Grav.* **8**, L109–14. [Chapter 8]

Hewitt, C.G. (1991c). Algebraic invariant curves in cosmological dynamical systems and exact solutions, *Gen. Rel. Grav.* **23**, 1363–83. [Chapter 9]

Hewitt, C.G. & Wainwright, J. (1990). Orthogonally transitive G_2 cosmologies, *Class. Quantum Grav.* **7**, 2295–316. [Introduction, Chapters 1, 12]

Hewitt, C.G. & Wainwright, J. (1992). Dynamical systems approach to tilted Bianchi cosmologies: irrotational models of type V, *Phys. Rev. D* **46**, 4242–52. [Chapters 8, 15]

Hewitt, C.G. & Wainwright, J. (1993). A dynamical systems approach to Bianchi cosmologies: orthogonal models of class B, *Class. Quantum Grav.* **10**, 99–124. [Chapters 4, 5, 7, 15]

Hewitt, C.G., Wainwright, J. & Glaum, M. (1991). Qualitative analysis of a class of inhomogeneous self–similar cosmological models: II. *Class. Quantum Grav.* **8**, 1505–18. [Chapter 12]

Hewitt, C.G., Wainwright, J. & Goode, S.W. (1988). Qualitative analysis of a class of inhomogeneous self–similar cosmological models, *Class. Quantum Grav.* **5**, 1313–27. [Chapter 12]

Hirsch, M.W. (1984). The dynamical systems approach to differential equations, *Bull. Amer. Math. Soc.* **11**, 1–63. [Introduction]

Hirsch, M.W. & Smale, S. (1974). *Differential Equations, Dynamical Systems, and Linear Algebra*. Academic Press. [Chapters 4, 14]

Hobill, D., Bernstein, D., Welge, M. & Simkins, D. (1991). The Mixmaster cosmology as a dynamical system, *Class. Quantum Grav.* **8**, 1155–71. [Chapter 11]

Hsu, L. & Wainwright, J. (1986). Self–similar spatially homogeneous cosmologies: orthogonal perfect fluid and vacuum solutions, *Class. Quantum Grav.* **3**, 1105–24. [Chapters 5, 9]

Isenberg, J. & Moncrief, V. (1990). Asymptotic behaviour of the gravitational field and the nature of singularities in Gowdy spacetimes, *Ann. Phys. NY* **199**, 84–122. [Chapters 1, 12, 15]

Israel, W. (1976). Nonstationary irreversible thermodynamics: a causal relativistic theory, *Ann. Phys. NY* **100**, 310–31. [Chapter 8]

Israel, W. & Stewart, J.M. (1979). Transient relativistic thermodynamics and kinetic theory, *Ann. Phys. NY* **118**, 341–72. [Chapter 8]

Jacobs, K.C. (1968). Spatially homogeneous and Euclidean cosmological models with shear, *Astrophys. J.* **153**, 661–78. [Chapter 9]

Jakobsen, P., Boksenberg, A., Deharveng, J.M., Greenfield, P., Jedrzejewski, R. & Paresce, F. (1994). Detection of intergalactic ionized helium absorption in a high redshift quasar, *Nature* **370**, 35–9 [Chapter 3]

Jantzen, R.T. (1984). Spatially homogeneous dynamics: a unified picture. In *Cosmology of the Early Universe*, eds. R. Ruffini and L.Z. Fang. World Scientific. This article is reprinted as Jantzen (1987). [Chapter 5]

Jantzen, R.T. (1986). Finite dimensional Einstein–Maxwell scalar field system, *Phys. Rev. D* **33**, 2121–35. [Chapter 10]

Jantzen, R.T. (1987). Spatially homogeneous dynamics: a unified picture. In *Gamow Cosmology*, eds. R. Ruffini and F. Melchiorri, Proc. Int. School of Physics E. Fermi, Course LXXXVI. North Holland . [Chapters 5, 8, 10, 11]

Jantzen, R.T. & Rosquist, K. (1986). Exact power law metrics in cosmology, *Class. Quantum Grav.* **3**, 281–309. [Chapters 5, 10]

Joseph, V. (1966). A spatially homogeneous gravitational field, *Proc. Cambr. Phil. Soc.* **62**, 87–9. [Chapter 9]

Kantowski, R. (1966). Some Relativistic Cosmological Models. Ph.D. thesis, University of Texas. [Chapter 9]

Kantowski, R. & Sachs, R.K. (1966). Some spatially homogeneous anisotropic relativisitic cosmological models, *J. Math. Phys.* **7**, 443–6. [Chapter 9]

Kasner, E. (1925). Solutions of Einstein's equations involving functions of only one variable, *Trans. Amer. Math. Soc.* **27**, 155–62. [Chapter 9]

Khalatnikov, I.M. & Lifshitz, E.M. (1970). General cosmological solution of the gravitational equations with a singularity in time, *Phys. Rev. Lett.* **24**, 76–9. [Chapter 11]

Khalatnikov, I.M. & Pokrovski, V.L. (1972). A contribution to the theory of homogeneous Einstein spaces. In *Magic Without Magic*, ed. J.R. Klauder. W.H. Freeman and Sons. [Chapter 10]

King, A.R. & Ellis, G.F.R. (1973). Tilted homogeneous cosmological models, *Commun. Math. Phys.* **31**, 209–42. [Chapters 1, 8]

Kirchgraber, U. & Stoffer, D. (1989). On the definition of chaos, *Z. angew. Math. Mech.* **69**, 175–85. [Chapter 11]

Kitada, Y. & Maeda, K. (1993). Cosmic no–hair theorem in homogeneous spacetimes I. Bianchi models, *Class. Quantum Grav.* **10**, 703–34. [Chapter 8]

Knudsen, C. (1994). Chaos without nonperiodicity, *Amer. Math. Monthly* **101**, 563–5. [Chapter 11]

Kofman, L. & Pogosyan, D. (1995). Dynamics of gravitational instability is nonlocal, *Astrophys. J.* **442**, 30–8. [Chapter 13]

Kolb, E.W. & Turner, M.S. (1990). *The Early Universe*. Addison Wesley. [Introduction, Chapter 3]

Kramer, D., Stephani, H., MacCallum, M., & Herlt, E. (1980). *Exact Solutions of Einstein's Field Equations*. Cambridge University Press. [Chapters 1, 9]

Krasinski, A. (1996). *Inhomogeneous Cosmological Models*. Cambridge University Press, in press (available as a preprint entitled *Physics in an inhomogeneous universe*). [Chapters 1, 2]

Kristian, J. & Sachs, R. K. (1966). Observations in cosmology, *Astrophys. J.* **143**, 379–99. [Chapter 2]

Kurki–Suonio, H. (1989). Dynamically inhomogeneous cosmic nucleosynthesis. In *Frontiers in Numerical Relativity*, eds. C.R. Evans, L.S. Finn and D.W. Hobill. Cambridge University Press. [Chapter 3]

Landau, L.D. & Lifshitz, E.M. (1975). *The Classical Theory of Fields*, 4th revised edition. Pergamon Press. [Chapter 11]

Lauer, T.R. & Postman, M. (1994). The motion of the local group with respect to the 15000 kilometer per second Abell cluster inertial frame, *Astrophys. J.* **425**, 418–38. [Chapter 3]

LeBlanc, V.G., Kerr, D. & Wainwright, J. (1995). Asymptotic states of magnetic Bianchi VI$_0$ cosmologies, *Class. Quantum Grav.* **12**, 513–41. [Chapters 4, 8, 15]

Lemaître, G. (1933). L'Univers en expansion, *Ann. Soc. Sci. Bruxelles A* **53**, 51–85. [Chapter 9]

Lesame, W.M., Dunsby, P.K.S. & Ellis, G.F.R. (1995). Integrability conditions for irrotational dust with a purely electric Weyl tensor: a tetrad analysis, *Phys. Rev. D* **52**, 3406–15. [Chapter 13]

Liang, E.P. (1976). Dynamics of primordial inhomogeneities in model universes, *Astrophys. J.* **204**, 235–50. [Chapter 1]

Lichtenberg, A.J. & Lieberman, M.A. (1983). *Regular and Stochastic Motion*. Springer–Verlag. [Chapter 11]

Lifshitz, E.M. & Khalatnikov, I.M. (1963). Investigations in relativistic cosmology, *Advances in Physics* **12**, 185–249. [Chapters 6, 9]

Lin, X–f. & Wald, R.M. (1989). Proof of the closed–universe–recollapse conjecture for diagonal Bianchi type IX cosmologies, *Phys. Rev. D.* **40**, 3280–6. [Chapter 8]

Longair, M.S. (1995). The physics of background radiation. In *The Deep Universe*, Saas–Fee Advanced Course 23, eds. B. Binggeli and R. Buser. Springer–Verlag. [Chapter 3]

Lorenz, E.N. (1963). Deterministic non–periodic flow, *J. Atmospheric Sci.* **20**, 130–41. [Chapter 11].

Lucchin, F. & Matarrese, S. (1985). Kinematical properties of generalized inflation, *Phys. Lett. B* **164**, 282–6. [Chapter 13]

Lukash, V.N. (1974). Some peculiarities in the evolution of homogeneous anisotropic cosmological models, *Sov. Astron.* **18**, 164–9. [Introduction, Chapter 5]

Lukash, V.N. (1975). Gravitational waves that conserve the homogeneity of space, *Sov. Phys. JETP* **40**, 792–9. [Chapter 9]

Lyth, D.H. & Mukherjee, M. (1988). Fluid flow description of density irregularities in the universe, *Phys. Rev. D* **38**, 485–9. [Chapter 14]

Lyth, D.H. & Stewart, E.D. (1990). The evolution of density perturbations in the universe, *Astrophys. J.* **361**, 343–53. [Chapter 14]

Ma, P.K–H. (1988). A Dynamical Systems Approach to the Oscillatory Singularity in Cosmology. M.Math. thesis, University of Waterloo, unpublished. [Chapters 6, 11]

Ma, P.K–H. & Wainwright, J. (1992). A dynamical systems approach to the oscillatory singularity in Bianchi cosmologies. In *Relativity Today*, ed. Z. Perjés. Nova Science Publishers. [Chapters 5, 6, 11]

Maartens, R. (1995). Dissipative cosmology, *Class. Quantum Grav.* **12**, 1455-65. [Chapter 8]

Maartens, R., Ellis, G.F.R. & Stoeger, W.J. (1995a). Limits on anisotropoy and inhomogeneity from the cosmic background radiation, *Phys. Rev. D* **51**, 1525-35. [Chapters 2, 3, 15]

Maartens, R., Ellis, G.F.R. & Stoeger, W. (1995b). Improved limits on anisotropy and inhomogeneity from the cosmic background radiation, *Phys. Rev. D* **51**, 5942-45. [Chapters 2, 3, 15]

MacCallum, M. (1973). Cosmological models from a geometric point of view. In *Cargèse Lectures in Physics*, Volume 6, ed. E. Schatzman. Gordon & Breach. [Introduction, Chapters 1, 11]

MacCallum, M.A.H. (1979). Anisotropic and inhomogeneous relativistic cosmologies. In *General Relativity*, eds. S.W. Hawking and W. Israel. Cambridge University Press. [Chapters 1, 10]

MacCallum, M.A.H. (1994). Relativistic cosmologies. In *Deterministic Chaos in General Relativity*, eds. D. Hobill, A. Burd and A. Coley. Plenum Press. [Chapter 1]

MacCallum, M.A.H & Ellis, G.F.R. (1970). A class of homogeneous cosmological models: II. Observations, *Comm. Math. Phys.* **19**, 31-64. [Chapter 1]

Maddox, S.J., Efstathiou, G., Sutherland, W.J. & Loveday, J. (1990). Galaxy correlations on large scales, *Mon. Not. R. Astron. Soc.* **242**, 43p-47p. [Chapter 3]

Madsen, M.S., Mimoso, J.P., Butcher, J.A. & Ellis, G.F.R. (1992). Evolution of the density parameter in inflationary cosmology re-examined, *Phys. Rev. D.* **46**, 1399-415. [Chapter 2]

Matarrese, S., Pantano, O. & Saez, D. (1993). General relativistic approach to the nonlinear evolution of collisionless matter, *Phys. Rev. D* **47**, 1311-23. [Chapter 13]

Matarrese, S., Pantano, O. & Saez, D. (1994a). General relativistic dynamics of irrotational dust: cosmological implications, *Phys. Rev. Lett.* **72**, 320-3. [Chapter 13]

Matarrese, S., Pantano, O. & Saez, D. (1994b). A relativistic approach to gravitational instability in the expanding universe: second order Lagrangian solutions, *Mon. Not. R. Astron. Soc.* **271**, 513-22. [Chapter 13]

Matravers, D.R., Vogel, D.L. & Madsen, M.S. (1984). Helium formation in a Bianchi type V cosmological model with tilt, *Class. Quantum Grav.* **1**, 407–14. [Chapter 3]

Matzner, R., Rothman, T. & Ellis, G.F.R. (1986). Conjecture on isotope production in the Bianchi cosmologies, *Phys. Rev. D.* **34**, 2926–33. [Chapter 3]

McIntosh, C.B.G. (1976). Homothetic motions in general relativity, *Gen. Rel. Grav.* **7**, 199–213. [Chapter 12]

Mészáros, A. (1994). On the anisotropy of the cosmic microwave background, *Astrophys. J.* **423**, 19–21. [Chapter 3]

Michalski, H. & Wainwright, J. (1975). Killing vector fields and the Einstein–Maxwell field equations in general relativity, *Gen. Rel. Grav.* **6**, 289–318. [Chapter 1]

Milnor, J. (1985) On the concept of attractor, *Commun. Math. Phys.* **99**, 177–95; **102**, 517–19. [Chapter 4]

Misner, C.W. (1967). Neutrino viscosity and the isotropy of primordial blackbody radiation, *Phys. Rev. Lett.* **19**, 533–5. [Chapter 15]

Misner, C.W. (1968). The isotropy of the universe, *Astrophys. J.* **151**, 431–57. [Chapter 15]

Misner, C.W. (1969a). Mixmaster universe, *Phys. Rev. Lett.* **22**, 1071–74. [Introduction, Chapters 5, 11]

Misner, C.W. (1969b). Quantum cosmology. I, *Phys. Rev.* **186**, 1319–27. [Introduction, Chapters 10, 11]

Misner, C.W. (1970). Classical and quantum dynamics of a closed universe. In *Relativity, Proceedings of the Relativity Conference in the Midwest*, eds. M. Carmeli, S.I. Fickler and L. Witten. Plenum Press. [Chapters 10, 11]

Misner, C.W. (1972). Minisuperspace. In *Magic Without Magic*, ed. J.R. Klauder. W.H. Freeman and Sons. [Chapters 10, 11]

Misner, C.W., Thorne, K.S. & Wheeler, J.A. (1973). *Gravitation.* W.H. Freeman and Sons. [Chapters 2, 8]

Narlikar, J.V. (1993). *Introduction to Cosmology*, 2nd edition. Cambridge University Press. [Introduction, Chapter 15]

Nemytskii, V.V. & Stepanov, V.V. (1960). *Qualitative Theory of Differential Equations.* Princeton University Press. [Chapter 4]

Netterfield, C.B., Jarosik, N., Page, L., Wilkinson, D. & Wollack, E. (1995). The anisotropy in the cosmic microwave background at degree angular scales, *Astrophys. J.* **445**, L69–72. [Chapter 3]

Newman, E.T. & Penrose, R. (1962). An approach to gravitational radiation by a method of spin coefficients, *J. Math. Phys.* **3**, 566–78. [Chapter 1]

Nilsson, U. & Uggla, C. (1996). Spatially self–similar locally rotationally symmetric perfect fluid models, (preprint), University of Stockholm. [Chapter 12]

Novikov, I.D. (1974). Isotropization of homogeneous cosmological models. In *Confrontation of Cosmological Theories with Observational Data*, ed. M.S. Longair. D. Reidel Company. [Chapter 15]

Nusser, A. & Dekel, A. (1993). Ω and the initial fluctuations from velocity and density fields, *Astrophys. J.* **405**, 437–48. [Chapter 3]

Oleson, M. (1971). A class of type [4] perfect fluid spacetimes, *J. Math. Phys.* **12**, 667–72. [Chapter 1]

Olson, D.W. (1978). Helium production and limits on the anisotropy of the universe, *Astrophys. J.* **219**, 777–80. [Chapter 3]

Padmanabhan, T. (1993). *Structure Formation in the Universe.* Cambridge University Press. [Chapter 14]

Panek, M. (1986). Large–scale microwave background fluctuations: gauge–invariant formalism, *Phys. Rev. D* **34**, 416–23. [Chapter 3]

Parker, T.S. & Chua, L.O. (1989). *Practical Numerical Algorithms for Chaotic Systems.* Springer–Verlag. [Chapter 11]

Partridge, R.N. (1994). The cosmic microwave radiation and cosmology, *Class. Quantum Grav.* **11**, A153–69. [Chapter 3]

Peebles, P.J.E. (1966). Primordial helium abundance and the primordial fireball, II. *Astrophys. J.* **146**, 542–52. [Chapter 3]

Peebles, P.J.E. (1993). *Principles of Physical Cosmology.* Princeton University Press. [Chapter 3]

Peebles, P.J.E., Schramm, D.N., Turner, E.L. & Kron, R.G. (1991). The case for the relativistic hot big–bang cosmology, *Nature* **352**, 769–76. [Introduction, Chapters 3, 15]

Peresetsky, A.A. (1985). Qualitative theory of open homogeneous cosmological models with motion of matter. In *Topics in Modern Mathematics*, Petrovskii Seminar No. 5, ed. O.A. Oleinik. Consultants Bureau. [Chapter 8]

Perko, L. (1991). *Differential Equations and Dynamical Systems.* Springer–Verlag. [Chapter 4]

Perko, T.E., Matzner, R.A. & Shepley, L.C. (1972). Galaxy formation in anisotropic cosmologies, *Phys. Rev. D* **6**, 969–83. [Chapter 14]

Persic, M. & Salucci, P. (1992). The baryon content of the universe, *Mon. Not. R. Astron. Soc.* **258**, 14p–18p. [Chapter 3]

Pierce M.J. *et al.* (1994). The Hubble constant and Virgo cluster distance from observations of Cepheid variables, *Nature* **371**, 385–9. [Chapter 3]

Raine, D.J. (1987). The quasi–isotropic universe. In *Gravitation in Astro-physics, Cargèse 1986*, eds. B. Carter and J.B. Hartle. Plenum Press. [Chapter 3]

Rees, M.J. & Sciama, D.W. (1968). Large–scale density inhomogeneity in the universe, *Nature* **217**, 511–16. [Chapter 3]

Rendall, A.D. (1996). The initial singularity in solutions of the Einstein–Vlasov system of Bianchi type I, *J. Math. Phys.* **37**, 438–51. [Chapter 8]

Robertson, H.P. (1935). Kinematics and world structure I, *Astrophys. J.* **82**, 284–301. [Chapter 1]

Robertson, H.P. (1936). Kinematics and world structure II and III, *Astrophys. J.* **83**, 187–201, 257–71. [Chapter 1]

Robinson, B. (1961). Relativistic universes with shear, *Proc. Nat. Acad. Sci.* **47**, 1852–57. [Chapter 9]

Robinson, I. & Trautman, A. (1962). Some spherical gravitational waves in general relativity, *Proc. Roy. Soc. London A* **265**, 463–73. [Chapter 9]

Rosquist, K. (1983). Exact rotating and expanding radiation–filled universe, *Phys. Lett. A* **97**, 145–6. [Chapter 8]

Rosquist, K. & Jantzen, R.T. (1985a). Spacetimes with a transitive similarity group, *Class. Quantum Grav.* **2**, L129–33. [Chapter 5]

Rosquist, K. & Jantzen, R.T. (1985b). Exact power law solutions of the Einstein equations, *Phys. Lett. A* **107**, 29–32. [Chapter 8]

Rosquist, K. & Jantzen R.T. (1988). Unified regularization of Bianchi cosmology, *Phys. Rep.* **166**, 89–124. [Chapter 5]

Rosquist, K., Uggla, C. & Jantzen, R.T. (1990). Extended dynamics and symmetries in perfect fluid Bianchi cosmologies, *Class. Quantum Grav.* **7**, 625–37. [Chapter 10]

Rothman, A. & Matzner, R. (1984). Nucleosynthesis in anisotropic cosmologies revisited, *Phys. Rev. D* **30**, 1649–68. [Chapter 3]

Ruban, V.A. (1977). Dynamics of anisotropic homogeneous generalizations of the Friedmann cosmological models, *Sov. Phys. JETP* **45**, 629–37. [Chapter 9]

Rubin, V. (1977). Is there evidence of anisotropy in the Hubble expansion? In *Proceedings of the 8th Texas Symposium on Relativistic Astrophysics, Ann. NY Acad. Sci.* **302**, 408–21. [Chapter 3]

Rugh, S.E. (1994). Chaos in the Einstein equations – characterization and importance. In *Deterministic Chaos in General Relativity*, eds. D. Hobill, A. Burd and A. Coley. Plenum Press. [Chapter 11]

Russ, H., Soffel, M.H., Xu, C. & Dunsby, P. K. S. (1993). A covariant and gauge–invariant formulation of the Sachs–Wolfe effect, *Phys. Rev. D* **48**, 4552–6. [Chapter 3]

Ryan, M.P. (1972). *Hamiltonian Cosmology*. Springer–Verlag. [Chapter 8]

Ryan, M.P. & Shepley, L.C. (1975). *Homogeneous Relativistic Cosmologies*. Princeton University Press. [Chapter 1]

Sachs, R.K. & Wolfe, A.M. (1967). Perturbations of a cosmological model and angular variations of the microwave background, *Astrophys. J.* **147**, 73–90. [Chapter 3]

Sandage, A. (1995). Practical cosmology: inventing the past. In *The Deep Universe*, Saas–Fee Advanced Course 23, eds. B. Binggeli and R. Buser. Springer–Verlag. [Chapter 2]

Sasselov, D. & Goldwirth, D. (1995). A new estimate of the uncertainties in the primordial helium abundance: new bounds on Ω baryons, *Astrophys. J.* **444**, L5–8. [Chapter 3]

Sciama, D.W. (1993). *Modern Cosmology and the Dark Matter Problem*. Cambridge University Press. [Chapter 3]

Scott, D., Silk, J. & White, M. (1995). From microwave anisotropies to cosmology, *Science* **268**, 829–35. [Chapter 3]

Senovilla, J.M.M. (1990). New class of inhomogeneous cosmological perfect fluid solutions without big–bang singularity, *Phys. Rev. Lett.* **64**, 2219–21. [Chapter 12]

Shinkai, H. & Maeda, K. (1994). Generality of inflation in a planar universe, *Phys. Rev. D* **49**, 6367–78. [Introduction, Chapter 1]

Sibirsky, K.S. (1975). *Introduction to Topological Dynamics*. Noordhoff. [Chapter 4]

Siklos, S.T.C. (1978). Occurrence of whimper singularities, *Commun. Math. Phys.* **58**, 255–72. [Chapter 1]

Siklos, S.T.C. (1981). Some Einstein spaces and their global properties, *J. Phys. A: Math. Gen.* **14**, 395–409. [Chapters 1, 9]

Siklos, S.T.C. (1984). Einstein's equations and some exact solutions. In *Relativistic Astrophysics and Cosmology*, Proc. 14th GIFT Int. Seminar, eds. X. Fustero and E. Verdaguer. World Scientific. [Chapters 1, 9]

Siklos, S.T.C. (1991). Stability of spatially homogeneous plane wave space–times: I, *Class. Quantum Grav.* **8**, 1587–604. [Chapter 7]

Siklos, S.T.C. (1996). Counting solutions of Einstein's equations, *J. Math. Phys.*, to appear. [Chapter 9]

Smale, S. (1967). Differentiable dynamical systems, *Bull. Amer. Math. Soc.* **73**, 747–817. [Chapter 11]

Smoot, G.F. & Steinhardt, P.J. (1993). Gravity's rainbow, *Class. Quantum Grav.* **10**, S19–32. [Chapter 3]

Stabell, R. & Refsdal, S. (1966). Classification of general relativistic world models, *Mon. Not. R. Astron. Soc.* **132**, 379–88. [Chapter 2]

Stewart, J.M. (1990). *Advanced General Relativity.* Cambridge University Press. [Chapter 1]

Stewart, J.M. & Ellis, G.F.R. (1968). Solutions of Einstein's equations for a fluid which exhibit local rotational symmetry, *J. Math. Phys.* **9**, 1072–82. [Chapter 1]

Stewart, J. M. & Walker, M. (1974). Perturbations of spacetimes in general relativity, *Proc. Roy. Soc. London A* **341**, 49–74. [Chapter 14]

Stoeger, W.R., Ellis, G.F.R. & Xu, C. (1994). Observational cosmology VI: the microwave background and the Sachs–Wolfe effect, *Phys. Rev. D* **49**, 1845–53. [Chapter 3]

Stoeger, W. Maartens, R. & Ellis, G.F.R. (1995). Proving almost–homogeneity of the universe: an almost–Ehlers, Geren and Sachs theorem, *Astrophys. J.* **443**, 1–5. [Chapters 2, 3]

Strauss, M.A. & Willick, J.A. (1995). The density and peculiar velocity fields of nearby galaxies, *Phys. Rep.* **261**, 271-431. [Chapter 3]

Symbalisty, E.M.D. & Schramm, D.W. (1981). Nucleocosmochronology, *Rep. Prog. in Phys.* **44**, 293–328. [Chapter 3]

Szekeres, P. (1975). A class of inhomogeneous cosmological models, *Comm. Math. Phys.* **41**, 55–64. [Chapters 12, 13]

Taub, A.H. (1951). Empty space–times admitting a three–parameter group of motions, *Ann. Math.* **53**, 472–90. [Chapters 5, 9, 10]

Temam, R. (1988a). *Infinite–Dimensional Dynamical Systems in Mechanics and Physics.* Springer–Verlag. [Introduction, Chapters 4, 12, 15]

Temam, R. (1988b). Dynamical systems in infinite dimensions. In *The Connection between Infinite–Dimensional and Finite–Dimensional Dynamical Systems*, eds. B. Nicolaenko, C. Foias and R. Temam. American Mathematical Society. [Introduction, Chapters 4, 12, 15]

Terrell, J. (1977). The luminosity distance relation in Friedmann cosmology, *Amer. J. Phys.* **45**, 869–70. [Chapter 2]

Thorne, K.S. (1967). Primordial element formation, primordial magnetic fields, and the isotropy of the universe, *Astrophys. J.* **148**, 51–68. [Chapters 1, 3]

Tully, R.B., Scaramella, R., Vettolani, G. & Zamorani, G. (1992). Possible geometric patterns in 0.1 c scale structure, *Astrophys. J.* **388**, 9–16. [Chapter 3]

Uggla, C. (1989). Asymptotic cosmological solutions: orthogonal Bianchi type II models, *Class. Quantum Grav.* **6**, 383–96. [Chapter 6]

Uggla, C. (1990). New exact perfect fluid solutions of Einstein's equations, *Class. Quantum Grav.* **7**, L171–4. [Chapter 9]

Uggla, C. (1992). Inhomogeneous self–similar cosmological models, *Class. Quantum Grav.* **9**, 2287–95. [Chapter 12]

Uggla, C., Jantzen, R.T. & Rosquist, K. (1995). Exact hypersurface–homogeneous solutions in cosmology and astrophysics, *Phys. Rev. D* **51**, 5522–57. [Chapters 9, 10]

Uggla, C., Jantzen, R.T., Rosquist, K. & von Zur–Mühlen, H. (1991). Remarks about late stage homogeneous cosmological dynamics, *Gen. Rel. Grav.* **23**, 947–66. [Chapter 10]

Uggla, C. & Rosquist, K. (1988). Asymptotic cosmological solutions: orthogonal Bianchi type I, II, III, IV, VI and VII models, *Class. Quantum Grav.* **5**, 767–84. [Chapter 7]

Uggla, C. & Rosquist, K. (1990). New exact perfect fluid solutions of Einstein's equations II, *Class. Quantum Grav.* **7**, L279–83. [Chapter 9]

Uggla, C. & von Zur–Müllen, H. (1990). Compactified and reduced dynamics for locally rotationally symmetric Bianchi type IX perfect fluid models, *Class. Quantum Grav.* **7**, 1365–85. [Chapter 8]

Van den Bergh, N. & Skea, J. (1992). Inhomogeneous perfect fluid cosmologies, *Class. Quantum Grav.* **9**, 527–32. [Chapter 12]

van den Hoogen, R.J. & Coley, A.A. (1995). Qualitative analysis of causal anisotropic viscous–fluid cosmological models, *Class. Quantum Grav.* **12**, 2335–54. [Chapter 8]

van Elst, H. & Ellis, G.F.R. (1996). The covariant approach to LRS perfect fluid spacetime geometries, *Class Quantum Grav.* **13**, 1099–127. [Chapter 1]

van Elst, H. & Uggla, C. (1996). General relativistic 1+3 orthonormal frame approach revisited, (preprint). [Chapter 1]

Vishik, M.I. (1992). *Asymptotic Behaviour of Solutions of Evolutionary Equations.* Cambridge University Press. [Chapter 4]

Wainwright, J. (1979). A classification scheme for non–rotating inhomogeneous cosmologies, *J. Phys. A: Math. Gen.* **12**, 2015–29. [Chapter 1]

Wainwright, J. (1981). Exact spatially inhomogeneous cosmologies, *J. Phys. A.: Math. Gen.* **12**, 1131–47. [Chapter 1]

Wainwright, J. (1983). A spatially homogeneous cosmological model with plane wave singularity, *Phys. Lett. A* **99**, 301–3. [Chapter 9]

Wainwright, J. (1984). Power law singularities in orthogonally spatially homogeneous cosmologies, *Gen. Rel. Grav.* **16**, 657–74. [Chapter 9]

Wainwright, J. (1985). Self–similar solutions of Einstein's equations. In

Galaxies, Axisymmetric Systems and Relativity, ed. M.A.H. MacCallum. Cambridge University Press. [Chapters 1, 5, 9]

Wainwright, J. (1988). On the asymptotic states of orthogonal spatially homogeneous cosmologies. In *Relativity Today*, Proceedings of the Second Hungarian Relativity Workshop, ed. Z. Perjes. World Scientific. [Chapter 5]

Wainwright, J. (1996). Relativistic cosmology. In *Proceedings of the 46th Scottish Universities Summer School in Physics*, eds. G. Hall and J. Pulham. Institute of Physics Publishing. [Chapters 2, 15]

Wainwright, J. & Anderson, P.J. (1984). Isotropic singularities and isotropization in a class of Bianchi type VI$_h$ cosmologies, *Gen. Rel. Grav.* **16**, 609–24. [Chapter 5]

Wainwright, J. & Goode, S.W. (1980). Some exact inhomogeneous cosmologies with equation of state $p = \gamma\mu$, *Phys. Rev.* D **25**, 1906–9. [Chapter 12]

Wainwright, J. & Hsu, L. (1989). A dynamical systems approach to Bianchi cosmologies: orthogonal models of class A, *Class. Quantum Grav.* **6**, 1409–31. [Chapters 4, 5, 6]]

Wainwright, J., Ince, W.C.W. & Marshman, B.J. (1979). Spatially homogeneous and inhomogeneous cosmologies with equation of state $p = \mu$, *Gen. Rel. Grav.* **10**, 259–71. [Chapter 9]

Wainwright, J. & Marshman, B.J. (1979). Some exact cosmological models with gravitational waves, *Phys. Lett.* A **72**, 275–6. [Chapter 9]

Wald, R.M. (1983). Asymptotic behaviour of homogeneous cosmological models in the presence of a positive cosmological constant, *Phys. Rev.* D **28**, 2118–20. [Chapter 8]

Wald, R. M. (1984). *General Relativity*. University of Chicago Press. [Chapters 1, 2, 10]

Walker, A.G. (1936). On Milne's theory of world structure, *Proc. London Math. Soc.* **42**, 90–127. [Chapter 1]

Weber, E. (1984). Kantowski–Sachs cosmological models approaching isotropy, *J. Math. Phys.* **25**, 3279–85. [Chapter 8]

Weber, E. (1986). Cosmologies with a non–interacting mixture of dust and radiation, *J. Math. Phys.* **27**, 1578–82. [Chapter 8]

Weber, E. (1987). Global qualitative study of Bianchi universes in the presence of a cosmological constant, *J. Math. Phys.* **28**, 1658–66. [Chapter 8]

Weinberg, S. (1972). *Gravitation and Cosmology*. John Wiley and Sons. [Chapter 2]

Weinberg, S. (1989). The cosmological constant problem, *Rev. Mod. Phys.* **61**, 1–23. [Chapter 13]

White, M., Scott, D. & Silk, J. (1994). Anisotropies in the cosmic microwave background, *Annual Rev. Astron. and Astrophys.* **32**, 319–70. [Chapter 3]

White, S.D.M., Navarro, J.F., Evrard, A.E. & Frenk, C.S. (1993). The baryon content of galaxy clusters: a challenge to cosmological orthodoxy, *Nature*, **366**, 429–33. [Chapter 3]

Wiggins, S. (1990). *Introduction to Applied Nonlinear Dynamical Systems and Chaos.* Springer–Verlag. [Chapters 4, 11]

Wolf, A., Swift, J.B., Swinney, H.L. & Vastano, J.A. (1985). Determining Lyapunov exponents from a time series, *Physica D* **16**, 285–317. [Chapter 11]

Woszczyna, A. (1992). Gauge–invariant cosmic structures – a dynamic systems approach *Phys. Rev. D* **45**, 1982–8. [Chapter 14]

Zel'dovich, Ya. B. (1970). Gravitational instability: an approximate theory for large density perturbations, *Astron. and Astrophys.* **5**, 84–9. [Chapter 13]

Zel'dovich, Ya. B., & Novikov, I. D. (1983). *The Structure and Evolution of the Universe.* University of Chicago Press. [Introduction, Chapters 5, 6, 11, 12, 15]

Subject index